:D LOGIC SERIES

37

Editor

.bbay, *Department of Computer Science, King's College, London, U.K.*

e†

ssistant

Department of Computer Science, King's College, London, U.K.

THE SERIES

ɔplied in an increasingly wide variety of disciplines, from the traditional philosophy and mathematics to the more recent disciplines of cognitive mputer science, artificial intelligence, and linguistics, leading to new vigor ent subject. Springer, through its Applied Logic Series, seeks to provide a utstanding books and research monographs in applied logic, and in doing so es the underlying unity and applicability of logic.

volumes: http://www.springer.com/series/5632

Hybrid Logic and its Proof-Theory

APPL

VOLUM

Managi

Dov M.

Co-Edit

Jon Barv

Editoria

Jane Spu

SCOPE
Logic is
subjects
science,
in this a
home fo
demonst

For furthe

Hybrid Logic and its Proof-Theory

by

Torben Braüner
Roskilde University, Denmark

Springer

Dr. Torben Braüner
Roskilde University
Programming, Logic and Intelligent Systems
Research Group (PLIS)
4000 Roskilde
Denmark
torben@ruc.dk

ISSN 1386-2790
ISBN 978-94-007-0001-7 e-ISBN 978-94-007-0002-4
DOI 10.1007/978-94-007-0002-4
Springer Dordrecht Heidelberg London New York

Library of Congress Control Number: 2010938944

Printed on acid-free paper

Springer is part of Springer Science+Business Media (www.springer.com)

Preface

This book is a collation of the research on hybrid logic and its proof-theory I have done over a number of years. To be more precise, the book presents a collection of research results originally published in the journal papers listed below. After each paper is an indication of where in the book the results of the paper are presented.

T. Braüner. Natural deduction for hybrid logic. *Journal of Logic and Computation*, 14:329–353, 2004a. Chapter 2.

T. Braüner. Axioms for classical, intuitionistic, and paraconsistent hybrid logic. *Journal of Logic, Language and Information*, 15:179–194, 2006. Chapters 2 and 8.

T. Bolander and T. Braüner Tableau-based decision procedures for hybrid logic. *Journal of Logic and Computation*, 16:737–763, 2006. Chapter 3.

T. Braüner. Two natural deduction systems for hybrid logic: A comparison. *Journal of Logic, Language and Information*, 13:1–23, 2004b. Chapter 4.

T. Braüner. Proof-theoretic functional completeness for the hybrid logics of everywhere and elsewhere. *Studia Logica*, 81:191–226, 2005c. Chapter 5.

T. Braüner. Natural deduction for first-order hybrid logic. *Journal of Logic, Language and Information*, 14:173–198, 2005b. Chapter 6.

T. Braüner. Adding intensional machinery to hybrid logic. *Journal of Logic and Computation*, 18:631–648, 2008. Chapter 7.

T. Braüner and V. de Paiva Intuitionistic hybrid logic. *Journal of Applied Logic*, 4:231–255, 2006. Chapter 8.

T. Braüner. Why does the proof-theory of hybrid logic work so well? *Journal of Applied Non-Classical Logics*, 17:521–543, 2007. Chapters 9 and 10.

The notation and terminology of the papers has been revised with the aim of giving a uniform presentation, and moreover, interdependencies have been pointed out. In some cases more substantial revisions as well as omissions have also taken place. Furthermore, new material has been added. In particular, some material from my article Braüner (2005a) in *The Stanford Encyclopedia of Philosophy* has been incorporated in Chapter 1 and some materialfrom my part of the chapter Braüner and

Ghilardi (2007) in *Handbook of Modal Logic* has been incorporated in Chapter 6. Material from the Chapters 1, 2, 3, 9, and 10 of this book is incorporated in my forthcoming chapter Braüner (2011) in *Handbook of Philosophical Logic*.

The present book is based on the thesis Braüner (2009) accepted in fulfillment of the requirements for the Danish higher doctorate dr.scient. (doctor scientiarum).

A Little Background on Hybrid Logic and Proof-Theory

Hybrid logics are obtained by adding further expressive power to ordinary modal logic. The history of hybrid logics goes back to the philosopher Arthur Prior's work in the 1960s. The most basic hybrid logic is obtained by adding nominals, which are propositional symbols of a new sort interpreted in a restricted way that enables reference to individual points in a Kripke model (where the points represent possible worlds, times, locations, epistemic states, states in a computer, or something else). Another addition is the satisfaction operator @, which enables the evaluation of formulas at particular points. It is notable that nominals and the satisfaction operator do not disturb the local character of the Kripke semantics. The extra expressive power is useful for many applications, for example, when reasoning about time one often wants to formulate a series of statements about what happens at specific times, and ordinary modal logic simply does not allow this.

The addition of hybrid-logical machinery increases the expressive power, but often decidability is retained. Hybrid logics are closely related to description logics, which are a family of decidable logics used for knowledge representation in Artificial Intelligence. In description logics the points in a Kripke model represent individuals in the specification of an ontology. At present, significant research effort is put into exploring the borderline between decidable and undecidable logics, one major reason being that decidability is important for computational applications.

The subject of proof-theory is the notion of proof and formal systems for representing proofs. There are a number of different types of proof systems. Some of the most important types are natural deduction systems, Gentzen sequent systems, tableau systems, and axiom systems. They are motivated in different ways: Proof systems of the first three types are suitable for actual reasoning. (Here the word "actual" has a broad meaning, not restricted to actual human reasoning. The logic does not care whether it is a human that carries out the reasoning, or the reasoning takes place in a computer, or in some other medium.) Axiom systems are usually not meant for actual reasoning, but are of a more foundational interest. When a decidable logic is considered, Gentzen and tableau systems have the desirable feature of often giving rise to decision procedures in a very direct way, therefore Gentzen and tableau systems lend themselves toward computer implementations. In fact, during the last couple of decades, tableau systems have become a highly active research area, involving basic research as well as practically applied work, for example tableau systems for high-speed theorem proving in description logics.

There is little consensus about proof-theory for ordinary modal logic, especially in connection with natural deduction systems, Gentzen sequent systems, and tableau systems. Many modal-logical proof systems lack important proof-theoretic properties and the relationships between proof systems for different modal logics are often unclear. In the quotation below Heinrich Wansing gives a succinct summary of the status of modal-logical proof-theory.

> Compared with the multitude of not only existing but also interesting axiomatically presentable normal modal propositional logics, the number of systems for which sequent calculus presentations (of some sort) are known is disappointingly small. In contrast to the axiomatic approach, the standard sequent-style proof theory for normal modal logics fails to be 'modular', and the very mechanism behind the small range of known possible variations is not very clear. (Wansing 1994, p. 128)

In the present book we shall demonstrate that hybrid-logical proof-theory remedies this lack of uniformity in modal-logical proof systems.

The Content of This Book

The main issue of this book is the proof-theory of hybrid logic. To be more precise: Natural deduction, Gentzen, tableau, and axiom systems for hybrid logic. We first deal with the propositional case, that is, we describe sound and complete natural deduction, Gentzen, and axiom systems for propositional hybrid logic. The natural deduction and Gentzen systems satisfy the requirements that such systems are expected to satisfy: The natural deduction system satisfies normalization, and normal derivations satisfy a version of the subformula property. The Gentzen system is cut-free and also the Gentzen derivations satisfy a version of the subformula property. Moreover, we give tableau-based decision procedures for two decidable fragments of hybrid logic, one of these being a decision procedure including the very expressive universal modality. After having dealt with the propositional case, we describe proof-theory for first-order hybrid logic, including intensional machinery. Furthermore, we describe proof-theory for intuitionistic hybrid logic.

Thus, we consider a spectrum of different versions of hybrid logic (propositional, first-order, intensional first-order, and intuitionistic) and a spectrum of different types of proof-systems for hybrid logic (natural deduction, Gentzen, tableau, and axiom systems). All these systems can be motivated independently, but the fact that the systems can be given corroborates the point of view that hybrid logic and hybrid-logical proof-theory is a natural enterprise. This line of thinking is expressed briefly and to the point in the following quotation by Nuel D. Belnap.

> It seems to be generally conceded that formal systems are natural or substantial if they can be looked at from several points of view. We tend to think of systems as artificial or *ad hoc* if most of their formal properties arise from some *one* notational system in terms of which they are described. (Anderson and Belnap 1975, p. 50)

Besides satisfying the above general requirements, hybrid-logical proof-theory furthermore satisfies the more concrete requirement that proof systems for wide classes of hybrid logics can be given in a uniform way, for example, natural deduction systems for a wide class of hybrid logics can be obtained in a uniform way by adding derivation rules as appropriate. This is simply not possible in connection with standard proof-theory for ordinary modal logic.

This leads us to the following question: Why does the proof-theory of hybrid logic behave so well compared to the proof-theory of ordinary modal logic? Before we give an answer to this question, we shall make two remarks: Firstly, we remark that the metalinguistic semantic, that is, model-theoretic, machinery of hybrid logic is internalized in the hybrid-logical object language (via satisfaction operators). Secondly, we remark that modal-logical rules for reasoning directly about models (called labelled rules) are proof-theoretically well-behaved, that is, they satisfy the proof-theoretic requirements such rules are expected to satisfy (but at the expense of making use of metalinguistic machinery). In the present book we demonstrate that the good behaviour of labelled rules is preserved by internalization. To be more precise, our natural deduction, Gentzen, and tableau rules for hybrid logic can be seen as internalized rules for reasoning directly about models, and what we do in the present book is that we provide a proof-theoretic analysis of the internalized rules, by which it is demonstrated that the internalized rules are proof-theoretically well-behaved.

The answer to the question above is accordingly that internalization of model-theoretic machinery in the object language enables us to give well-behaved proof-theory for hybrid logic. So model-theory is a prerequisite for our proof-theoretic analysis—in this sense the present book has proof-theory as well as model-theory as its starting point.

Acknowledgements

I have benefited greatly from discussions and collaboration with a number of researchers.

First I would like to thank Peter Øhrstrøm for introducing me to Arthur Prior's work and for including me in his Prior project at Aalborg University back in 1997. Also thanks to Per Hasle for conversations on Prior's work and many other subjects. Prior's work was my first contact with what now is known as hybrid logic.

My work on the proof-theory of hybrid logic has benefited in particular from discussions with Carlos Areces, Patrick Blackburn, Thomas Bolander, Balder ten Cate, Melvin Fitting, Valentin Goranko, Jens Ulrik Hansen, Maarten Marx, Valeria de Paiva, and Jørgen Villadsen. Special thanks to Thomas for fruitful collaboration in connection with the projects HyLoMOL (Hybrid Logic Meets Other Logics, 2005–2008) and HYLOCORE (Hybrid Logic, Computation, and Reasoning Methods, 2009–2012). I am grateful to Patrick for many conversations whenever I visited France or he visited Denmark. Thanks to Thomas Müller for discussions on some of the more philosophical material in this book and thanks to Sara L. Uckelman for helpful comments on a preliminary version of the book.

I would like to thank the evaluation committee—Stig Andur Pedersen, Patrick Blackburn, and Melvin Fitting—for many useful comments and suggestions on my dr.scient. thesis Braüner (2009) on which the present book is based.

I wish to acknowledge the financial support received from The Danish Natural Science Research Council as funding for the above mentioned projects HyLoMOL and HYLOCORE.

Thanks to my wife Anne and our children, Sara, Lea, Thomas, and Maria, for support and for being there.

Roskilde, Denmark Torben Braüner

Contents

Chapter 1
Introduction to Hybrid Logic

In this chapter we give the basics of hybrid logic. The chapter is structured as follows. In the first section of the chapter we give an informal motivation of hybrid logic. In the second section we give the formal syntax and semantics and we give translations forwards and backwards between hybrid logic and first-order logic. In the third section we discuss the work of Arthur Prior and describe how hybrid logic has its origin in his work. In the fourth section we outline the development of hybrid logic since Prior.

1.1 Informal Motivation

The term "hybrid logic" covers a number of logics obtained by adding further expressive power to ordinary modal logic.[1] The history of what now is known as hybrid logic goes back to Arthur Prior's work in the 1960s, which we shall come back to in Section 1.3. The term "hybrid logic" was coined by Patrick Blackburn and Jerry Seligman in their paper Blackburn and Seligman (1995). The most basic hybrid logic is obtained by adding nominals, which are propositional symbols of a new sort interpreted in a restricted way that enables reference to individual points in a Kripke model. In what follows we shall give a more detailed explanation.

In the standard Kripke semantics for modal logic, the truth-value of a formula is relative to points in a set, that is, a formula is evaluated "locally" at a point. Usually, the points are taken to represent possible worlds, times, locations, epistemic states, states in a computer, or something else. Thus, in the Kripke semantics, a propositional symbol might have different truth-values at different points. This allows us to formalize natural language statements whose truth-values are relative to, for example times, like the statement

it is raining

[1] This should not be confused with the term "hybrid systems" which in computer science is used for systems that combine discrete and continuous features.

T. Braüner, *Hybrid Logic and its Proof-Theory*, Applied Logic Series 37,
DOI 10.1007/978-94-007-0002-4_1, © Springer Science+Business Media B.V. 2011

which has clearly different truth-values at different times. Such statements can be formalized in ordinary modal logic using ordinary propositional symbols. Now, certain natural language statements are true at exactly one time, possible world, or something else. An example is the statement

it is five o'clock May 10th 2007

which is true at the time five o'clock May 10th 2007, but false at all other times. While the first kind of statement can be formalized in ordinary modal logic, the second kind of statement cannot, the reason being that there is only one sort of propositional symbol available, namely ordinary propositional symbols, which are not restricted to being true at exactly one point in the Kripke semantics.

A major motivation for hybrid logic is to add further expressive power to ordinary modal logic with the aim of being able to formalize the second kind of statement. This is obtained by adding to ordinary modal logic a second sort of propositional symbol called a nominal such that in the Kripke semantics each nominal is true at exactly one point. In other words, a nominal is interpreted with the restriction that the set of points at which it is true is a singleton set, not an arbitrary set. A natural language statement of the second kind (like the example statement with the time five o'clock May 10th 2007) is then formalized using a nominal, not an ordinary propositional symbol (which is used to formalize the example statement with rainy weather). The fact that a nominal is true at exactly one point implies that a nominal can be considered a term referring to a point, for example, if a is a nominal that stands for "it is five o'clock May 10th 2007", then the nominal a can be considered a term referring to the time five o'clock May 10th 2007.[2] Thus, in hybrid logic a

[2] Considering a nominal as a symbol that refers to something is not the only way to view nominals. Two different views on nominals can be identified in the works of Arthur Prior, as is clear from the quotation below where Prior discusses the addition of nominals to a temporal version of modal logic called tense logic.

> We might ... equate the instant a with a conjunction of all those propositions which would ordinarily be said to be true at that instant, or we might equate it with some proposition which would ordinarily be said to be true at that instant only, and so could serve as an index of it. (Hasle et al. 2003, p. 124)

In the second half of the sentence, the nominal a is viewed as a proposition that can serve as an index of an instant, which is clearly in line with considering a nominal as a symbol that refers to an instant. On the other hand, in the first half of the sentence, the nominal a is viewed as a description of the content of an instant. The alternative view on nominals expressed in the first half of the sentence quoted above can also be found in a number of other places in Prior's works, for example the following.

> The essential trick is to treat the instant variables as a special sort of *propositional* variables, by identifying an 'instant' with the totality of what would ordinarily be said to be true *at* that instant, ... (Hasle et al. 2003, p. 141)

See the discussion of Prior's work in Section 1.3 of the present book, in particular Footnote 6 of that section. Moreover, see the discussion in Blackburn (2006), the last paragraph of page 353, including Footnote 7, and the first complete paragraph of page 362, in particular Footnote 11. Incidentally, note that the description of the content of an instant as the conjunction of all propositions true at that instant is similar to a maximal consistent set of formulas.

term is a specific sort of propositional symbol whereas in first-order logic it is an argument to a predicate.

Most hybrid logics involve further additional machinery than nominals. There is a number of options for adding further machinery; here we shall consider a kind of operator called satisfaction operators. The motivation for adding satisfaction operators is to be able to formalize a statement being true at a particular time, possible world, or something else. For example, we want to be able to formalize that the statement "it is raining" is true at the time five o'clock May 10th 2007, that is, that

> at five o'clock May 10th 2007, it is raining.

This is formalized by the formula $@_a p$ where the nominal a stands for "it is five o'clock May 10th 2007" as above and where p is an ordinary propositional symbol that stands for "it is raining". It is the part $@_a$ of the formula $@_a p$ that is called a satisfaction operator. In general, if a is a nominal and ϕ is an arbitrary formula, then a new formula $@_a \phi$ can be built (in some literature the notation $a : \phi$ is used instead of $@_a \phi$). A formula of the form $@_a \phi$ is called a satisfaction statement. The satisfaction statement $@_a \phi$ expresses that the formula ϕ is true at one particular point, namely the point to which the nominal a refers.

To sum up, we have now added further expressive power to ordinary modal logic in the form of nominals and satisfaction operators. Informally, the nominal a has the truth-condition

> a is true relative to a point w
> if and only if
> the reference of a is identical to w

and the satisfaction statement $@_a \phi$ has the truth-condition

> $@_a \phi$ is true relative to a point w
> if and only if
> ϕ is true relative to the reference of a

Observe that actually the point w does not matter in the truth-condition for $@_a \phi$ since the satisfaction operator $@_a$ moves the point of evaluation to the reference of a whatever the identity of w. Note that the addition of nominals and satisfaction operators does not disturb the local character of the Kripke semantics: The truth-value of a formula is still relative to points in a set and the added machinery only involves reference to particular points, not all points in the set.

It is worth noting that nominals together with satisfaction operators allow us to express that two points are identical: If the nominals a and b refer to the points u and v, then the formula $@_a b$ expresses that u and v are identical. The following line of reasoning shows why.

> $@_a b$ is true relative to a point w
> if and only if
> b is true relative to the reference of a
> if and only if
> b is true relative to u
> if and only if

the reference of b is identical to u
if and only if
v is identical to u

The identity relation on a set has the well-known properties reflexivity, symmetry, and transitivity, which is reflected in the fact that the formulas

$$@_a a$$
$$@_a b \rightarrow @_b a$$
$$(@_a b \wedge @_b c) \rightarrow @_a c$$

are valid formulas of hybrid logic. To see that these hybrid-logical formulas correspond to the properties reflexivity, symmetry, and transitivity, read $@_a b$ as $a = b$ etc. Also the formula

$$(@_a b \wedge @_a \phi) \rightarrow @_b \phi$$

is valid. This hybrid-logical formula corresponds to the standard rule called rule of replacement. Reflexivity and replacement are in the natural deduction system of Section 2.2 directly formulated as the rules (*Ref*) and (*Nom*1), see Figure 2.3 of that section (and see the discussion of the side-condition on (*Nom*1) in Section 2.2.2). As remarked in Section 2.2, natural deduction rules corresponding to symmetry and transitivity are derivable from (*Ref*) and (*Nom*1).

Beside nominals and satisfaction operators, in what follows we shall consider the binders $\forall$ and $\downarrow$, which allow us to build formulas $\forall a\phi$ and $\downarrow a\phi$. The binders bind nominals to points in two different ways: The $\forall$ binder quantifies over all points analogous to the standard first-order universal quantifier, that is, $\forall a\phi$ is true relative to w if and only if whatever point the nominal a refers to, ϕ is true relative to w. The $\downarrow$ binder binds a nominal to the point of evaluation, that is, $\downarrow a\phi$ is true relative to w if and only if ϕ is true relative to w when a refers to w. It turns out that the $\downarrow$ binder is definable in terms of $\forall$.

Above we noted that nominals and satisfaction operators do not disturb the local character of the Kripke semantics. Also the $\downarrow$ binder leaves the local character of the semantics undisturbed since this binder just binds a nominal to the point of evaluation. Things are more complicated with the $\forall$ binder. This binder has a non-local character in the sense that it involves reference to all points in the Kripke semantics. Moreover, together with nominals and satisfaction operators, the $\forall$ binder gives rise to non-local expressivity in the form of full first-order expressive power (which we shall show in Section 1.2.1). However, the $\forall$ binder does not give rise to full first-order expressive power just together with nominals, that is, in the absence of satisfaction operators (or some similar machinery). Thus, it is really the interaction between the $\forall$ binder and satisfaction operators that gives rise to full first-order expressive power, and hence, non-local expressivity.[3]

[3] In fact, Blackburn and Seligman (1995) give a result (Proposition 4.5 on p. 264) indicating that the $\forall$ binder has a surprisingly local character when it is not accompanied by satisfaction operators or some similar machinery. Informally, this result says that the $\forall$ binder is then insensitive to the information at points outside the submodel generated by the point of evaluation, that is, it cannot detect the truth-values of formulas at such points.

To conclude, extending ordinary modal logic with hybrid-logical machinery (disregarding the extreme case involving both $\forall$ and satisfaction operators), gives us a more expressive logic without sacrificing the local character of the Kripke semantics.[4]

1.2 Formal Syntax and Semantics

In what follows we give the formal syntax and semantics of hybrid logic. In many cases we will adopt the terminology of Blackburn et al. (2001) and Areces et al. (2001). The hybrid logic we consider is obtained by adding a second sort of propositional symbol, called *nominals*, to ordinary modal logic, that is, propositional logic extended with a modal operator $\Box$.[5] It is assumed that a set of ordinary propositional symbols and a countably infinite set of nominals are given. The sets are assumed to be disjoint. The metavariables $p, q, r, \ldots$ range over ordinary propositional symbols and $a, b, c, \ldots$ range over nominals. Besides nominals, an operator $@_a$ called a *satisfaction operator* is added for each nominal a. Sometimes the operator $@_a$ is called an *at operator*. Moreover, we shall consider the *binders* $\forall$ and $\downarrow$. The formulas of hybrid modal logic are defined by the grammar

$$S ::= p \mid a \mid S \wedge S \mid S \rightarrow S \mid \bot \mid \Box S \mid @_a S \mid \forall a S \mid \downarrow a S$$

where p ranges over ordinary propositional symbols and a ranges over nominals. In what follows, the metavariables $\phi, \psi, \theta, \ldots$ range over formulas. Formulas of the form $@_a\phi$ are called *satisfaction statements*, cf. Blackburn (2000a). The notions of free and bound occurrences of nominals are defined as in first-order logic with the addition that the free nominal occurrences in $@_a\phi$ are the free nominal occurrences in ϕ together with the occurrence of a, and moreover, the free nominal occurrences in $\downarrow a\phi$ are the free nominal occurrences in ϕ except for occurrences of a. Also, if $\overline{a}$ is a list of pairwise distinct nominals and $\overline{c}$ is a list of nominals of the same length as $\overline{a}$, then $\psi[\overline{c}/\overline{a}]$ is the formula ψ where the nominals $\overline{c}$ have been simultaneously substituted for all free occurrences of the nominals $\overline{a}$. If a nominal a_i in $\overline{a}$ occurs free in ψ within the scope of $\forall c_i$ or $\downarrow c_i$, then the nominal c_i in ψ is renamed as appropriate (this can be done since there are infinitely many nominals). The connectives negation, nullary conjunction, disjunction, and biimpliction are defined by the conventions that $\neg\phi$ is an abbreviation for $\phi \rightarrow \bot$, $\top$ is an abbreviation for $\neg\bot$, $\phi \vee \psi$ is an abbreviation for $\neg(\neg\phi \wedge \neg\psi)$, and $\phi \leftrightarrow \psi$ is an abbreviation

[4] Further discussion of this point can be found in a number of places, notably Blackburn (2006). This paper also discusses hybrid-logical versions of *bisimulations*, which give a mathematical way to illustrate the local character of the Kripke semantics. See also Simons (2006) which discusses a number of logics of location involving what we here call satisfaction operators.

[5] All results in the present book can be generalized to cover an arbitrary, finite number of modal operators, but in the interest of simplicity, we shall stick to one modal operator unless otherwise is specified.

for $(\phi \to \psi) \wedge (\psi \to \phi)$. Similarly, $\Diamond\phi$ is an abbreviation for $\neg\Box\neg\phi$ and $\exists a\phi$ is an abbreviation for $\neg\forall a\neg\phi$.

We now define models and frames.

Definition 1.1. A *model* for hybrid logic is a tuple $(W, R, \{V_w\}_{w\in W})$ where

1. W is a non-empty set;
2. R is a binary relation on W; and
3. for each w, V_w is a function that to each ordinary propositional symbol assigns an element of $\{0, 1\}$.

The pair (W, R) is called a *frame* and the model is said to be *based* on this frame. The elements of W are called *worlds* and the relation R is called the *accessibility relation*. A propositional symbol p is said to be *true* at w if $V_w(p) = 1$ and it is said to be *false* at w if $V_w(p) = 0$.

Note that a model for hybrid logic is the same as a model for ordinary modal logic. To give an extremely simple example of a model, we let $W = \{w, v\}$ and $R = \{(w, v)\}$, and moreover, we let $V_w(p) = 0$ and $V_v(p) = 1$. All other propositional symbols than p are ignored. This model can be depicted as

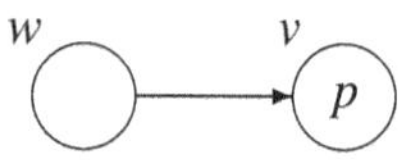

where circles represent worlds and an arrow indicates that two worlds are related by the accessibility relation. A propositional symbol in a circle means that the symbol is true and the absence of a propositional symbol means that it is false.

Given a model $\mathfrak{M} = (W, R, \{V_w\}_{w\in W})$, an *assignment* is a function g that to each nominal assigns an element of W. Given assignments g' and g, $g' \stackrel{a}{\sim} g$ means that g' agrees with g on all nominals save possibly a. The relation $\mathfrak{M}, g, w \models \phi$ is defined by induction, where g is an assignment, w is an element of W, and ϕ is a formula.

$$
\begin{aligned}
\mathfrak{M}, g, w &\models p \text{ iff } V_w(p) = 1 \\
\mathfrak{M}, g, w &\models a \text{ iff } w = g(a) \\
\mathfrak{M}, g, w &\models \phi \wedge \psi \text{ iff } \mathfrak{M}, g, w \models \phi \text{ and } \mathfrak{M}, g, w \models \psi \\
\mathfrak{M}, g, w &\models \phi \to \psi \text{ iff } \mathfrak{M}, g, w \models \phi \text{ implies } \mathfrak{M}, g, w \models \psi \\
\mathfrak{M}, g, w &\models \bot \text{ iff falsum} \\
\mathfrak{M}, g, w &\models \Box\phi \text{ iff for any } v \in W \text{ such that } wRv,\ \mathfrak{M}, g, v \models \phi \\
\mathfrak{M}, g, w &\models @_a\phi \text{ iff } \mathfrak{M}, g, g(a) \models \phi \\
\mathfrak{M}, g, w &\models \forall a\phi \text{ iff for any } g' \stackrel{a}{\sim} g,\ \mathfrak{M}, g', w \models \phi \\
\mathfrak{M}, g, w &\models \downarrow a\phi \text{ iff } \mathfrak{M}, g', w \models \phi \text{ where } g' \stackrel{a}{\sim} g \text{ and } g'(a) = w
\end{aligned}
$$

A formula ϕ is said to be *true* at w if $\mathfrak{M}, g, w \models \phi$; otherwise it is said to be *false* at w. By convention $\mathfrak{M}, g \models \phi$ means $\mathfrak{M}, g, w \models \phi$ for every element w of W and $\mathfrak{M} \models \phi$ means $\mathfrak{M}, g \models \phi$ for every assignment g. A formula ϕ is *valid* in a frame if and only if $\mathfrak{M} \models \phi$ for any model $\mathfrak{M}$ that is based on the frame. A formula ϕ is *valid* in a class of frames if and only if ϕ is valid in any frame in the class of frames in question. A formula ϕ is *valid* if and only if ϕ is valid in the class of all frames.

Now, let $\mathcal{O} \subseteq \{\downarrow, \forall\}$. In what follows $\mathcal{H}(\mathcal{O})$ denotes the fragment of hybrid logic in which the only binders are the binders in the set $\mathcal{O}$. If $\mathcal{O} = \emptyset$, then we simply write $\mathcal{H}$, and if $\mathcal{O} = \{\downarrow\}$, then we write $\mathcal{H}(\downarrow)$, etc. It is assumed that the set $\mathcal{O}$ of binders is fixed.

Note that $\downarrow$ is definable in terms of $\forall$ since the formula $\downarrow a\phi \leftrightarrow \forall a(a \rightarrow \phi)$ is valid. The fact that hybridizing ordinary modal logic actually does give more expressive power can for example be seen by considering the formula $\downarrow c\Box\neg c$. It is straightforward to check that this formula is valid in a frame if and only if the frame is irreflexive. Thus, irreflexivity can be expressed by a hybrid-logical formula, but it is well known that it cannot be expressed by any formula of ordinary modal logic. Irreflexivity can actually be expressed just by adding nominals to ordinary modal logic, namely by the formula $c \rightarrow \Box\neg c$. It is clear that if a frame is irreflexive, then $c \rightarrow \Box\neg c$ is valid in the frame. On the other hand, if $c \rightarrow \Box\neg c$ is valid in a frame, then the frame is irreflexive: Let (W,R) be a frame in which $c \rightarrow \Box\neg c$ is valid and let w be an element of W, then $\mathfrak{M}, g, w \models c \rightarrow \Box\neg c$ where $\mathfrak{M}$ is an arbitrarily chosen model based on (W,R) and g is an arbitrarily chosen assignment such that $g(c) = w$, and from this it follows that wRw is false. Hence, the formula $c \rightarrow \Box\neg c$ expresses irreflexivity. Other examples of properties expressible in hybrid logic, but not in ordinary modal logic, are asymmetry (expressed by $c \rightarrow \Box\neg\Diamond c$), antisymmetry (expressed by $c \rightarrow \Box(\Diamond c \rightarrow c)$), and universality (expressed by $\Diamond c$).

1.2.1 Translation into First-Order Logic

Hybrid logic can be translated into first-order logic with equality and (a fragment of) first-order logic with equality can be translated back into (a fragment of) hybrid logic. The translation from hybrid logic into first-order logic we consider in this section is an extension of the well-known *standard translation* from modal logic into first-order logic, see Areces et al. (2001) and van Benthem (1983).

The first-order language under consideration has a 1-place predicate symbol corresponding to each ordinary propositional symbol of modal logic, a 2-place predicate symbol corresponding to the modality, and a 2-place predicate symbol corresponding to equality. The language does not have constant or function symbols. It is assumed that a countably infinite set of first-order variables is given. The metavariables a, b, c, ... range over first-order variables. There are no function symbols or constants. So the formulas of the first-order language we consider are defined by the grammar

$$S ::= p^*(a) \mid R(a,b) \mid a = b \mid S \wedge S \mid S \rightarrow S \mid \bot \mid \forall a S$$

where p ranges over ordinary propositional symbols of hybrid logic, and a and b range over first-order variables. Note that according to the grammar above, for each ordinary propositional symbol p of the modal language there is a corresponding 1-place predicate symbol p^* in the first-order language. The predicate symbol p^* will be interpreted such that it relativises the interpretation of the corresponding modal propositional symbol p to worlds. In the grammar above, R is a designated predicate symbol which will be interpreted using the accessibility relation (with the same name). In what follows, we shall identify first-order variables with nominals of hybrid logic. Note in this connection that the set of metavariables ranging over first-order variables is identical to the set of metavariables ranging over nominals. Free and bound occurrences of variables are defined as usual for first-order logic. Also, $\psi[c/a]$ is the formula ψ where the variable c has been substituted for all free occurrences of the variable a. As usual, if the variable a occurs free in ψ within the scope of $\forall c$, then the variable c in ψ is renamed as appropriate. The connectives $\neg$, $\top$, $\vee$, $\leftrightarrow$, and $\exists$ are defined in one of the usual ways.

We first translate the hybrid logic $\mathcal{H}(\downarrow,\forall)$ into first-order logic with equality. It is assumed that two nominals a and b are given which do not occur in the formulas to be translated. The translations ST_a and ST_b are defined by mutual induction. We just give the translation ST_a. Recall that nominals and first-order variables are identified.

$$\begin{aligned}
ST_a(p) &= p^*(a)\\
ST_a(c) &= a = c\\
ST_a(\phi \wedge \psi) &= ST_a(\phi) \wedge ST_a(\psi)\\
ST_a(\phi \rightarrow \psi) &= ST_a(\phi) \rightarrow ST_a(\psi)\\
ST_a(\bot) &= \bot\\
ST_a(\Box\phi) &= \forall b(R(a,b) \rightarrow ST_b(\phi))\\
ST_a(@_c\phi) &= ST_a(\phi)[c/a]\\
ST_a(\forall c\phi) &= \forall c ST_a(\phi)\\
ST_a(\downarrow c\phi) &= ST_a(\phi)[a/c]
\end{aligned}$$

The definition of ST_b is obtained by exchanging a and b. As an example, we demonstrate step by step how the hybrid-logical formula $\downarrow c\Box\neg c$ is translated into a first-order formula:

$$\begin{aligned}
ST_a(\downarrow c\Box\neg c) &= ST_a(\Box\neg c)[a/c]\\
&= \forall b(R(a,b) \rightarrow ST_b(\neg c))[a/c]\\
&= \forall b(R(a,b) \rightarrow \neg ST_b(c))[a/c]\\
&= \forall b(R(a,b) \rightarrow \neg b = c)[a/c]\\
&= \forall b(R(a,b) \rightarrow \neg b = a)
\end{aligned}$$

The resulting first-order formula is equivalent to $\neg R(a,a)$ which shows that $\downarrow c\Box\neg c$ indeed does correspond to the accessibility relation being irreflexive, cf. above. What has been done in the translation is that the semantics of hybrid logic has been formalized in terms of first-order logic; note how each clause in the translation formalizes a clause in the definition of the semantics, that is, the relation $\mathfrak{M}, g, w \models \phi$.

The translation ST_a is truth-preserving. To state this formally, we make use of the well-known observation that a model for hybrid logic can be considered as a model for first-order logic and vice versa.

Definition 1.2. Given a model $\mathfrak{M} = (W, R, \{V_w\}_{w \in W})$ for hybrid logic, a model $\mathfrak{M}^* = (W, V^*)$ for first-order logic is defined by letting

- $V^*(p^*) = \{w \mid V_w(p) = 1\}$ and
- $V^*(R) = R$.

It is straightforward to see that the map $(\cdot)^*$ which maps $\mathfrak{M}$ to $\mathfrak{M}^*$ is bijective. Moreover, an assignment in the sense of classical hybrid logic can be considered as an assignment in the sense of classical first-order logic and vice versa.

Given a model $\mathfrak{M}$ for first-order logic, the relation $\mathfrak{M}, g \models \phi$ is defined by induction in the standard way, where g is an assignment and ϕ is a first-order formula.

$$\begin{array}{rl}
\mathfrak{M}, g \models p^*(a) & \text{iff } g(a) \in V(p^*) \\
\mathfrak{M}, g \models R(a,b) & \text{iff } g(a)V(R)g(b) \\
\mathfrak{M}, g \models a = b & \text{iff } g(a) = g(b) \\
\mathfrak{M}, g \models \phi \wedge \psi & \text{iff } \mathfrak{M}, g \models \phi \text{ and } \mathfrak{M}, g \models \psi \\
\mathfrak{M}, g \models \phi \rightarrow \psi & \text{iff } \mathfrak{M}, g \models \phi \text{ implies } \mathfrak{M}, g \models \psi \\
\mathfrak{M}, g \models \bot & \text{iff falsum} \\
\mathfrak{M}, g \models \forall a \phi & \text{iff for any } g' \overset{a}{\sim} g,\ \mathfrak{M}, g' \models \phi
\end{array}$$

The formula ϕ is said to be *true* if $\mathfrak{M}, g \models \phi$; otherwise it is said to be *false*. By convention $\mathfrak{M} \models \phi$ means $\mathfrak{M}, g \models \phi$ for every assignment g. We shall later make use of the first-order semantics in connection with the interpretation of geometric theories.

It can now be stated formally that the translation is truth-preserving.

Proposition 1.1. *Let $\mathfrak{M}$ be a model for hybrid logic and let ϕ be a hybrid-logical formula in which the nominals a and b do not occur. For any assignment g, it is the case that $\mathfrak{M}, g, g(a) \models \phi$ if and only if $\mathfrak{M}^*, g \models ST_a(\phi)$ (and the same for ST_b).*

Proof. Induction on the structure of ϕ.

Thus, hybrid logic, considered as a language for talking about models, has the same expressive power as the fragment of first-order logic obtained by taking the image of hybrid logic under the translation ST_a.

First-order logic with equality can be translated into the hybrid logic $\mathcal{H}(\forall)$ by the translation HT given below.

$$\begin{array}{rcl}
HT(p^*(a)) & = & @_a p \\
HT(R(a,c)) & = & @_a \Diamond c \\
HT(a = c) & = & @_a c \\
HT(\phi \wedge \psi) & = & HT(\phi) \wedge HT(\psi) \\
HT(\phi \rightarrow \psi) & = & HT(\phi) \rightarrow HT(\psi) \\
HT(\bot) & = & \bot \\
HT(\forall a \phi) & = & \forall a HT(\phi)
\end{array}$$

The translation HT is truth-preserving.

Proposition 1.2. *Let $\mathfrak{M}$ be a model for hybrid logic. For any first-order formula ϕ and any assignment g, it is the case that $\mathfrak{M}^*, g \models \phi$ if and only if $\mathfrak{M}, g \models HT(\phi)$.*

Proof. Induction on the structure of ϕ.

Thus, in the sense above the hybrid logic $\mathcal{H}(\forall)$ has the same expressive power as first-order logic with equality. It is implicit in the proposition above that the first-order formula ϕ is a formula of the first-order language defined by the grammar given earlier in the present section. The history of the above observations goes back to the work of Arthur Prior, which we shall come back to in the next section.

In a way similar to the above translation, a fragment of first-order logic with equality which is called the *bounded fragment* can be translated into the hybrid logic $\mathcal{H}(\downarrow)$. This was pointed out in Areces et al. (2001). The bounded fragment is obtained from the above grammar for first-order logic by replacing the clause $\forall aS$ by the new clause $\forall c(R(a,c) \rightarrow S$ where it is required that the variables a and c are distinct. In Areces et al. (2001) a number of independent semantic characterizations of the bounded fragment are given. A translation from the bounded fragment to the hybrid logic $\mathcal{H}(\downarrow)$ can be obtained by replacing the last clause in the translation HT above by the following.

$$HT(\forall c(R(a,c) \rightarrow \phi)) = @_a\Box \downarrow cHT(\phi)$$

It is straightforward to check that Proposition 1.2 still holds, hence, the hybrid logic $\mathcal{H}(\downarrow)$ has the same expressive power as the bounded fragment of first-order logic (note that for any formula ϕ of $\mathcal{H}(\downarrow)$, the formula $ST_a(\phi)$ is in the bounded fragment).

1.3 The Origin of Hybrid Logic in Prior's Work

The history of hybrid logic goes back to Arthur Prior's hybrid tense logic, which is a hybridized version of ordinary tense logic. Arthur Prior (1914–1969) is usually considered the founding father of modern temporal logic, his main contribution being the formal logic of tenses. In his memorial paper on Prior, A.J.P. Kenny (1970) summed up Prior's life and work as follows.

> Prior's greatest scholarly achievement was undoubtedly the creation and development of tense-logic. But his research and reflection on this topic led him to elaborate, piece by piece, a whole metaphysical system of an individual and characteristic stamp. He had many different interests at different periods of his life, but from different angles he constantly returned to the same central and unchanging themes. Throughout his life, for instance, he worked away at the knot of problems surrounding determinism: first as a predestinarian theologian, then as a moral philosopher, finally as a metaphysician and logician. (Kenny 1970, p. 348)

Prior's reflections on determinism and other issues related to the philosophy of time were a major motivation for his formulation of tense logic. With the aim of discussing tense logic and hybrid tense logic further, we shall give a formal definition of hybrid tense logic: The language of hybrid tense logic is simply the language of hybrid logic defined above except that there are two modal operators, namely G and H, instead of the single modal operator $\Box$. The two new modal operators are called *tense operators*. The semantics of hybrid tense logic is the semantics of hybrid logic, cf. earlier, with the clause for $\Box$ replaced by clauses for the tense operators G and H.

$$\mathfrak{M}, g, w \models G\phi \text{ iff for any } v \in W \text{ such that } wRv,\ \mathfrak{M}, g, v \models \phi$$
$$\mathfrak{M}, g, w \models H\phi \text{ iff for any } v \in W \text{ such that } vRw,\ \mathfrak{M}, g, v \models \phi$$

Thus, there are now two modal operators, namely one that "looks forwards" along the accessibility relation R and one that "looks backwards". In tense logic the elements of the set W are called *moments* or *instants* and the accesibility relation R is now also called the *earlier-later relation*.

It is straightforward to modify the translations ST_a and HT in the previous section such that translations are obtained between a tense-logical version of $\mathscr{H}(\forall)$ and first-order logic with equality. The first-order logic under consideration is what Prior called *first-order earlier-later logic*. Given the translations, it follows that Prior's first-order earlier-later logic has the same expressive power as the tense-logical version of $\mathscr{H}(\forall)$, that is, hybrid tense logic.

Now, Prior introduced hybrid tense logic in connection with what he called four grades of tense-logical involvement. The four grades were presented in Prior (1968), chapter XI (also chapter XI in the new edition, Hasle et al. (2003)). Moreover, see Prior (1967), chapter V.6 and appendix B.3-4. For a more general discussion of the four grades, see the posthumously published book Fine and Prior (1977). The stages progress from pure first-order earlier-later logic to what can be regarded as a pure tense logic, where the second grade is a "neutral" logic encompassing first-order earlier-later logic and tense logic on the same footing. The motivation for Prior's four grades of tense-logical involvement was philosophical. Prior considered instants to be "artificial" entities which due to their abstractness should not be taken as primitive concepts.

> ...my desire to sweep 'instants' under the metaphysical table is not prompted by any worries about their punctual or dimensionless character but purely by their abstractness. ...'instants' as literal objects, or as cross-sections of a literal object, go along with a picture of 'time' as a literal object, a sort of snake which either eats its tail or doesn't, either has ends or doesn't, either is made of separate segments or isn't; and this picture I think we must drop. (Prior 1967, p. 189)

Given the explicit reference to instants in first-order earlier-later logic, Prior found that first-order earlier-later logic gives rise to undesired ontological import. Instead of first-order earlier-later logic, he preferred tense logic.

> Some of us at least would prefer to see 'instants', and the 'time-series' which they are supposed to constitute, as mere logical constructions out of tensed facts. (Hasle et al. 2003, p. 120)

This is why Prior's goal was to extend tense logic such that it could be considered as encompassing first-order earlier-later logic. Technically, the goal was to extend tense logic such that first-order earlier-later logic could be translated into it. It was with this goal in mind Prior introduced what he called *instant-propositions*.

> What I shall call the third grade of tense-logical involvement consists in treating the instant-variables a, b, c, etc. as also representing propositions. (Hasle et al. 2003, p. 124)

In the context of modal logic, Prior called such propositions *possible-world-propositions*. Of course, this is what we here call nominals. Prior also introduced the binder $\forall$ and what we here call satisfaction operators (he used the notation $T(a,\phi)$ instead of $@_a\phi$ for satisfaction operators). The extended tense-logic thus obtained is the logic he called third grade tense logic, hence, the third grade tense logic is identical to the tense-logical version of $\mathcal{H}(\forall)$, hybrid tense logic, which has the same expressive power as first-order earlier-later logic, as remarked above.

Prior gave an alternative, but equivalent, formulation of the third grade tense logic in which the satisfaction operator is replaced by a modal operator A called the *universal* modality (some authors call it the *global* modality). The universal modality has a fixed interpretation: The truth-condition is that a formula $A\phi$ is true (at any world) if and only if the formula ϕ is true at all worlds. Thus, the universal modality is interpreted using the universal binary relation. Formally, the clause for the satisfaction operator in the semantics is replaced by a clause for the modal operator A.

$$\mathfrak{M},g,w \models A\phi \text{ iff for any } v \in W,\, \mathfrak{M},g,v \models \phi.$$

Thus, besides the tense operators G and H, the language under consideration here also contains the modal operator A. The two formulations of the third degree are equivalent since the satisfaction operator and the universal modality are interdefinable in the presence of nominals and the $\forall$ binder, this being the case as the formulas $A\phi \leftrightarrow \forall a(@_a\phi)$ and $@_a\phi \leftrightarrow A(a \rightarrow \phi)$ are valid.

Prior's fourth grade tense logic is obtained from the third grade tense logic by replacing the satisfaction operator (or the universal modality in the alternative formulation of the third grade) by a defined modal operator L such that

$$\mathfrak{M},g,w \models L\phi \text{ iff for any } v \in W \text{ such that } wR^*v,\, \mathfrak{M},g,v \models \phi$$

where the binary relation R^* is the reflective, symmetric, and transitive closure of the earlier-later relation R. Prior considered two ways to define the operator L in what he took to be purely tense-logical terms. In the first case he allowed what amounts to infinite conjunctions of formulas. If infinite conjunctions are allowed, the operator L can be defined by the conventions that

$$L\phi = L^0\phi \wedge L^1\phi \wedge \ldots$$

and

$$\begin{aligned} L^0\phi &= \phi \\ L^{n+1}\phi &= GL^n\phi \wedge HL^n\phi \end{aligned}$$

Note that for any given natural number k, $L^k\phi$ is a formula in the object language (which does not involve natural numbers). For example, if $k = 1$ and $\phi = p$, then $L^1\phi = Gp \wedge Hp$. In the second case Prior assumed time to have a structure making $L\phi$ equivalent to

$$L^0\phi \wedge L^1\phi \wedge \ldots \wedge L^k\phi$$

for some fixed natural number k whereby infinite conjunctions are avoided. If for example time is linear, that is, transitive, backwards linear, and forwards linear, then $k = 1$ will do. If time is branching, that is, transitive and backwards linear, then $k = 2$ will do. In whichever way the operator L is defined, the fourth grade tense logic has the same expressive power as first-order earlier-later logic if it is assumed that the time-series is unique, that is, if it is assumed that any two instants are connected by some number of steps in either direction along the earlier-later relation R. For Prior it was natural to assume that the time-series is unique, as is witnessed by the following quotation.

> For is not the question as to whether 'our' time-series (whatever its structure) is unique, a genuine one? I would urge the following consideration against saying that it is, or at all events against saying it too hurriedly: It is only if we have a more-or-less 'Platonistic' conception of what a time-series is, that we can raise this question. If, as I would contend, it is only by tensed statements that we can give the cash-value of assertions which purport to be about 'time', the question as to whether there are or could be unconnected time-series is a senseless one. We think we can give it a sense because it is as easy to draw unconnected lines and networks as it is to draw connected ones; but these diagrams cannot represent *time*, as they cannot be translated into the basic non-figurative temporal language. (Prior 1967, pp. 198–199)

The reason why the fourth grade tense logic has first-order expressive power when the time-series is unique, is that the fourth-grade modality L then has the same effect as the universal modality A which is used in (the alternative formulation of) the third-grade logic, and the third-grade logic has first-order expressive power, as we argued above. This is discussed in more detail in Braüner (2002) by the present author.

To sum up, Prior obtained tense logics having the same expressive power as first-order earlier-later logic, namely the third and fourth grade tense logics, by adding to ordinary tense logic further expressive power in the form of hybrid-logical machinery (and in the case of the fourth grade tense logic by making appropriate assumptions about the structure of time, including an assumption that the time-series is unique). So Prior clearly reached his technical goal. Prior also found that he reached his philosophical goal, namely that of avoiding an ontology including instants.

> The 'entities' which we 'countenance' in our 'ontology' ... depend on what variables we take seriously as individual variables in a first-order theory, i.e. as subjects of predicates rather than as *assertibilia* which may be qualified by modalities. If we prefer to handle instant-variables, for example, or person-variables, as subjects of predicates, then we may be taken to believe in the existence of instants, or of persons. If, on the other hand, we prefer to treat either of these as *propositional* variables, i.e. as arguments of truth-functions and of modal functions, then we may be taken as *not* believing in the existence of instants, etc. (they don't exist; rather, they are or are not the case). (Hasle et al. 2003, p. 220)

However, it has been debated whether or not Prior managed to avoid an instant ontology. We shall return to this later in Section 1.3.1 (where we also return to the person-variables mentioned in the quotation above).

The discussion on Prior's third grade tense logic and first-order earlier-later logic is closely related to the discussion on two different conceptions of time, namely the *A-series* and *B-series* conceptions, a terminology introduced in 1908 by the philosopher McTaggart. According to the A-series conception, also called the *dynamic* view, the past, present, and future tenses are primitive concepts from which other temporal concepts, in particular instants and the earlier-later relation, are to be derived. On the other hand, according to the B-series conception, also called the *static* view, instants and the earlier-later relation are primitive. The A-series conception embodies the local way in which human beings experience the flow of time whereas the B-series conception embodies a Gods-eye-view of time, where time is a sequence of objectively and tenselessly existing instants. It is notable that representations of both the A-series and B-series conceptions can be found in natural language (the A-series conception in the form of tense inflection of verbs and the B-series conception in particular in the form of nominal constructions like "five o'clock May 10th 2007"). Of course, first-order earlier-later logic is associated with the B-series conception and Prior's third grade tense logic is associated with the A-series conception, which was Prior's own view, as succintly expressed in the following quotation.

> So far, then, as I have anything that you could call a philosophical creed, its first article is this: I believe in the reality of the distinction between past, present, and future. I believe that what we see as a progress of events *is* a progress of events, a *coming to pass* of one thing after another, and not just a timeless tapestry with everything stuck there for good and all. (Prior 1996, p. 47)

The discussion of A-series and B-series is reflected in discussions of time in Artificial Intelligence, see Galton (2006). The paper Blackburn (2006) discusses all the above issues as well as a number of other issues in hybrid logic and their origin in Prior's work. The above issues are also discussed in many papers of the collection Copeland (1996), in particular in Richard Sylvan (1996). See Øhrstrøm and Hasle (1993), the book Øhrstrøm and Hasle (1995), and the handbook chapters Øhrstrøm and Hasle (2005a,b) for general accounts of Prior's work. See also the encyclopedia article Copeland (2007). A very recent assessment of Prior's philosophical and logical views can be found in Müller (2007).

1.3.1 Did Prior Reach His Philosophical Goal?

It has been debated whether Prior reached his philosophical goal with the third and fourth grade logics, namely that of avoiding an ontology including instants.

According to one criticism, the ontological import of the third and fourth grade logics is the same as the ontological import of first-order earlier-later logic since the third and fourth grade logics involve what are considered direct analogies to first-order primitives, in particular, nominals are considered a direct analogy to first-order variables and the $\forall$ binder is considered a direct analogy to the first-order

$\forall$ quantifier.[6] Such a criticism can be found in Sylvan (1996).[7] Note that this is a philosophical, not a technical, discussion. The technical, to be precise, mathematical, result that first-order earlier-later logic and the third grade logic (as well as the fourth grade logic in the light of appropriate assumptions on the structure of time) have the same expressive power, in the sense that there are truth-preserving translations in both directions between the logics, does not itself give an answer to the philosophical question as to whether the logics have the same ontological import.

Clearly, Prior's view on logic differs in a number of ways from the views held by most contemporary logicians, in particular logicians inclined towards model-theory. A criticism from the perspective of contemporary model-theory has been raised by Blackburn (2006).

> If the fundamental unit of logical modeling is a formal language *together with* a set-theoretical interpretation, then it makes little sense to claim, for example, that first-order logic automatically brings greater ontological commitment than (say) propositional modal logic. Under the model-theoretic conception, both make use of the same set-theoretic structures, so their ontological commitments are at least prima facie identical. Perhaps arguments could be mounted (based, perhaps, on the fact that modal logic is decidable and has the finite model property) that modal logic commits us to less. But such arguments would have to be carefully constructed. In the light of modern correspondence theory, simple knockdown arguments based on the presence or absence of explicit quantifiers in the object language are unconvincing. (Blackburn 2006, pp. 358–359)

Another criticism raised by Blackburn (2006) has to do with a logic Prior called *egocentric logic*. We now briefly describe this logic. Egocentric logic is technically the same as the third grade tense logic, but the points in the Kripke semantics are now taken to represent persons, not instants, and the accessibility relation relates two

[6] It appears that this criticism presupposes a view on nominals according to which a nominal is a symbol that refers to something, like a first-order variable does. As remarked in Footnote 2 in Section 1.1, there is an alternative view on nominals according to which a nominal is viewed as a description of the content of an instant. It is not clear whether the criticism applies if this alternative view on nominals is adopted.

[7] Sylvan actually argues that it is not necessary to reduce first-order earlier-later logic (the B-series conception of time) to tense logic (the A-series conception), or vice versa. Sylvan points out that Prior regarded tense-logical postulates as being capable of giving the meaning of statements like 'time is continuous' and 'time is infinite both ways', cf. Prior (1967, p. 74). To this Sylvan responds as follows.

> Time is an item, a theoretical object, which bears both the tensed and the temporally ordered properties which the item in question genuinely has. ...
>
> Part of the elegance of such a simple characterization of Time is that it neatly decouples the stable sense of 'time' ... from various vexed issues as to exactly which properties the item genuinely has (and so from what Time is 'really' like). Whichever it should have, under evolving or under alternative theories, the item can remain abstractly one and the same. Naturally, tight coupling remains between the item and its properties; but it is not a meaning connection, it is a theory-dependent linkage. (Sylvan 1996, p. 114)

Sylvan sees Prior's goal to reduce the B-series talk to A-series talk as part of a more general, and in Sylvan's view overdeveloped, reductionist inclination of analytic philosophers, which also encompasses philosophers having the converse reduction as a goal, that is, having the goal of reducing A-series to B-series, cf. Sylvan (1996, p. 112).

persons if and only if the second person is taller than the first one. Of course, just as the third grade tense logic is a modal-logical counterpart to first-order earlier-later logic, egocentric logic is a modal-logical counterpart to a first-order logic which technically is the same as first-order earlier-later logic, but where the points in a model are taken to represent persons. What was observed by Prior is that egocentric logic is just an instance of a general relationship: Any first-order logic has a modal-logical counterpart, whatever signature the first-order logic has and whatever the points in a model are taken to represent. Thus, in this sense there is nothing special about tense logic. But Prior considered tense logic to have a priviledged status that distinguishes it from other logics, in particular egocentric logic.

> Tense logic is for me, if I may use the phrase, *metaphysically fundamental*, and not just an artificially torn-off fragment of the first-order theory of the earlier-later relation. Egocentric logic is a different matter; I find it hard to believe that individuals really are just propositions of a certain sort, or just 'points of view', or that the real world of individuals is just a logical construction out of such points of view. (Hasle et al. 2003, p. 232)

Thus, as the quotation indicates, Prior considered tense logic to have a special philosophical status, but in the sense described above, there is nothing special about tense logic. This calls for an explanation. Prior concluded the following.

> So far as I can see, there is nothing philosophically disreputable in saying that (i) persons just *are* genuine individuals, so that their figuring as individual variables in a first-order theory needs no explaining (*this* first-order theory being, on the contrary, the only way of giving sense to its 'modal' counterpart), whereas (ii) instants are *not* genuine individuals, so that *their* figuring as values of individual variables *does* need explaining, and it is the related 'modal' logic (tense logic) which gives the first-order theory what sense it has. (Hasle et al. 2003, pp. 219–220)

However, Prior's conclusion is critized in Blackburn's paper for being unsatisfactorily justified, which is in line with the other criticism expressed in the above quotation from Blackburn's paper.

1.4 The Development Since Prior

Below we outline the development of hybrid logic since Prior. We shall present a selection of works rather than trying to be encyclopaedic. See the handbook chapter Areces and ten Cate (2007) for a detailed overview.

The first completely rigorous definition of hybrid logic was given by Robert Bull which in 1970 appeared in a special issue of the journal *Theoria* in memory of Prior. Bull introduces a third sort of propositional symbols where a propositional symbol is assumed to be true exactly at one branch ("course of events") in a branching time model. This idea of sorting propositional symbols according to restrictions on their interpretations has later been developed further by a number of authors, see section 5 of Blackburn and Tzakova (1999) as well as Section 9 of Blackburn (2000a) for discussions. The idea of sorting is also discussed in the unpublished manuscript Goranko (2000).

The hybrid logical machinery originally invented by Prior in the late 1960s was reinvented in the 1980s by Solomon Passy and Tinko Tinchev from Bulgaria, see Passy and Tinchev (1985, 1991). Rather than ordinary modal logic, this work took place in connection with the much more expressive Propositional Dynamic Logic.

A major contribution in the 1990s was the introduction of the $\downarrow$ binder by Valentin Goranko, see Goranko (1994, 1996).[8] Since then, hybrid logic with the $\downarrow$ binder has been extensively studied by a number of people, notably Patrick Blackburn and his co-authors, for example, in Blackburn and Seligman (1995) it is shown that this logic does not have the finite model property and that the logic is undecidable.[9] Also, various expressivity results are given in Blackburn and Seligman (1995). See Areces et al. (2001) for a number of model-theoretic aspects. A very comprehensive study of the model-theory of hybrid logic is the PhD thesis of ten Cate (2004).

Also the weaker hybrid logic obtained by omitting both of the binders $\downarrow$ and $\forall$ has been the subject of extensive exploration. An early work on the binder-free hybrid logic (but including the very expressive universal modality) is the paper by Gargov and Goranko (1993). It turns out that the binder-free logic and a number of variants of it are decidable. In Areces et al. (1999), a number of complexity results are given for hybrid modal and tense logics over various classes of frames, for example arbitrary, transitive, linear, and branching. It is remarkable that the satisfiability problem of the binder-free hybrid logic over arbitrary frames is solvable in polynomial space (PSPACE), which is the same as the complexity of satisfiability in ordinary modal logic. Thus, hybridizing ordinary modal logic gives more expressive power, but the complexity stays the same.

It is remarkable that first-order hybrid logic offers precisely the features needed to prove interpolation theorems.[10] While interpolation fails in a number of well-known first-order modal logics, their hybridized counterparts have this property, see Areces et al. (2003) as well as Blackburn and Marx (2003). The first paper gives a model-theoretic proof of interpolation whereas the second paper gives an algorithm for calculating interpolants based on a tableau system.[11] In the first paper interpolation is proved to hold for any bounded fragment definable class of skeletons, with either varying, increasing, decreasing, or constant domains (see Definition 6.1 in Section 6.1).

---

[8] A variation of the $\downarrow$ binder (called the "freeze" quantifier) was actually introduced already in 1989 in connection with real-time logics, see Alur and Henzinger (1989). See also the survey Alur and Henzinger (1992). The $\downarrow$ binder and the freeze quantifier were discovered independently of each other.

[9] To prove these results, Blackburn and Seligman (1995) introduce a proof technique called the spy-point technique, which later has been used in many other connections.

[10] The interpolation theorem for propositional logic says that for any valid formula $\phi \rightarrow \psi$ there exists a formula θ containing only the common propositional symbols of ϕ and ψ such that the formulas $\phi \rightarrow \theta$ and $\theta \rightarrow \psi$ are valid. Interpolation theorems for other logics are formulated in an analogous fashion.

[11] An unexplored line of work is to find out whether interpolation can be proved in other ways, for example using proof systems like the linear reasoning systems which in Fitting (1984) are used to prove interpolation for some particular propositional and first-order modal logics.

A number of papers have dealt with axioms for hybrid logic, for example Blackburn (1993), Blackburn and Tzakova (1999), and Blackburn and ten Cate (2006). Blackburn and Tzakova (1999) gives an axiom system for hybrid logic and shows the remarkable result that if the axiom system is extended with a set of additional axioms which are *pure* formulas (that is, formulas where all propositional symbols are nominals), then the extended axiom system is complete with respect to the class of frames validating the axioms in question.[12] Pure formulas correspond to first-order conditions on the accessibility relation (cf. the translation ST_a in Section 1.2.1), so axiom systems for new hybrid logics with first-order conditions on the accessibility relation can be obtained in a uniform way simply by adding axioms as appropriate. So, if for example the formula $\downarrow c\Box\neg c$ is added as an axiom, then the resulting system is complete with respect to irreflexive frames, cf. Section 1.2. Blackburn and ten Cate (2006) investigate orthodox proof-rules (which are proof-rules without side-conditions) in axiom systems, and it is shown that if one requires extended completeness using pure formulas, then unorthodox proof-rules are indispensable in axiom systems for binder-free hybrid logic. However, an axiom system can be given only involving orthodox proof-rules for the stronger hybrid logic including the $\downarrow$ binder. Another axiom system for hybrid logic is given in Braüner (2006) by the present author (see Section 2.5 of this book) and an axiom system for first-order hybrid logic is given in Braüner (2005b) also by the present author (see Section 6.3). In Gabbay and Malod (2002) an axiom system is given for a logic similar to hybrid logic, obtained by extending ordinary modal logic with first-order machinery for naming worlds.

Besides giving an axiom system for standard classical hybrid logic, Braüner (2006) also gives axiom systems for intuitionistic and paraconsistent hybrid logic (see Sections 8.3 and 8.4). The paper ten Cate and Litak (2007) gives hybrid-logical axiom systems that are sound and complete with respect to topological semantics, that is, generalisations of the standard Kripke semantics where the modal operator is interpreted in terms of a topology on the set of possible worlds. Strictly speaking, the topological semantics only generalize Kripke semantics where the frames are reflexive and transitive, but ten Cate and Litak (2007) also consider an even more general kind of semantics called neighbourhood semantics which generalizes all Kripke semantics. Topological semantics is interesting for a number of reasons, one being that it is applicable for spatial reasoning (topological spaces are abstractions from metric spaces which in turn are abstractions from Euclidean space). A major reason for the interest in neighbourhood semantics is that it does not validate the formula $\Box(\phi \rightarrow \psi) \rightarrow (\Box\phi \rightarrow \Box\psi)$ and nor does it validate the standard modal-logical rule called necessitation, that is, from ϕ derive $\Box\phi$ (the Kripke semantics validates both, but for some applications this is undesirable, for example, if the modal operator represents an agent's knowledge, then these two validities together imply that the agent is logical omniscient, that is, the agent's knowledge is closed under logical consequence, which at least for human agents is implausible). Further work on topological semantics for hybrid logic can be found in Sustretov (2009).

[12] See ten Cate (2004) for semantic characterizations of frame classes definable by pure hybrid-logical formulas.

Work in resolution calculi and model-checking for hybrid logic is in the early phases, see Areces et al. (2001) and Areces and Heguiabehere (2002) for resolution calculi and see Franceschet and de Rijke (2006) and Lange (2009) for results on model-checking.

Tableau, Gentzen, and natural deduction style proof-theory for hybrid logic work very well compared to ordinary modal logic. Usually, when a modal tableau, Gentzen, or natural deduction system is given, it is for one particular modal logic and it has turned out to be problematic to formulate such systems for modal logics in a uniform way without introducing metalinguistic machinery. This can be remedied by hybridization, that is, hybridization of modal logics enables the formulation of uniform tableau, Gentzen, and natural deduction systems for wide classes of logics. Blackburn (2000a) introduces a tableau system for hybrid logic that has this desirable feature: Analogous to the axiom systems of Blackburn and Tzakova (1999) and Blackburn and ten Cate (2006), completeness is preserved if the tableau system is extended with a set of pure axioms, that is, a set of pure formulas that are allowed to be added to a tableau during the tableau construction. See Hansen (2007) for another tableau system for hybrid logic.

The tableau system of Blackburn (2000a) is the basis for a decision procedure for the binder-free fragment of hybrid logic given in Bolander and Braüner (2006) by the present author together with Thomas Bolander (see Chapter 3). The tableau-based decision procedures of Bolander and Braüner (2006) have been further developed in Bolander and Blackburn (2007, 2009). Cerrito (2010) present another tableau-based decision procedure for hybrid logic. Other decision procedures for hybrid logics, which also are based on proof-theory, are given in Kaminski and Smolka (2007, 2009). The procedures of these two papers are based on the higher-order formulation of hybrid logic (involving the simply typed λ-calculus) given in Hardt and Smolka (2006).

Jens Ulrik Hansen, Thomas Bolander, and the present author (2008) give a tableau-based decision procedure for many-valued hybrid logic, that is, hybrid logic where the two-valued classical logic basis has been generalized to a many-valued logic basis involving a truth-value space having the structure of a finite Heyting algebra (this many-valued hybrid logic can also be seen as a hybridized version of the many-valued modal logic given in Fitting (1992a,b, 1995)). Hansen (2010) gives a tableau-based decision procedure for a hybridized version of a dynamic epistemic logic called public annoucement logic.

Natural deduction style proof-theory of propositional and first-order hybrid logic has been explored in Braüner (2004a, 2005b) by the present author (see Sections 2.2 and 6.2). Braüner (2004a) also gives a Gentzen system for hybrid logic (see Section 2.4). These natural deduction and Gentzen systems can be extended with additional proof-rules corresponding to first-order conditions on the accessibility relations expressed by geometric theories; this is analogous to extending tableau and axiom systems with pure axioms.[13] The present author togetherwith Valeria de

---

[13] Like frame classes definable by pure formulas, frame classes definable by geometric theories can be given a semantic characterization, see the remark at the end of Section 2.2.1.

Paiva (2006) gives a natural deduction system for intuitionistic hybrid logic (see Section 8.2). In the context of situation theory, Gentzen and natural deduction systems for logics similar to hybrid logics were explored in the early 1990s by Jerry Seligman, see the overview in Seligman (2001). In Braüner (2004b), a natural deduction system given in Seligman (1997) is compared to the system of Braüner (2004a) (see Chapter 4).

The fact that hybridization of modal logics enables the formulation of uniform tableau, Gentzen, and natural deduction systems for wide classes of logics is discussed in detail in Braüner (2007) by the present author (see Chapters 9 and 10).

The development of hybrid logic is only outlined above, in particular, we have only outlined hybrid-logical proof-theory. The proof-theory of hybrid logic will be the main issue in what follows.

Chapter 2
Proof-Theory of Propositional Hybrid Logic

In this chapter we introduce the proof-theory of propositional hybrid logic. The chapter is structured as follows. In the first section of the chapter we sketch the basics of natural deduction systems and in the second section we introduce a natural deduction system for hybrid logic. In the third section we sketch the basics of Gentzen systems and in the fourth section we introduce a Gentzen system corresponding to the natural deduction system for hybrid logic. In the fifth section we give an axiom system for hybrid logic. The natural deduction system and the Gentzen system are taken from Braüner (2004a) whereas the axiom system is taken from Braüner (2006).

2.1 The Basics of Natural Deduction Systems

Before giving our hybrid-logical natural deduction system, we shall sketch the basics of natural deduction and fix terminology for later use.

Natural deduction style derivation rules for ordinary classical first-order logic were originally introduced by Gerhard Gentzen (1969) and later on developed much further by Dag Prawitz (1965, 1971). See Troelstra and Schwichtenberg (1996) for a general introduction to natural deduction systems. With reference to Gentzen's work, Prawitz made the following remarks on the significance of natural deduction.

> ...the essential logical content of intuitive logical operations that can be formulated in the languages considered can be understood as composed of the atomic inferences isolated by Gentzen. It is in this sense that we may understand the terminology *natural* deduction.
>
> Nevertheless, Gentzen's systems are also natural in the more superficial sense of corresponding rather well to informal practices; in other words, the structure of informal proofs are often preserved rather well when formalized within the systems of natural deduction. (Prawitz 1971, p. 245)

The method of reasoning in natural deduction systems is called "forwards" reasoning: When you want to find a derivation of a certain formula you start with the rules and try to build a derivation of the formula you have in mind. This is contrary to

T. Braüner, *Hybrid Logic and its Proof-Theory*, Applied Logic Series 37,
DOI 10.1007/978-94-007-0002-4_2, © Springer Science+Business Media B.V. 2011

tableau systems which are backward reasoning systems since you explicitly start with a particular formula and try to build a proof of it using tableau rules, cf. Section 3.1.

A *derivation* in a natural deduction system has the form of a finite tree where the nodes are labelled with formulas such that for any formula occurrence ϕ in the derivation, either ϕ is a leaf of the derivation or the immediate successors of ϕ in the derivation are the premises of a rule-instance which has ϕ as the conclusion. In what follows, the metavariables π, τ, ... range over derivations. A formula occurrence that is a leaf but is not the conclusion of a rule-instance with zero premises, is called an *assumption* of the derivation. The root of a derivation is called the *end-formula* of the derivation. All assumptions are annotated with numbers. An assumption is either *undischarged* or *discharged*. If an assumption is discharged, then it is discharged at one particular rule-instance and this is indicated by annotating the assumption and the rule-instance with identical numbers. We shall often omit this information when no confusion can occur. A rule-instance annotated with some number discharges all undischarged assumptions that are above it and are annotated with the number in question, and moreover, are occurrences of a formula determined by the rule-instance.

Two assumptions in a derivation belong to the same *parcel* if they are annotated with the same number and are occurrences of the same formula, and moreover, either are both undischarged or have both been discharged at the same rule-instance. Thus, in this terminology rules discharge parcels.[1] We shall make use of the standard notations

$$\begin{array}{c} [\phi^r] \\ \vdots\, \pi \\ \psi \end{array} \qquad\qquad \begin{array}{c} (\phi^r) \\ \vdots\, \pi \\ \psi \end{array} \qquad\qquad \begin{array}{c} \vdots\, \tau \\ \phi \\ \vdots\, \pi \\ \psi \end{array}$$

which from left to right mean (i) a derivation π where ψ is the end-formula and $[\phi^r]$ is the parcel consisting of all undischarged assumptions that have the form ϕ^r; (ii) a derivation π where ψ is the end-formula and (ϕ^r) is a single undischarged assumption of the form ϕ^r; and (iii) a derivation π where ψ is the end-formula and a derivation τ with end-formula ϕ has been substituted for all the undischarged assumptions indicated by either $[\phi^r]$ or (ϕ^r). A derivation in a natural deduction system is generated by a set of derivation rules from derivations consisting of a single undischarged assumption.

We shall make use of the following conventions. The metavariables Γ, Δ, ... range over sets of formulas. A derivation π is called a *derivation of* ϕ if the end-formula of π is an occurrence of ϕ, and moreover, π is called a *derivation from* Γ if each undischarged assumption in π is an occurrence of a formula in Γ (note

[1] Instead of annotating assumptions and rule-instances with numbers, the discharging of parcels could have been recorded by placing a list of the undischarged parcels at every stage in the derivation. That is, instead of a formula ϕ in a derivation, we have a sequent $\psi_1, \ldots, \psi_m \vdash \phi$ where the formulas in the list $\psi_1, \ldots, \psi_m$ correspond to the undischarged parcels at the stage in question. Such sequents should not be confused with sequents in a Gentzen system.

that numbers annotating undischarged assumptions are ignored). If there exists a derivation of ϕ from the empty set $\emptyset$, then we shall simply say that ϕ is *derivable*.

A characteristic feature of natural deduction is that there are two different kinds of rules for each connective; there are rules called introduction rules which introduce a connective (that is, the connective occurs in the conclusion of the rule, but not in the premises) and there are rules called elimination rules which eliminate a connective (the connective occurs in a premiss of the rule, but not in the conclusion). Introduction rules traditionally have names in the form $(\ldots I \ldots)$, and similarly, elimination rules traditionally have names in the form $(\ldots E \ldots)$. For an instructive and important example, see the standard natural deduction rules for propositional logic in Figure 2.1. Note that the rule $(\bot 1)$ is neither an introduction rule nor an elimination rule (recall that $\neg\phi$ is an abbreviation for $\phi \to \bot$). Below is a sample derivation to illustrate how natural deduction derivations are presented.

$$\dfrac{\dfrac{\dfrac{p \wedge (p \to q)^1}{p \to q}\,(\wedge E2) \qquad \dfrac{p \wedge (p \to q)^1}{p}\,(\wedge E1)}{q}\,(\to E)}{(p \wedge (p \to q)) \to q}\,(\to I)^1$$

The derivation above is a derivation of $(p \wedge (p \to q)) \to q$ from $\emptyset$.

$$\dfrac{\phi \quad \psi}{\phi \wedge \psi}\,(\wedge I) \qquad \dfrac{\phi \wedge \psi}{\phi}\,(\wedge E1) \qquad \dfrac{\phi \wedge \psi}{\psi}\,(\wedge E2)$$

$$\dfrac{\begin{matrix}[\phi] \\ \vdots \\ \psi\end{matrix}}{\phi \to \psi}\,(\to I) \qquad \dfrac{\phi \to \psi \quad \phi}{\psi}\,(\to E)$$

$$\dfrac{\begin{matrix}[\neg\phi] \\ \vdots \\ \bot\end{matrix}}{\phi}\,(\bot 1)^*$$

$*$ ϕ is a propositional symbol.

Fig. 2.1 Natural deduction rules for propositional logic

The introduction and elimination rules for a connective are expected to satisfy Dag Prawitz' *inversion principle*.

> ... a proof of the conclusion of an elimination is already "contained" in the proofs of the premises when the major premise is inferred by introduction. (Prawitz 1971, pp. 246–247)

(The major premise of an elimination rule is the premise that exhibits the connective being eliminated.) The history of the inversion principle goes back to Prawitz (1965). In the above formulation of the inversion principle it is not made explicit what is meant by requiring that some derivations (called proofs by Prawitz) "contain" a derivation of a certain formula, but it means that a derivation of the formula in question can be obtained by composition of derivations, that is, by substitution of derivations for undischarged assumptions. In the case of first-order logic, not only substitution of derivations for undischarged assumptions is allowed, but also substitution of variables for variables in derivations.

The standard rules for the connective $\wedge$ in Figure 2.1 are an instructive example of introduction and elimination rules that satisfy the inversion principle: If the premiss of an instance of $(\wedge E1)$ is the conclusion of an introduction rule, necessarily the rule $(\wedge I)$, then the conclusion of $(\wedge E1)$ already occurs as premiss of the instance of $(\wedge I)$ (of course, the case of $(\wedge E2)$ is analogous). Note that composition of derivations is not needed here.

The inversion principle refers to a particular kind of formula occurrence in a derivation, namely a formula occurrence which is both introduced by an introduction rule and eliminated by an elimination rule. Such a formula occurrence is called a maximum formula. According to the inversion principle, a maximum formula can be considered a "detour" in the derivation, and it follows from the principle that the maximum formula can be removed by rewriting the derivation. This rewrite process is formalized in a kind of rewrite rules that are called proper reduction rules. For example, the introduction and elimination rules for the connectives $\wedge$ and $\rightarrow$ in Figure 2.1 give rise to the following proper reduction rules.

$(\wedge I)$ followed by $(\wedge E1)$ (analogously in the case of $(\wedge E2)$)

$$\dfrac{\dfrac{\begin{array}{c}\vdots\,\pi_1\\ \phi\end{array}\qquad\begin{array}{c}\vdots\,\pi_2\\ \psi\end{array}}{\phi\wedge\psi}}{\phi}\quad\rightsquigarrow\quad\begin{array}{c}\vdots\,\pi_1\\ \phi\end{array}$$

$(\rightarrow I)$ followed by $(\rightarrow E)$

$$\dfrac{\dfrac{\begin{array}{c}[\phi]\\ \vdots\,\pi_1\\ \psi\end{array}}{\phi\rightarrow\psi}\qquad\begin{array}{c}\vdots\,\pi_2\\ \phi\end{array}}{\psi}\quad\rightsquigarrow\quad\begin{array}{c}\vdots\,\pi_2\\ \phi\\ \vdots\,\pi_1\\ \psi\end{array}$$

Some natural deduction systems also involve other kinds of reduction rules than proper reduction rules. A derivation is called normal if no reduction rules can be applied to it, and for most natural deduction systems a normalization theorem can be proved which says that any derivation can be rewritten to such a normal derivation by repeated applications of reductions. In most natural deduction systems, normal derivations satisfy the subformula property which says that any formula in a derivation, save possibly some exceptions, is a subformula of the end-formula or one of the undischarged assumptions.

As is clear form the above, the inversion principle can be seen as a prerequisite for formulating a normalization theorem since such a theorem is relative to a set of reduction rules. See in particular Chapter 10 for further discussion of the inversion principle. The inversion principle is also discussed in Section 5.3.

2.2 Natural Deduction for Propositional Hybrid Logic

In this section we shall give a sound and complete natural deduction system for the hybrid logic $\mathscr{H}(\mathscr{O})$, where $\mathscr{O}$ is any subset of the set of binders $\{\forall, \downarrow\}$. Moreover, we shall show how to extend the natural deduction system with additional derivation rules corresponding to first-order conditions on the accessibility relation. The conditions we consider are expressed by geometric theories. Different geometric theories give rise to different hybrid logics, so natural deduction systems for new hybrid logics can be obtained in a uniform way simply by adding derivation rules as appropriate. Furthermore, we prove a normalization theorem which says that any derivation can be rewritten to a normal derivation by repeated applications of reduction steps. Normal derivations satisfy a version of the subformula property called the quasi-subformula property.

The derivation rules for the natural deduction system are given in Figures 2.2 and 2.3. All formulas in the rules are satisfaction statements (recall that a satisfaction statement is a formula of the form $@_a\phi$). Our natural deduction system for $\mathscr{H}(\mathscr{O})$ is obtained from the rules given in Figures 2.2 and 2.3 by leaving out the rules for the binders that are not in the set $\mathscr{O}$. The system thus obtained will be denoted $\mathbf{N}_{\mathscr{H}(\mathscr{O})}$. So, for example, the system $\mathbf{N}_{\mathscr{H}(\forall)}$ is obtained by leaving out the rules $(\downarrow I)$ and $(\downarrow E)$. Note that the rules $(\bot 1)$ and $(\bot 2)$ are neither introduction rules nor elimination rules. We shall discuss the side-condition on the rules $(\bot 1)$ and $(Nom1)$ in Section 2.2.2.

Below we give some examples of derivations in the system $\mathbf{N}_{\mathscr{H}}$. The end-formula of the first example derivation is the standard modal axiom K prefixed by a satisfaction operator (it is assumed that the nominal a in the derivation is new).

$$\dfrac{\dfrac{\dfrac{\dfrac{\dfrac{@_b\Box(\phi \rightarrow \psi)^3 \quad @_b\Diamond a^1}{@_a(\phi \rightarrow \psi)}(\Box E) \qquad \dfrac{@_b\Box\phi^2 \quad @_b\Diamond a^1}{@_a\phi}(\Box E)}{@_a\psi}(\rightarrow E)}{@_b\Box\psi}(\Box I)^1}{@_b(\Box\phi \rightarrow \Box\psi)}(\rightarrow I)^2}{@_b(\Box(\phi \rightarrow \psi) \rightarrow (\Box\phi \rightarrow \Box\psi))}(\rightarrow I)^3$$

$$\dfrac{@_a\phi \qquad @_a\psi}{@_a(\phi\wedge\psi)}\;(\wedge I) \qquad\qquad \dfrac{@_a(\phi\wedge\psi)}{@_a\phi}\;(\wedge E1) \qquad \dfrac{@_a(\phi\wedge\psi)}{@_a\psi}\;(\wedge E2)$$

$$\dfrac{\begin{array}{c}[@_a\phi]\\ \vdots\\ @_a\psi\end{array}}{@_a(\phi\rightarrow\psi)}\;(\rightarrow I) \qquad\qquad \dfrac{@_a(\phi\rightarrow\psi) \qquad @_a\phi}{@_a\psi}\;(\rightarrow E)$$

$$\dfrac{\begin{array}{c}[@_a\neg\phi]\\ \vdots\\ @_a\bot\end{array}}{@_a\phi}\;(\bot 1)^{*} \qquad\qquad \dfrac{@_a\bot}{@_c\bot}\;(\bot 2)$$

$$\dfrac{@_a\phi}{@_c@_a\phi}\;(@I) \qquad\qquad \dfrac{@_c@_a\phi}{@_a\phi}\;(@E)$$

$$\dfrac{\begin{array}{c}[@_a\Diamond c]\\ \vdots\\ @_c\phi\end{array}}{@_a\Box\phi}\;(\Box I)^{\star} \qquad\qquad \dfrac{@_a\Box\phi \qquad @_a\Diamond e}{@_e\phi}\;(\Box E)$$

$$\dfrac{@_a\phi[c/b]}{@_a\forall b\phi}\;(\forall I)^{\dagger} \qquad\qquad \dfrac{@_a\forall b\phi}{@_a\phi[e/b]}\;(\forall E)$$

$$\dfrac{\begin{array}{c}[@_a c]\\ \vdots\\ @_c\phi[c/b]\end{array}}{@_a\downarrow b\phi}\;(\downarrow I)^{\ddagger} \qquad\qquad \dfrac{@_a\downarrow b\phi \qquad @_a e}{@_e\phi[e/b]}\;(\downarrow E)$$

$*$ ϕ is a propositional symbol (ordinary or a nominal).
$\star$ c does not occur free in $@_a\Box\phi$ or in any undischarged assumptions other than the specified occurrences of $@_a\Diamond c$.
$\dagger$ c does not occur free in $@_a\forall b\phi$ or in any undischarged assumptions.
$\ddagger$ c does not occur free in $@_a\downarrow b\phi$ or in any undischarged assumptions other than the specified occurrences of $@_a c$.

Fig. 2.2 Natural deduction rules for connectives

$$\dfrac{}{@_a a}\;(Ref) \qquad\qquad \dfrac{@_a c \qquad @_a\phi}{@_c\phi}\;(Nom1)^{*} \qquad\qquad \dfrac{@_a c \qquad @_a\Diamond b}{@_c\Diamond b}\;(Nom2)$$

$*$ ϕ is a propositional symbol (ordinary or a nominal).

Fig. 2.3 Natural deduction rules for nominals

The end-formulas of the following two example derivations are hybrid-logical axioms from Areces et al. (2001) prefixed by satisfaction operators. The axiom below is the modal axiom K for satisfaction operators.

$$\dfrac{\dfrac{\dfrac{\dfrac{\dfrac{@_b @_a(\phi \to \psi)^2}{@_a(\phi \to \psi)}\,(@E) \qquad \dfrac{@_b @_a \phi^1}{@_a \phi}\,(@E)}{@_a \psi}\,(\to E)}{@_b @_a \psi}\,(@I)}{@_b(@_a \phi \to @_a \psi)}\,(\to I)^1}{@_b(@_a(\phi \to \psi) \to (@_a \phi \to @_a \psi))}\,(\to I)^2$$

Compare the example derivation above with the Gentzen derivation of the same formula given in Section 2.4. The axiom below is called the *Bridge* axiom.

$$\dfrac{\dfrac{\dfrac{\dfrac{\dfrac{@_a \Box \neg \phi^1 \qquad \dfrac{@_a(\Diamond c \wedge @_c \phi)^2}{@_a \Diamond c}\,(\wedge E1)}{@_c \neg \phi}\,(\Box E) \qquad \dfrac{\dfrac{@_a(\Diamond c \wedge @_c \phi)^2}{@_a @_c \phi}\,(\wedge E2)}{@_c \phi}\,(@E)}{@_c \bot}\,(\to E)}{@_a \bot}\,(\bot 2)}{@_a \Diamond \phi}\,(\to I)^1}{@_a((\Diamond c \wedge @_c \phi) \to \Diamond \phi)}\,(\to I)^2$$

Remark: In the rule $(Nom1)$ nominals are treated as a second sort of propositional symbol on a par with ordinary propositional symbols. The rule $(Nom1)$ can be split up in the pair of rules

$$\dfrac{@_a c \qquad @_a p}{@_c p} \qquad\qquad \dfrac{@_a c \qquad @_a b}{@_c b}$$

which makes clear the role of nominals in equational reasoning, since $(Nom1)$ restricted to nominals (the right-hand-side rule above) together with (Ref) can be used to derive rules corresponding to equality being symmetric and transitive, namely

$$\dfrac{@_a c}{@_c a} \qquad\qquad \dfrac{@_a b \qquad @_b c}{@_a c}$$

and conversely, the rules corresponding to equality being symmetric and transitive can be used to derive the restriction of $(Nom1)$ to nominals.

The natural deduction system $\mathbf{N}_{\mathscr{H}(\mathscr{O})}$ corresponds to the class of all frames, that is, the class of frames where no conditions are imposed on the accessibility conditions (this is formalized in the soundness and completeness results, Theorems 2.1 and 2.2). Hence, it is a hybrid version of the standard modal logic K.

2.2.1 Conditions on the Accessibility Relation

In what follows we shall consider natural deduction systems obtained by extending $\mathbf{N}_{\mathscr{H}(\mathscr{O})}$ with additional derivation rules corresponding to first-order conditions on the accessibility relations. The conditions we consider are expressed by geometric theories. A first-order formula is *geometric* if it is built out of atomic formulas of the forms $R(a,c)$ and $a = c$ using only the connectives $\bot$, $\wedge$, $\vee$, and $\exists$.

In general, geometric formulas corresponds to affirmative assertions, that is, assertions which can be affirmed if they are true. For example, the assertion "Some ravens are black" is affirmative but "All ravens are black" is not. If some ravens are black, then there is a raven which is black, and finding it affirms the assertion. On the other hand, if all ravens are black, then it cannot be affirmed, it is, for example, impossible to check all past and future ravens. The point is here that all the connectives used in geometric formulas preserve the property of being an affirmative assertion, this is actually the case even if infinite disjunctions are allowed (which they are not here). The geometric logic terminology stems from general topology (finite conjunctions and infinite disjunctions correspond to the open sets of a topology being closed under finite intersections and infinite unions) which traditionally is motivated by Euclidean space. See Vickers (1988, 1993) for more information on geometric logic.

In what follows, the metavariables S_k and S_{jk} range over atomic first-order formulas of the forms $R(a,c)$ and $a = c$. Using the translation HT given in Section 1.2.1, atomic formulas of the mentioned forms are translated into hybrid logic as follows.

$$\begin{aligned} HT(R(a,c)) &= @_a \Diamond c \\ HT(a = c) &= @_a c \end{aligned}$$

A *geometric theory* is a finite set of closed first-order formulas, each having the form $\forall \overline{a}(\phi \rightarrow \psi)$, where the formulas ϕ and ψ are geometric, $\overline{a}$ is a list $a_1, \ldots, a_l$ of variables, and $\forall \overline{a}$ is an abbreviation for $\forall a_1 \ldots \forall a_l$. It can be proved, cf. Simpson (1994), that any geometric theory is equivalent to a *basic geometric theory* which is a geometric theory in which each formula has the form

$$(*) \qquad \forall \overline{a}((S_1 \wedge \cdots \wedge S_n) \rightarrow \exists \overline{c} \bigvee_{j=1}^{m} (S_{j1} \wedge \cdots \wedge S_{jn_j}))$$

where $n, m \geq 0$ and $n_1, \ldots, n_m \geq 1$. For simplicity, we assume that the variables in the list $\overline{a}$ are pairwise distinct, that the variables in $\overline{c}$ are pairwise distinct, and that no variable occurs in both $\overline{c}$ and $\overline{a}$. A sample of formulas of the form $(*)$ displayed above is given in Figure 2.4. Note that such a formula is a Horn clause if $\overline{c}$ is empty,

$m = 1$, and $n_m = 1$. Thus, the first two formulas in Figure 2.4 are Horn clauses. Also, note that the third formula in Figure 2.4 is identical to $\forall a \neg R(a,a)$ (recall that $\neg\phi$ is an abbreviation for $\phi \rightarrow \bot$).

1. Symmetry	$\forall a \forall c(R(a,c) \rightarrow R(c,a))$
2. Antisymmetry	$\forall a \forall c((R(a,c) \wedge R(c,a)) \rightarrow a = c)$
3. Irreflexivity	$\forall a(R(a,a) \rightarrow \bot)$
4. Directedness	$\forall a \forall b \forall c((R(a,b) \wedge R(a,c)) \rightarrow \exists d(R(b,d) \wedge R(c,d)))$

Fig. 2.4 A sample of conditions on the accessibility relation

In what follows, the metavariables s_k and s_{jk} range over hybrid-logical formulas of the forms $@_a \Diamond c$ and $@_a c$. It turns out that basic geometric theories correspond to straightforward natural deduction rules for hybrid logic: With a formula θ of the form displayed above, we associate the natural deduction rule (R_θ) given in Figure 2.5, where s_k is of the form $HT(S_k)$ and s_{jk} is of the form $HT(S_{jk})$. If $\theta_1, \ldots, \theta_4$ are formulas of the forms given in Figure 2.4, then the associated natural deduction rules $(R_{\theta_1}), \ldots, (R_{\theta_4})$ are the rules in Figure 2.6. Note that the rule (R_{θ_3}) has zero non-relational premises. Now, let **T** be any basic geometric theory. The natural deduction system obtained by extending $\mathbf{N}_{\mathcal{H}(\mathcal{O})}$ with the set of rules $\{(R_\theta) \mid \theta \in \mathbf{T}\}$ will be denoted $\mathbf{N}_{\mathcal{H}(\mathcal{O})} + \mathbf{T}$. We shall assume that we are working with a fixed basic geometric theory **T** unless otherwise specified.

$$\dfrac{s_1 \quad \ldots \quad s_n \qquad \begin{matrix}[s_{11}]\ldots[s_{1n_1}]\\ \vdots \\ \phi\end{matrix} \qquad \ldots \qquad \begin{matrix}[s_{m1}]\ldots[s_{mn_m}]\\ \vdots \\ \phi\end{matrix}}{\phi}\ (R_\theta)^*$$

$*$ None of the nominals in $\overline{c}$ occur free in ϕ or in any of the undischarged assumptions other than the specified occurrences of s_{jk}. (Recall that nominals are identified with first-order variables and that $\overline{c}$ are the first-order variables existentially quantified over in the formula θ.)

Fig. 2.5 Natural deduction rules for geometric theories

It is straightforward to check that if a formula θ of the form displayed above is a Horn clause, then the rule (R_θ) given in Figure 2.5 can be replaced by the simpler rule below (which we have called (R_θ) too).

$$\dfrac{s_1 \quad \ldots \quad s_n}{s_{11}}\ (R_\theta)$$

$$\dfrac{@_a\Diamond c \qquad \begin{matrix}[@_c\Diamond a]\\ \vdots \\ \phi\end{matrix}}{\phi}\,(R_{\theta_1}) \qquad\qquad \dfrac{@_a\Diamond c \qquad @_c\Diamond a \qquad \begin{matrix}[@_a c]\\ \vdots \\ \phi\end{matrix}}{\phi}\,(R_{\theta_2})$$

$$\dfrac{@_a\Diamond a}{\phi}\,(R_{\theta_3}) \qquad\qquad \dfrac{@_a\Diamond b \qquad @_a\Diamond c \qquad \begin{matrix}[@_b\Diamond d][@_c\Diamond d]\\ \vdots \\ \phi\end{matrix}}{\phi}\,(R_{\theta_4})^*$$

$*$ d does not occur free in ϕ or in any undischarged assumptions other than the specified occurrences of $@_c\Diamond d$ and $@_b\Diamond d$.

Fig. 2.6 Rules corresponding to conditions on the accessibility relation

Natural deduction rules corresponding to Horn clauses were discussed already in Prawitz (1971).

Remark: A semantic characterization of frame classes definable by geometric theories can be found in Chang and Keisler (1990, p. 322), Exercise 5.2.24.[2] We briefly summarize this exercise. A *homomorphism* from a frame (W,R) to a frame (W',R') is a function f from W to W' such that for any $w,v \in W$, if wRv then $f(w)R'f(v)$. A *direct system* is a sequence of frames $\mathfrak{F}_1, \mathfrak{F}_2, \ldots$ together with a sequence of functions $f_1, f_2, \ldots$ such that for each i, the function f_i is a homomorphism from $\mathfrak{F}_i$ to $\mathfrak{F}_{i+1}$. There is a natural way to define a limit frame, called a *direct limit*, for a direct system, the exact definition can be found in the above mentioned exercise. It can be shown that there exists a direct limit for any direct system and that this direct limit is unique up to isomorphism. Now, according to the exercise, a closed first-order formula ϕ is equivalent to the conjunction of the formulas in a geometric theory if and only if ϕ is preserved under direct limits, that is, whenever each of the frames $\mathfrak{F}_1, \mathfrak{F}_2, \ldots$ in a direct system validates ϕ, the direct limit also validates ϕ. It follows that a class of frames definable by a closed first-order formula is definable by a geometric theory if and only if the class in question is closed under direct limits.

2.2.2 *Some Admissible Rules*

Below we shall prove a small proposition regarding some admissible rules. A rule is *admissible* in a natural deduction system if, for every derivation of a formula ϕ from a set of formulas Γ involving the rule in question, there exists a derivation of ϕ from Γ not involving the rule. We first need a convention: The *degree* of a formula

[2] This was pointed out to the author by Balder ten Cate (personal communication).

is the number of occurrences of non-nullary connectives in it. Thus, for example, the degree of the formula $@_c(p \wedge q)$ is 2.

Proposition 2.1. *The rules*

$$
\dfrac{\begin{array}{c}[@_a\neg\phi]\\ \vdots \\ @_a\bot\end{array}}{@_a\phi}\,(\bot)
\qquad\qquad
\dfrac{@_a c \quad @_a\phi}{@_c\phi}\,(Nom)
$$

are admissible in $\mathbf{N}_{\mathscr{H}(\mathscr{O})}+\mathbf{T}$.

Proof. The proof that $(\bot)$ is admissible is along the lines of a similar proof for ordinary classical first-order logic given in Prawitz (1965). Let π be a derivation in which the highest degree of the conclusion of an instance of the rule $(\bot)$ is d. If $d = 0$, then it is straightforward to modify π such that it does not contain any instances of $(\bot)$. If $d > 0$, then there exists an instance of $(\bot)$ with conclusion ψ with degree d such that no conclusion of an instance of $(\bot)$ that stands above ψ is with degree d. It is now straightforward to modify π such that the instance of $(\bot)$ with conclusion ψ disappears and such that the conclusions of the new instances of $(\bot)$ that arise from this modification have degrees less than d. We only cover one case, namely the case where ψ is of the form $@_a@_b\theta$. In this case the left-hand-side derivation below is replaced by the right-hand-side derivation.

$$
\dfrac{\begin{array}{c}[@_a\neg @_b\theta]\\ \vdots\,\pi \\ @_a\bot\end{array}}{@_a@_b\theta}\,(\bot)
\qquad\qquad
\dfrac{\dfrac{\dfrac{\begin{array}{c}\dfrac{\dfrac{\dfrac{\dfrac{@_a@_b\theta^1}{@_b\theta}\,(@E) \quad @_b\neg\theta^2}{@_b\bot}\,(\rightarrow E)}{@_a\bot}\,(\bot 2)}{@_a\neg @_b\theta}\,(\rightarrow I)^1\\ \vdots\,\pi \\ @_a\bot\end{array}}{@_b\bot}\,(\bot 2)}{@_b\theta}\,(\bot)^2}{@_a@_b\theta}\,(@I)
$$

We are thus done by induction. The proof that (Nom) is admissible is analogous to the proof that $(\bot)$ is admissible. We only cover the case where the instance of (Nom) has a conclusion of the form $@_a(\psi \wedge \theta)$. In this case the left-hand-side derivation below is replaced by the right-hand-side derivation.

$$\dfrac{\begin{array}{c}\vdots\,\pi_1\\ @_a c\end{array} \qquad \begin{array}{c}\vdots\,\pi_2\\ @_a(\psi\wedge\theta)\end{array}}{@_c(\psi\wedge\theta)}\,(Nom) \qquad \dfrac{\dfrac{\begin{array}{c}\vdots\,\pi_1\\ @_a c\end{array} \qquad \dfrac{\begin{array}{c}\vdots\,\pi_2\\ @_a(\psi\wedge\theta)\end{array}}{@_a\psi}\,(\wedge E1)}{@_c\psi}\,(Nom) \qquad \dfrac{\begin{array}{c}\vdots\,\pi_1\\ @_a c\end{array} \qquad \dfrac{\begin{array}{c}\vdots\,\pi_2\\ @_a(\psi\wedge\theta)\end{array}}{@_a\theta}\,(\wedge E2)}{@_c\theta}\,(Nom)}{@_c(\psi\wedge\theta)}\,(\wedge I)$$

Again, we are done by induction.

Note in the proposition above that ϕ can be any formula; not just a propositional symbol. Thus, the rule $(\bot)$ generalizes the rule $(\bot 1)$ whereas (Nom) generalizes $(Nom1)$ (and the rule $(Nom2)$ as well). The side-conditions on the rules $(\bot 1)$ and $(Nom1)$ enable us to prove a normalization theorem (Theorem 2.3) such that normal derivations satisfy a version of the subformula property called the quasi-subformula property (Theorem 2.4). In the case with $(\bot 1)$, it is well-known from the literature that the subformula property does not hold without the side-condition, cf. Prawitz (1965, 1971). We shall return to the subformula property later.

2.2.3 Soundness and Completeness

The aim of this section is to prove soundness and completeness of the natural deduction system for propositional hybrid logic. We shall need the standard substitution lemma below.

Lemma 2.1. *(Substitution lemma) Let $\mathfrak{M}$ be a model and let ψ be a formula. For any world w and any assignments g and g' such that $g(a) = g'(c)$ and $g \overset{a}{\sim} g'$, it is the case that $\mathfrak{M}, g, w \models \psi$ if and only if $\mathfrak{M}, g', w \models \psi[c/a]$.*

Proof. Straightforward induction on the structure of ψ.

Recall that we are working with a fixed basic geometric theory $\mathbf{T}$. A model $\mathfrak{M}$ for hybrid logic is called a $\mathbf{T}$-*model* if and only if $\mathfrak{M}^* \models \theta$ for every formula $\theta \in \mathbf{T}$ (recall that $\mathfrak{M}^*$ is the first-order model corresponding to the hybrid-logical model $\mathfrak{M}$). Remark: Being a $\mathbf{T}$-model is really a property of the frame on which the model is based, the reason being that the formulas in $\mathbf{T}$ do not contain predicate symbols besides R and $=$.

Theorem 2.1. *(Soundness) Let ψ be a satisfaction statement and let Γ be a set of satisfaction statements. The first statement below implies the second statement.*

1. *ψ is derivable from Γ in $\mathbf{N}_{\mathcal{H}(\mathcal{O})} + \mathbf{T}$.*
2. *For any $\mathbf{T}$-model $\mathfrak{M}$ and any assignment g, if, for any formula $\theta \in \Gamma$, $\mathfrak{M}, g \models \theta$, then $\mathfrak{M}, g \models \psi$.*

Proof. Induction on the structure of the derivation of ψ. We only cover the case where ψ is the conclusion of an instance of the rule $(\downarrow I)$ (see Figure 2.2). Let $\mathfrak{M}$ be a **T**-model and g an assignment such that for any formula $\theta \in \Gamma$, $\mathfrak{M}, g \models \theta$. Let g' be the assignment such that $g' \stackrel{c}{\sim} g$ and $g'(c) = g(a)$, and moreover, let $\Gamma' \subseteq \Gamma$ be the set of undischarged assumptions in the derivation of $@_a \downarrow b\phi$ (again, see Figure 2.2). Then for any formula $\theta \in \Gamma'$, $\mathfrak{M}, g' \models \theta$ since c does not occur free in any of the formulas in Γ' according to the side-condition on the rule. So $\mathfrak{M}, g' \models @_c\phi[c/b]$ by induction as $\mathfrak{M}, g' \models @_a c$. But then $\mathfrak{M}, g', g'(c) \models \phi[c/b]$ and hence it is the case that $\mathfrak{M}, g'', g'(c) \models \phi$ by Lemma 2.1 where g'' is the assignment such that $g'' \stackrel{b}{\sim} g'$ and $g''(b) = g'(c)$. Therefore $\mathfrak{M}, g', g'(c) \models \downarrow b\phi$ which implies that $\mathfrak{M}, g \models @_a \downarrow b\phi$ as c does not occur free in $\downarrow b\theta$ according to the side-condition.

In what follows, we shall prove completeness. The proof we give is similar to the completeness proof in Blackburn (2000a). However, we use maximal consistent sets instead of Hintikka sets. Also, our proof is in some ways similar to the completeness proof in Basin et al. (1997).

Definition 2.1. A set of satisfaction statements Γ in $\mathscr{H}(\mathscr{O})$ is $\mathbf{N}_{\mathscr{H}(\mathscr{O})} + \mathbf{T}$-*inconsistent* if and only if $@_a\bot$ is derivable from Γ in $\mathbf{N}_{\mathscr{H}(\mathscr{O})} + \mathbf{T}$ for some nominal a and Γ is $\mathbf{N}_{\mathscr{H}(\mathscr{O})} + \mathbf{T}$-*consistent* if and only if Γ is not $\mathbf{N}_{\mathscr{H}(\mathscr{O})} + \mathbf{T}$-inconsistent. Moreover, Γ is *maximal* $\mathbf{N}_{\mathscr{H}(\mathscr{O})} + \mathbf{T}$-*consistent* if and only if Γ is $\mathbf{N}_{\mathscr{H}(\mathscr{O})} + \mathbf{T}$-consistent and any set of satisfaction statements in $\mathscr{H}(\mathscr{O})$ that properly extends Γ is $\mathbf{N}_{\mathscr{H}(\mathscr{O})} + \mathbf{T}$-inconsistent.

We shall frequently omit the reference to $\mathscr{H}(\mathscr{O})$ and $\mathbf{N}_{\mathscr{H}(\mathscr{O})} + \mathbf{T}$ where no confusion can occur. The definition above leads us to the lemma below.

Lemma 2.2. *If a set of satisfaction statements Γ is consistent, then for every satisfaction statement $@_a\phi$, either $\Gamma \cup \{@_a\phi\}$ is consistent or $\Gamma \cup \{@_a\neg\phi\}$ is consistent.*

Proof. Straightforward.

The Lindenbaum lemma below is similar to the Lindenbaum lemma in Basin et al. (1997).

Lemma 2.3. *(Lindenbaum lemma) Let $\overline{\mathscr{H}(\mathscr{O})}$ be the hybrid logic obtained by extending the set of nominals in $\mathscr{H}(\mathscr{O})$ with a countably infinite set of new nominals. Let $\phi_1, \phi_2, \phi_3, \ldots$ be an enumeration of all satisfaction statements in $\overline{\mathscr{H}(\mathscr{O})}$. For every $\mathbf{N}_{\mathscr{H}(\mathscr{O})} + \mathbf{T}$-consistent set of satisfaction statements Γ, a maximal $\mathbf{N}_{\overline{\mathscr{H}(\mathscr{O})}} + \mathbf{T}$-consistent set of satisfaction statements $\Gamma^* \supseteq \Gamma$ is defined as follows. Firstly, Γ^0 is defined to be Γ. Secondly, Γ^{n+1} is defined by induction. If $\Gamma^n \cup \{\phi_{n+1}\}$ is $\mathbf{N}_{\overline{\mathscr{H}(\mathscr{O})}} + \mathbf{T}$-inconsistent, then Γ^{n+1} is defined to be Γ^n. Otherwise Γ^{n+1} is defined to be*

1. *$\Gamma^n \cup \{\phi_{n+1}, @_b\psi, @_a\Diamond b\}$ if ϕ_{n+1} is of the form $@_a\Diamond\psi$;*
2. *$\Gamma^n \cup \{\phi_{n+1}, @_b\psi[b/c], @_ab\}$ if ϕ_{n+1} is of the form $@_a \downarrow c\psi$;*
3. *$\Gamma^n \cup \{\phi_{n+1}, @_a\psi[b/c]\}$ if ϕ_{n+1} is of the form $@_a\exists c\psi$;*
4. *$\Gamma^n \cup \{\phi_{n+1}, @_e \bigvee_{j=1}^m (s_{j1} \wedge \cdots \wedge s_{jn_j})[\overline{d}, \overline{b}/\overline{a}, \overline{c}]\}$ if there exists a formula in $\mathbf{T}$ of the form $\forall\overline{a}((S_1 \wedge \cdots \wedge S_n) \rightarrow \exists\overline{c} \bigvee_{j=1}^m (S_{j1} \wedge \cdots \wedge S_{jn_j}))$ such that $m \geq 1$ and $\phi_{n+1} = @_e(s_1 \wedge \cdots \wedge s_n)[\overline{d}/\overline{a}]$ for some nominals $\overline{d}$ and e; and*

5. $\Gamma^n \cup \{\phi_{n+1}\}$ *if none of the clauses above apply.*

In clause 1, 2, and 3, b is a new nominal that does not occur in Γ^n or ϕ_{n+1}, and similarly, in clause 4, $\overline{b}$ is a list of new nominals such that none of the nominals in $\overline{b}$ occur in Γ^n or ϕ_{n+1}. Finally, Γ^* is defined to be $\cup_{n\geq 0}\Gamma^n$.

Proof. Firstly, Γ^0 is $\mathbf{N}_{\mathscr{H}(\mathscr{O})}+\mathbf{T}$-consistent by definition and hence also $\mathbf{N}_{\overline{\mathscr{H}(\mathscr{O})}}+\mathbf{T}$-consistent. Secondly, to check that the consistency of Γ^n implies the consistency of Γ^{n+1}, we need to check the first four clauses in the definition of Γ^{n+1}.

- If ϕ_{n+1} is of the form $@_a\Diamond\psi$, then assume conversely that $@_f\bot$ is derivable from $\Gamma^n \cup \{\phi_{n+1}, @_b\psi, @_a\Diamond b\}$. Then $@_b\neg\psi$ is derivable from $\Gamma^n \cup \{\phi_{n+1}, @_a\Diamond b\}$ wherefore $@_a\Box\neg\psi$ is derivable from $\Gamma^n \cup \{\phi_{n+1}\}$ by the rule $(\Box I)$. But then $@_a\bot$ is derivable from $\Gamma^n \cup \{\phi_{n+1}\}$ as $\phi_{n+1} = @_a\neg\Box\neg\psi$.
- If ϕ_{n+1} is of the form $@_a \downarrow c\psi$, then assume conversely that $@_f\bot$ is derivable from $\Gamma^n \cup \{\phi_{n+1}, @_b\psi[b/c], @_ab\}$. Then $@_b\neg\psi[b/c]$ is derivable from $\Gamma^n \cup \{\phi_{n+1}, @_ab\}$ wherefore $@_a \downarrow c\neg\psi$ is derivable from $\Gamma^n \cup \{\phi_{n+1}\}$ by the rule $(\downarrow I)$. But $@_a(\downarrow c\neg\psi \rightarrow \neg\downarrow c\psi)$ is derivable, thus $@_a\neg\downarrow c\psi$ and hence also $@_a\bot$ is derivable from $\Gamma^n \cup \{\phi_{n+1}\}$.
- The case involving $\exists$ is similar to the case involving $\Diamond$.
- If there exists a formula $\forall\overline{a}((S_1 \wedge \cdots \wedge S_n) \rightarrow \exists\overline{c}\vee_{j=1}^{m}(S_{j1} \wedge \cdots \wedge S_{jn_j}))$ in $\mathbf{T}$ such that $m \geq 1$ and $\phi_{n+1} = @_e(s_1 \wedge \cdots \wedge s_n)[\overline{d}/\overline{a}]$ for some nominals $\overline{d}$ and e, then assume conversely that the formula $@_f\bot$ is derivable from the set $\Gamma^n \cup \{\phi_{n+1}, @_e\vee_{j=1}^{m}(s_{j1} \wedge \cdots \wedge s_{jn_j})[\overline{d},\overline{b}/\overline{a},\overline{c}]\}$. Then it is the case that the formula $@_e\wedge_{j=1}^{m}\neg(s_{j1} \wedge \cdots \wedge s_{jn_j})[\overline{d},\overline{b}/\overline{a},\overline{c}]$ is derivable from $\Gamma^n \cup \{\phi_{n+1}\}$, and hence, $@_e\bot$ is derivable from $\Gamma^n \cup \{\phi_{n+1}, s_{j1}[\overline{d},\overline{b}/\overline{a},\overline{c}],\ldots,s_{jn_j}[\overline{d},\overline{b}/\overline{a},\overline{c}]\}$ for any j where $1 \leq j \leq m$. But $s_1[\overline{d}/\overline{a}], \ldots, s_n[\overline{d}/\overline{a}]$ are derivable from $\Gamma^n \cup \{\phi_{n+1}\}$. Therefore $@_e\bot$ is derivable from $\Gamma^n \cup \{\phi_{n+1}\}$ by the rule (R_θ).

We conclude that each Γ^n is consistent which trivially implies that Γ^* is consistent. We now just need to prove that Γ^* is maximal consistent. Assume conversely that there exists a satisfaction statement $@_a\phi$ such that $@_a\phi \notin \Gamma^*$ as well as $@_a\neg\phi \notin \Gamma^*$, cf. Lemma 2.2. Then $\phi_p \notin \Gamma^p$ and $\phi_q \notin \Gamma^q$ where $\phi_p = @_a\phi$ and $\phi_q = @_a\neg\phi$. So $\Gamma^{p-1} \cup \{\phi_p\}$ is inconsistent and so is $\Gamma^{q-1} \cup \{\phi_q\}$. If $p < q$, then $\Gamma^{p-1} \subseteq \Gamma^{q-1}$ wherefore $\Gamma^{q-1} \cup \{\phi_p\}$ is inconsistent. Thus, Γ^{q-1} is inconsistent by Lemma 2.2. The argument is analogous if $q < p$.

If $\forall \notin \mathscr{O}$, then ϕ_{n+1} in the lemma above can obviously not be of the form $@_a\exists c\psi$, so the parts of the definition of Γ^{n+1} involving that case are superfluous. An analogous remark applies if $\downarrow\notin \mathscr{O}$. Below we shall define a canonical model. First a small lemma.

Lemma 2.4. *Let Δ be a maximal $\mathbf{N}_{\overline{\mathscr{H}(\mathscr{O})}}+\mathbf{T}$-consistent set of satisfaction statements. Let $\sim_\Delta$ be the binary relation on the set of nominals of $\overline{\mathscr{H}(\mathscr{O})}$ defined by the convention that $a \sim_\Delta a'$ if and only if $@_aa' \in \Delta$. Then the relation $\sim_\Delta$ is an equivalence relation with the following properties.*

1. If $a \sim_\Delta a'$, $c \sim_\Delta c'$, and $@_a\Diamond c \in \Delta$, then $@_{a'}\Diamond c' \in \Delta$.
2. If $a \sim_\Delta a'$ and $@_a p \in \Delta$, then $@_{a'} p \in \Delta$.

Proof. It follows straightforwardly from Lemma 2.2 and the rules (Ref) and $(Nom1)$ that $\sim_\Delta$ is reflexive, symmetric, and transitive. The first mentioned property follows from Lemma 2.2, the rule $(Nom2)$, and the observation that $@_{a'}\Diamond c'$ is derivable from $\{@_{a'}\Diamond c, @_c c'\}$. The second property follows from Lemma 2.2 and the rule $(Nom1)$.

Given a nominal a, we let $[a]$ denote the equivalence class of a with respect to $\sim_\Delta$. We now define a canonical model.

Definition 2.2. (Canonical model) Let Δ be a maximal $\mathbf{N}_{\overline{\mathscr{H}(\mathscr{O})}} + \mathbf{T}$-consistent set of satisfaction statements. A model $\mathfrak{M}^\Delta = (W^\Delta, R_1^\Delta, \ldots, R_m^\Delta, \{V_w^\Delta\}_{w \in W^\Delta})$ and an assignment g^Δ for $\mathfrak{M}^\Delta$ are defined as follows.

- $W^\Delta = \{[a] \mid a \text{ is a nominal of } \overline{\mathscr{H}(\mathscr{O})}\}$.
- $R^\Delta = \{([a],[c]) \mid @_a\Diamond c \in \Delta\}$.
- $V_{[a]}^\Delta(p) = \begin{cases} 1 \text{ if } @_a p \in \Delta. \\ 0 \text{ otherwise.} \end{cases}$
- $g^\Delta(a) = [a]$.

Note that the first property of $\sim_\Delta$ mentioned in Lemma 2.4 implies that R^Δ is well-defined, and similarly, the second property implies that V_w^Δ is well-defined. Given the Lindenbaum lemma and the definition of a canonical model, we just need one small lemma before we are ready to prove a truth lemma.

Lemma 2.5. *Let ϕ be a satisfaction statement of the hybrid logic $\mathscr{H}(\mathscr{O})$, and let c and b be nominals such that b does not occur in ϕ. Let ϕ' be ϕ where each occurrence of c that is not free has been replaced by b. Then ϕ' is derivable from $\{\phi\}$ and ϕ is derivable from $\{\phi'\}$ in $\mathbf{N}_{\mathscr{H}(\mathscr{O})} + \emptyset$.*

Proof. Induction on the degree of ϕ.

For example, the satisfaction statement $@_d\forall b(@_b p)$ of $\mathscr{H}(\forall)$ is derivable from $\{@_d\forall c(@_c p)\}$. Now the truth lemma.

Lemma 2.6. *(Truth lemma) Let Γ be a $\mathbf{N}_{\mathscr{H}(\mathscr{O})} + \mathbf{T}$-consistent set of satisfaction statements. Then for any satisfaction statement $@_a\phi$, it is the case that $@_a\phi \in \Gamma^*$ if and only if $\mathfrak{M}^{\Gamma^*}, g^{\Gamma^*}, [a] \models \phi$.*

Proof. Induction on the degree of ϕ. We only consider two cases; the other cases are similar.

The first case is where ϕ is of the form $\Box\theta$. Assume that $@_a\Box\theta \in \Gamma^*$. We then have to prove that $\mathfrak{M}^{\Gamma^*}, g^{\Gamma^*}, [c] \models \theta$ for any nominal c such that $[a]R^{\Gamma^*}[c]$, that is, such that $@_a\Diamond c \in \Gamma^*$. But $@_a\Diamond c \in \Gamma^*$ implies $@_c\theta \in \Gamma^*$ by the rule $(\Box E)$ and this implies $\mathfrak{M}^{\Gamma^*}, g^{\Gamma^*}, [c] \models \theta$ by induction. On the other hand, assume that $\mathfrak{M}^{\Gamma^*}, g^{\Gamma^*}, [a] \models \Box\theta$, that is, $\mathfrak{M}^{\Gamma^*}, g^{\Gamma^*}, [c] \models \theta$ for any nominal c such that $@_a\Diamond c \in \Gamma^*$. Now, if $@_a\neg\Box\theta \in \Gamma^*$, then also $@_a\Diamond\neg\theta \in \Gamma^*$ as $@_a(\neg\Box\theta \rightarrow \Diamond\neg\theta)$ is derivable. Therefore by definition of Γ^*, there exists a nominal b such that

$@_b\neg\theta \in \Gamma^*$ and $@_a\Diamond b \in \Gamma^*$. But then $\mathfrak{M}^{\Gamma^*}, g^{\Gamma^*}, [b] \models \theta$ by assumption and hence $@_b\theta \in \Gamma^*$ by induction. Thus, we conclude that $@_a\neg\Box\theta \notin \Gamma^*$ and hence $@_a\Box\theta \in \Gamma^*$ by Lemma 2.2.

The second case we consider is where ϕ is of the form $\downarrow c\theta$. Assume that we have $@_a \downarrow c\theta \in \Gamma^*$. We then have to prove that $\mathfrak{M}^{\Gamma^*}, g, [a] \models \theta$ where $g \overset{c}{\sim} g^{\Gamma^*}$ and $g(c) = [a]$. Let θ' be θ where each occurrence of a that is not free has been replaced by some nominal that does not occur in $@_a\theta$. Then $@_a \downarrow c\theta' \in \Gamma^*$ as $@_a(\downarrow c\theta \to \downarrow c\theta')$ is derivable by Lemma 2.5. So $@_a\theta'[a/c] \in \Gamma^*$ by the rules $(\downarrow E)$ and (Ref). By induction we get $\mathfrak{M}^{\Gamma^*}, g^{\Gamma^*}, [a] \models \theta'[a/c]$ and therefore $\mathfrak{M}^{\Gamma^*}, g, [a] \models \theta'$ by Lemma 2.1. But $@_a(\theta' \to \theta)$ is derivable by Lemma 2.5 and therefore valid by Theorem 2.1, so $\mathfrak{M}^{\Gamma^*}, g, [a] \models \theta$. On the other hand, assume that $\mathfrak{M}^{\Gamma^*}, g^{\Gamma^*}, [a] \models \downarrow c\theta$. If $@_a\neg \downarrow c\theta \in \Gamma^*$, then also $@_a \downarrow c\neg\theta \in \Gamma^*$ as $@_a(\neg \downarrow c\theta \to \downarrow c\neg\theta)$ is derivable. Therefore by definition of Γ^*, there exists a nominal b such that $@_b\neg\theta[b/c] \in \Gamma^*$ and $@_a b \in \Gamma^*$. Now, let $g \overset{c}{\sim} g^{\Gamma^*}$ such that $g(c) = [a]$. Then by assumption $\mathfrak{M}^{\Gamma^*}, g, [a] \models \theta$ and hence $\mathfrak{M}^{\Gamma^*}, g^{\Gamma^*}, [a] \models \theta[b/c]$ by Lemma 2.1 as $[a] = [b]$ since $a \sim_{\Gamma^*} b$. Therefore $@_b\theta[b/c] \in \Gamma^*$ by induction. We conclude that $@_a\neg \downarrow c\theta \notin \Gamma^*$ and hence $@_a \downarrow c\theta \in \Gamma^*$ by Lemma 2.2.

The treatment of $\downarrow$ in the lemma above is similar to the treatment of the binder $\exists$ in the truth lemma of Blackburn and Tzakova (1998). Now we need only one lemma before we can prove completeness.

Lemma 2.7. *Let Γ be a $\mathbf{N}_{\mathscr{H}(\mathscr{O})} + \mathbf{T}$-consistent set of satisfaction statements. Then the canonical model $\mathfrak{M}^{\Gamma^*}$ is a $\mathbf{T}$-model.*

Proof. If $\theta \in \mathbf{T}$, then θ has the form $\forall\overline{a}((S_1 \wedge \cdots \wedge S_n) \to \exists\overline{c} \bigvee_{j=1}^{m} (S_{j1} \wedge \cdots \wedge S_{jn_j}))$ where $\overline{a} = a_1, \ldots, a_l$. Assume g is an assignment for a hybrid-logical $\mathfrak{M}^{\Gamma^*}$ such that $(\mathfrak{M}^{\Gamma^*})^*, g \models S_1, \ldots, (\mathfrak{M}^{\Gamma^*})^*, g \models S_n$. Let $g(a_1) = [d_1], \cdots, g(a_l) = [d_l]$. Then it is the case that $s_1[\overline{d}/\overline{a}], \cdots, s_n[\overline{d}/\overline{a}] \in \Gamma^*$ by the definition of a canonical model. If it is the case that $m \geq 1$, then by definition of Γ^* there exists a list of nominals $\overline{b}$ such that $@_e \bigvee_{j=1}^{m} (s_{j1} \wedge \cdots \wedge s_{jn_j})[\overline{d}, \overline{b}/\overline{a}, \overline{c}] \in \Gamma^*$ since $@_e(s_1 \wedge \cdots \wedge s_n)[\overline{d}/\overline{a}] \in \Gamma^*$ where e is an arbitrary nominal. Therefore $@_e(s_{j1} \wedge \cdots \wedge s_{jn_j})[\overline{d}, \overline{b}/\overline{a}, \overline{c}] \in \Gamma^*$ and hence $s_{j1}[\overline{d}, \overline{b}/\overline{a}, \overline{c}], \ldots, s_{jn_j}[\overline{d}, \overline{b}/\overline{a}, \overline{c}] \in \Gamma^*$ for some j where $1 \leq j \leq m$. But then it follows from the definition of a canonical model that $(\mathfrak{M}^{\Gamma^*})^*, g \models \exists\overline{c} \bigvee_{j=1}^{m} (S_{j1} \wedge \cdots \wedge S_{jn_j})$. On the other hand, if $m = 0$, then $@_e\bot \in \Gamma^*$ by the rule (R_θ) which contradicts the consistency of Γ^*.

Now the completeness theorem.

Theorem 2.2. *(Completeness) Let ψ be a satisfaction statement and let Γ be a set of satisfaction statements. The second statement below implies the first statement.*

1. *ψ is derivable from Γ in $\mathbf{N}_{\mathscr{H}(\mathscr{O})} + \mathbf{T}$.*
2. *For any $\mathbf{T}$-model $\mathfrak{M}$ and any assignment g, if, for any formula $\theta \in \Gamma$, $\mathfrak{M}, g \models \theta$, then $\mathfrak{M}, g \models \psi$.*

Proof. We are done if Γ is inconsistent, cf. Proposition 2.1. So assume that Γ is consistent. Now, assume that ψ is not derivable from Γ and let $\psi = @_a\phi$. Then $\Gamma \cup \{@_a\neg\phi\}$ is consistent. Let $\Delta = (\Gamma \cup \{@_a\neg\phi\})^*$, cf. Lemma 2.3, and consider the model $\mathfrak{M}^\Delta$ and the assignment g^Δ. By Lemma 2.6, $\mathfrak{M}^\Delta, g^\Delta \models \theta$ for any formula $\theta \in \Gamma$, and also $\mathfrak{M}^\Delta, g^\Delta \models @_a\neg\phi$. But this contradicts the second statement in the theorem since $\mathfrak{M}^\Delta$ is a **T**-model by Lemma 2.7.

2.2.4 Normalization

In this section we give reduction rules for the natural deduction system $\mathbf{N}_{\mathscr{H}(\mathscr{O})} + \mathbf{T}$ and we prove a normalization theorem. First some conventions. If a premise of a rule has the form $@_ac$ or $@_a\Diamond c$, then it is called a *relational premise*, and similarly, if the conclusion of a rule has the form $@_ac$ or $@_a\Diamond c$, then it is called a *relational conclusion*. Moreover, if an assumption discharged by a rule has the form $@_ac$ or $@_a\Diamond c$, then it is called a *relationally discharged assumption*. The premise of the form $@_a\phi$ in the rule $(\rightarrow E)$ is called *minor*. A premise of an elimination rule that is neither minor nor relational is called *major*. Note that the notion of a relational premise is defined in terms of rules; not rule-instances. A similar remark applies to the other notions above. Thus, a formula occurrence in a derivation might be of the form $@_a\Diamond c$ and also be the major premise of an instance of $(\rightarrow E)$. Note that the premises $s_1, \ldots, s_n$ in a (R_θ) rule are relational and that all the assumptions discharged by such a rule are relationally discharged.

A *maximum formula* in a derivation is a formula occurrence that is both the conclusion of an introduction rule and the major premise of an elimination rule. Maximum formulas can be removed by applying *proper reductions*. The rules for proper reductions are given below. We consider each case in turn. In what follows, we let $\pi[\overline{c}/\overline{a}]$ be the derivation π where each formula occurrence ψ has been replaced by $\psi[\overline{c}/\overline{a}]$.

$(\wedge I)$ followed by $(\wedge E1)$ (analogously in the case involving $(\wedge E2)$)

$$\dfrac{\dfrac{\begin{array}{c}\vdots\,\pi_1\\ @_a\phi\end{array}\qquad\begin{array}{c}\vdots\,\pi_2\\ @_a\psi\end{array}}{@_a(\phi\wedge\psi)}}{@_a\phi} \quad\rightsquigarrow\quad \begin{array}{c}\vdots\,\pi_1\\ @_a\phi\end{array}$$

$(\rightarrow I)$ followed by $(\rightarrow E)$

$$\dfrac{\dfrac{\begin{matrix}[@_a\phi]\\ \vdots\ \pi_1\\ @_a\psi\end{matrix}}{@_a(\phi\rightarrow\psi)} \qquad \begin{matrix}\vdots\ \pi_2\\ @_a\phi\end{matrix}}{@_a\psi} \quad\rightsquigarrow\quad \begin{matrix}\vdots\ \pi_2\\ @_a\phi\\ \vdots\ \pi_1\\ @_a\psi\end{matrix}$$

$(@I)$ followed by $(@E)$

$$\dfrac{\dfrac{\begin{matrix}\vdots\ \pi\\ @_a\phi\end{matrix}}{@_c@_a\phi}}{@_a\phi} \quad\rightsquigarrow\quad \begin{matrix}\vdots\ \pi\\ @_a\phi\end{matrix}$$

$(\Box I)$ followed by $(\Box E)$

$$\dfrac{\dfrac{\begin{matrix}[@_a\Diamond c]\\ \vdots\ \pi_1\\ @_c\phi\end{matrix}}{@_a\Box\phi} \qquad \begin{matrix}\vdots\ \pi_2\\ @_a\Diamond e\end{matrix}}{@_e\phi} \quad\rightsquigarrow\quad \begin{matrix}\vdots\ \pi_2\\ @_a\Diamond e\\ \vdots\ \pi_1[e/c]\\ @_e\phi\end{matrix}$$

$(\downarrow I)$ followed by $(\downarrow E)$

$$\dfrac{\dfrac{\begin{matrix}[@_a c]\\ \vdots\ \pi_1\\ @_c\phi[c/b]\end{matrix}}{@_a\downarrow b\phi} \qquad \begin{matrix}\vdots\ \pi_2\\ @_a e\end{matrix}}{@_e\phi[e/b]} \quad\rightsquigarrow\quad \begin{matrix}\vdots\ \pi_2\\ @_a e\\ \vdots\ \pi_1[e/c]\\ @_e\phi[e/b]\end{matrix}$$

$(\forall I)$ followed by $(\forall E)$

$$\dfrac{\dfrac{\begin{matrix}\vdots\ \pi\\ @_a\phi[c/b]\end{matrix}}{@_a\forall b\phi}}{@_a\phi[e/b]} \quad\rightsquigarrow\quad \begin{matrix}\vdots\ \pi[e/c]\\ @_a\phi[e/b]\end{matrix}$$

We also need reduction rules in connection with the (R_θ) derivation rules corresponding to geometric theories. A *permutable formula* in a derivation is a formula occurrence that is both the conclusion of a (R_θ) rule and the major premise of an elimination rule. Permutable formulas in a derivation can be removed by applying *permutative reductions*. The rule for permutative reductions is as follows in the case where the elimination rule has two premises.

$$\dfrac{\dfrac{\begin{matrix}\vdots\,\tau_1\\ s_1\end{matrix}\quad\cdots\quad\begin{matrix}\vdots\,\tau_n\\ s_n\end{matrix}\qquad\begin{matrix}[s_{11}]\ldots[s_{1n_1}]\\ \vdots\,\pi_1\\ \phi\end{matrix}\quad\cdots\quad\begin{matrix}[s_{m1}]\ldots[s_{mn_m}]\\ \vdots\,\pi_m\\ \phi\end{matrix}}{\phi}\qquad\begin{matrix}\vdots\,\pi\\ \theta\end{matrix}}{\psi}$$

$$\rightsquigarrow$$

$$\dfrac{\begin{matrix}\vdots\,\tau_1\\ s_1\end{matrix}\quad\cdots\quad\begin{matrix}\vdots\,\tau_n\\ s_n\end{matrix}\qquad\dfrac{\begin{matrix}[s_{11}[\overline{b}/\overline{c}]]\ldots[s_{1n_1}[\overline{b}/\overline{c}]]\\ \vdots\,\pi_1[\overline{b}/\overline{c}]\\ \phi\end{matrix}\qquad\begin{matrix}\vdots\,\pi\\ \theta\end{matrix}}{\psi}\quad\cdots\quad\dfrac{\begin{matrix}[s_{m1}[\overline{b}/\overline{c}]]\ldots[s_{mn_m}[\overline{b}/\overline{c}]]\\ \vdots\,\pi_m[\overline{b}/\overline{c}]\\ \phi\end{matrix}\qquad\begin{matrix}\vdots\,\pi\\ \theta\end{matrix}}{\psi}}{\psi}$$

The nominals in the list $\overline{b}$ are pairwise distinct and new. Note that according to the side-condition on the rules (R_θ), cf. Figure 2.5, none of the nominals in $\overline{c}$ occur free in ϕ, hence, it is ensured that the formula $\phi[\overline{b}/\overline{c}]$ is identical to ϕ. This remark also applies to the undischarged assumptions in the derivations of ϕ. The case where the elimination rule has only one premise is obtained by deleting all instances of the derivation π from the reduction rule above.

A derivation is *normal* if it contains no maximum or permutable formula. In what follows we shall prove a normalization theorem which says that any derivation can be rewritten to a normal derivation by repeated applications of reductions. To this end we need a number of definitions and lemmas. In the case of ordinary classical first-order logic, it is always possible to select reductions such that applying a reduction to a maximum formula only generates new maximum formulas having a lower degree than the original one. The technique used in the standard normalization proof for first-order logic (originally given in Prawitz (1965)) is based on this property. The natural deduction system considered here does not have this property since the reduction rule for $\Box$ might generate new maximum formulas of the form $@_a\Diamond e$, that is, maximum formulas that do not necessarily have a lower degree than the original one (here we ignore permutable formulas). Thus, the standard technique does not work directly here. This problem is solved by using what we have called the $\Box$-graph of a derivation to systematically control the application of reductions to new maximum formulas like $@_a\Diamond e$.

Definition 2.3. The $\Box$-*graph* of a derivation π is the binary relation on the set of formula occurrences of π which is defined as follows. A pair of formula occurrences (ϕ, ψ) is an element of the $\Box$-graph of π if and only if it satisfies one of the following conditions.

1. ϕ is of the form $@_a\Box\neg c$, ψ is of the form $@_a\Diamond e$, and ϕ is either the major premise of an instance of $(\Box E)$ which has ψ as the relational premise or ϕ is the minor premise of an instance of $(\rightarrow E)$ which has ψ as the major premise.
2. ϕ is of the form $@_a\Diamond e$, ψ is of the form $@_a\Box\neg c$, and ϕ is either an assumption discharged at an instance of $(\Box I)$ which has ψ as the conclusion or ϕ is the conclusion of an instance of $(\rightarrow I)$ at which ψ is discharged.
3. ϕ and ψ are both of the form $@_a\Box\neg c$ and ϕ is a non-relational premise of a (R_θ) rule which has ψ as the conclusion.
4. ϕ and ψ are both of the form $@_a\Diamond c$ and ψ is a non-relational premise of a (R_θ) rule which has ϕ as the conclusion.

(Recall that the formulas $@_a\Diamond e$ and $@_a(\Box\neg e \rightarrow \bot)$ are identical.) Note that the $\Box$-graph of π is a relation on the set of formula occurrences of π; not the set of formulas occurring in π. Also, note that it follows from the definition above that every formula occurrence in a $\Box$-graph is of the form $@_a\Box\neg c$ or $@_a\Diamond c$.

Lemma 2.8. *The $\Box$-graph of a derivation π does not contain cycles.*

Proof. Induction on the structure of π. There are four cases to check according to the definition of a $\Box$-graph. The first case has a subcase for each of the rules $(\Box E)$ and $(\rightarrow E)$. We consider the first subcase where π has the form

$$\dfrac{\begin{array}{cc} \vdots\,\tau & \vdots\,\sigma \\ @_a\Box\neg c & @_a\Diamond e \end{array}}{@_e\neg c}\,(\Box E)$$

Now, the $\Box$-graph of π is the union of the $\Box$-graph of τ, and the $\Box$-graph of σ, and

$$\{(@_a\Box\neg c, @_a\Diamond e)\}.$$

By induction the $\Box$-graphs of τ and σ do not contain cycles. If the $\Box$-graph of π has a cycle, then it contains both of the formula occurrences $@_a\Box\neg c$ and $@_a\Diamond e$ indicated above since the $\Box$-graphs of τ and σ do not have common nodes (as the derivations τ and σ do not have common nodes). But this cannot be the case, again since the $\Box$-graphs of τ and σ do not have common nodes. The second subcase of the first case as well as the second, third, and fourth cases are similar.

Definition 2.4. A *stubborn formula* in a derivation π is a maximum or permutable formula of the form $@_a\Box\neg c$ or $@_a\Diamond c$ and the *potential* of a stubborn formula in π is the maximal length of a chain in the $\Box$-graph of π that contains the stubborn formula.

Note that the notion of potential in the definition above is well-defined as Lemma 2.8 implies that the set of lengths of chains in the $\Box$-graph of π is bounded.

Lemma 2.9. *Let π be a derivation where all stubborn formulas have potential less than or equal to d. Assume that ϕ is a stubborn formula with potential d such that no formula occurrence above a minor or relational premise of the rule instance of* *which ϕ is a major premise, is stubborn and with potential d. Let π' be the derivation obtained by applying the appropriate reduction such that ϕ is removed. Then all stubborn formulas in π' have potential less than or equal to d, and moreover, the number of stubborn formulas with potential d in π' is less than the number of stubborn formulas with potential d in π.*

Proof. There are four cases to check: If ϕ is a maximum formula, then it is either the conclusion of a $(\Box I)$ rule and the major premise of a $(\Box E)$ rule or it is the conclusion of a $(\rightarrow I)$ rule and the major premise of a $(\rightarrow E)$ rule. If ϕ is a permutable formula, then it is the conclusion of a (R_θ) rule and the major premise of either a $(\Box E)$ rule or a $(\rightarrow E)$ rule. We consider the first case where π and π' have the forms below.

$$\dfrac{\dfrac{\begin{array}{c}[@_a\Diamond d]\\ \vdots\ \pi_1\\ @_d\neg c\end{array}}{@_a\Box\neg c}\,(\Box I) \qquad \begin{array}{c}\vdots\ \pi_2\\ @_a\Diamond e\end{array}}{\begin{array}{c}@_e\neg c\\ \vdots\ \tau\\ \psi\end{array}}\,(\Box E) \qquad\qquad \begin{array}{c}\vdots\ \pi_2\\ @_a\Diamond e\\ \vdots\ \pi_1[e/d]\\ @_e\neg c\\ \vdots\ \tau\\ \psi\end{array}$$

Note that any formula occurrence in π', except the indicated occurrences of $@_a\Diamond e$ and $@_e\neg c$, in an obvious way can be mapped to a formula occurrence in π. Let f be the map thus defined (note that f need not be injective as the instance of $(\Box I)$ in π might discharge more than one occurrence of $@_a\Diamond d$). Using the map f, a map from the $\Box$-graph of π' to the $\Box$-graph of π is defined as follows. An element (ξ,χ) of the $\Box$-graph of π', where the formula occurrences ξ and χ both are in the domain of f, is mapped to $(f(\xi),f(\chi))$, which straightforwardly can be shown to be an element of the $\Box$-graph of π (observe that no assumption in π_2 is discharged at a rule-instance in $\pi_1[e/d]$). An element (ξ,χ), where ξ is one of the indicated occurrences of $@_a\Diamond e$, is mapped to $(\xi',f(\chi))$, where ξ' is the relational premise of the instance of $(\Box E)$, and an element (ξ,χ), where χ is one of the indicated occurrences of $@_a\Diamond e$, is mapped to $(f(\xi),\chi')$, where χ' is the assumption discharged by the instance of $(\Box I)$ corresponding to the occurrence of $@_a\Diamond e$ in question. By using the map from the $\Box$-graph of π' to the $\Box$-graph of π, any chain in the $\Box$-graph of π' that does not contain any of the indicated occurrences of $@_a\Diamond e$ can in an obvious way be mapped to a chain in the $\Box$-graph of π of the same length which does not contain the indicated occurrences of $@_a\Diamond d$, $@_a\Box\neg c$, and $@_a\Diamond e$, and similarly, any chain in the $\Box$-graph of π' that contains one of the indicated occurrences of $@_a\Diamond d$ can in an obvious way be mapped to a chain in the $\Box$-graph of π of greater length which contains the mentioned formula occurrences. The conclusions of the lemma follow straightforwardly. The other three cases are similar.

Definition 2.5. A *segment* in a derivation π is a non-empty list $\phi_1,\ldots,\phi_n$ of formula occurrences in π with the following properties.

1. ϕ_1 is not the conclusion of a (R_θ) rule with more than zero non-relational premises.
2. For each $i < n$, ϕ_i is a non-relational premise of a (R_θ) rule which has ϕ_{i+1} as the conclusion.
3. ϕ_n is not a non-relational premise of a (R_θ) rule.

The *length* of a segment is the number of formula occurrences in the segment.

The following lemma is along the lines of a similar result for ordinary classical first-order logic given in Prawitz (1965).

Lemma 2.10. *By repeated applications of proper and permutative reductions, any derivation π can be rewritten to a derivation π' in which each maximum or permutable formula is stubborn. If all maximum or permutable formulas in π are of the form $@_a\neg c$, then the derivation π' is normal.*

Proof. To any derivation π we assign the pair (d,k) of non-negative integers where d is the maximal degree of a non-stubborn maximum or permutable formula in π, or 0 if there is no such formula occurrence, and k is the sum of the lengths of segments in π in which the last formula occurrence is a non-stubborn maximum or permutable formula with degree d (note that a list of formula occurrences with only one element is a segment if the one and only formula occurrence in the list is a maximum formula). The proof is by induction on such pairs equipped with the lexicographic order. Let π be a derivation to which a pair (d,k) is assigned such that $d > 0$. It is straightforward that there exists a non-stubborn maximum or permutable formula ϕ with degree d such that no non-stubborn maximum or permutable formula above a minor or relational premise of the rule instance of which ϕ is the major premise is with degree d. (Consider an arbitrary formula occurrence θ in the set of non-stubborn maximum or permutable formulas with degree d. We are done if θ satisfies the mentioned criterium. Otherwise, consider instead a non-stubborn maximum or permutable formula with degree d such that it is above a minor or relational premise of the rule instance of which θ is major premise. This step is repeated until a formula occurrence is found that satisfies the mentioned criterium.) Let π' be the derivation obtained by applying the appropriate reduction rule such that ϕ is removed. Then it is straightforward to check that the pair (d',k') assigned to π' is less than (d,k) in the lexicographic order. Moreover, it is straightforward to check that if all maximum or permutable formulas in π are of the form $@_a\neg c$, then all maximum or permutable formulas in π' are also of the form $@_a\neg c$.

We are now ready to prove the normalization theorem.

Theorem 2.3. *(Normalization) Any derivation in $\mathbf{N}_{\mathscr{H}(\mathscr{O})} + \mathbf{T}$ can be rewritten to a normal derivation by repeated applications of proper and permutative reductions.*

Proof. The first step of the theorem is to prove that any derivation can be rewritten to a derivation in which each maximum or permutable formula is stubborn. This follows from Lemma 2.10.

The second step of the theorem is to prove that any derivation which is the result of the first step can be rewritten to a derivation in which all maximum or permutable formulas are of the form $@_a \neg c$ (thus, all stubborn formulas have been removed). The second step is similar to Lemma 2.10. To any derivation π in which each non-stubborn maximum or permutable formula is of the form $@_a \neg c$, we assign the pair (d,k) of non-negative integers where d is the maximal potential of a stubborn formula in π or 0 if there is no such formula occurrence and k is the number of stubborn formulas in π with potential d. Let π be a derivation to which a pair (d,k) is assigned such that $d > 0$. It is straightforward that there exists a stubborn formula ϕ with potential d such that no formula occurrence above a minor or relational premise of the rule instance of which ϕ is a major premise is stubborn and with potential d. Let π' be the derivation obtained by applying the appropriate reduction such that ϕ is removed. Then by inspecting the reduction rules it is trivial to check that each maximum or permutable formula in π' either is of the form $@_a \neg c$ or is stubborn, and moreover, by Lemma 2.9 the pair (d',k') assigned to π' is less than (d,k) in the lexicographic order. Thus, by induction we obtain a derivation in which each maximum or permutable formula is of the form $@_a \neg c$.

The third step of the theorem is to prove that any derivation which is the result of the second step can be rewritten to a normal derivation. This follows from Lemma 2.10.

Note that our notion of normalization involves permutative reductions which is unusual for a classical natural deduction system. Intuitionistic systems, on the other hand, generally involve permutative reductions in connection with derivation rules for the connectives $\perp$, $\vee$, and $\exists$.

2.2.5 The Form of Normal Derivations

Below we shall adapt an important definition from Prawitz (1965) to hybrid logic.

Definition 2.6. A *branch* in a derivation π is a non-empty list $\phi_1, \ldots, \phi_n$ of formula occurrences in π with the following properties.

1. For each $i < n$, ϕ_i stands immediately above ϕ_{i+1}.
2. ϕ_1 is an assumption, or a relational conclusion, or the conclusion of a (R_θ) rule with zero non-relational premises.
3. ϕ_n is either the end-formula of π or a minor or relational premise.
4. For each $i < n$, ϕ_i is not a minor or relational premise.

Note that ϕ_1 in the definition above might be a discharged assumption.

Lemma 2.11. *Any formula occurrence in a derivation π belongs to a branch in π.*

Proof. Induction on the structure of π.

The definition of a branch leads us to the lemma below which says that a branch in a normal derivation can be split into three parts: An analytical part in which formulas are broken down into their components by successive applications of the elimination rules, a minimum part in which an instance of the rule $(\bot 1)$ may occur, and a synthetical part in which formulas are put together by successive applications of the introduction rules. See Prawitz (1971).

Lemma 2.12. *Let $\beta = \phi_1, \ldots, \phi_n$ be a branch in a normal derivation. Then there exists a formula occurrence ϕ_i in β, called the* minimum formula *in β, such that*

1. *for each $j < i$, ϕ_j is the major premise of an elimination rule, or the non-relational premise of an instance of $(Nom1)$, or the premise of an instance of the rule $(\bot 2)$, or a non-relational premise of an instance of a (R_θ) rule;*
2. *if $i \neq n$, then ϕ_i is a premise of an introduction rule or the premise of an instance of the rule $(\bot 1)$; and*
3. *for each j, where $i < j < n$, ϕ_j is a premise of an introduction rule, or the non-relational premise of an instance of $(Nom1)$, or a non-relational premise of an instance of a (R_θ) rule.*

Proof. Let ϕ_i be the first formula occurrence in β which is not the major premise of an elimination rule, is not the non-relational premise of an instance of $(Nom1)$, and is not the premise of an instance of the rule $(\bot 2)$, and is not a non-relational premise of an instance of a (R_θ) rule (such a formula occurrence exists in β as ϕ_n satisfies the mentioned criterium). We are done if $i = n$. Otherwise ϕ_i is a premise of an introduction rule or the premise of an instance of the rule $(\bot 1)$ (by inspection of the rules and the definition of a branch). If ϕ_i is the premise of an instance of the rule $(\bot 1)$, then ϕ_{i+1} has the form $@_a\psi$ where ψ is a propositional symbol. Therefore each ϕ_j, where $i < j < n$, is a premise of an introduction rule, or the non-relational premise of an instance of $(Nom1)$, or a non-relational premise of an instance of a (R_θ) rule (by inspection of the rules, the definition of a branch, and normality of π). Similarly, if ϕ_i is a premise of an introduction rule, then each ϕ_j, where $i < j < n$, is a premise of an introduction rule or a non-relational premise of an instance of a (R_θ) rule.

The lemma above is more technically involved than the corresponding result in Prawitz (1965), the reason being the disturbing effect of $(Nom1)$, $(\bot 2)$, and the (R_θ) rules. In the theorem below we make use of the following definition.

Definition 2.7. The notion of a *subformula* is defined by the conventions that

- ϕ is a subformula of ϕ;
- if $\psi \wedge \theta$ or $\psi \rightarrow \theta$ is a subformula of ϕ, then so are ψ and θ;
- if $@_a\psi$ or $\Box\psi$ is a subformula of ϕ, then so is ψ; and
- if $\downarrow a\psi$ or $\forall a\psi$ is a subformula of ϕ, then so is $\psi[c/a]$ for any nominal c.

A formula $@_a\phi$ is a *quasi-subformula* of a formula $@_c\psi$ if and only if ϕ is a subformula of ψ.

Now comes the theorem which says that normal derivations satisfy a version of the subformula property.

Theorem 2.4. *(Quasi-subformula property) Let Γ be a set of satisfaction statements and let π be a normal derivation of ϕ from Γ in $\mathbf{N}_{\mathcal{H}(\mathcal{O})} + \mathbf{T}$. Moreover, let θ be a formula occurrence in π such that*

1. *θ is not an assumption discharged by an instance of the rule $(\bot 1)$ where the discharged assumption is the major premise of an instance of $(\rightarrow E)$;*
2. *θ is not an occurrence of $@_a\bot$ in a branch whose first formula is an assumption discharged by an instance of the rule $(\bot 1)$ where the discharged assumption is the major premise of an instance of $(\rightarrow E)$; and*
3. *θ is not an occurrence of $@_a\bot$ in a branch whose first formula is the conclusion of a (R_θ) rule with zero non-relational premises.*

Then θ is a quasi-subformula of ϕ, or of some formula in Γ, or of some relational premise, or of some relational conclusion, or of some relationally discharged assumption.

Proof. First a small convention: The *order* of a branch in π is the number of formula occurrences in π which stand below the last formula occurrence of the branch. Now consider a branch $\beta = \phi_1, \ldots, \phi_n$ in π of order p. By induction we can assume that the theorem holds for all formula occurrences in branches of order less than p. Note that it follows from Lemma 2.12 that any formula occurrence ϕ_j such that $j \leq i$, where ϕ_i minimum formula in β, is a quasi-subformula of ϕ_1, and similarly, any ϕ_j such that $j > i$ is a quasi-subformula of ϕ_n.

We first consider ϕ_n. We are done if ϕ_n is the end-formula ϕ or a relational premise. Otherwise ϕ_n is the minor premise of an instance of $(\rightarrow E)$. If the major premise of this instance of $(\rightarrow E)$ is not an assumption discharged by an instance of the rule $(\bot 1)$, then we are done by induction as the major premise belongs to a branch of order less than p. If the major premise of the instance of $(\rightarrow E)$ in question is an assumption discharged by an instance of the rule $(\bot 1)$, then we are done by induction as the conclusion of this instance of $(\bot 1)$, which has the same form as ϕ_n, belongs to a branch of order less than p.

We now consider ϕ_1. We are done if ϕ_1 is an undischarged assumption, or a relationally discharged assumption, or a relational conclusion. If ϕ_1 is the conclusion of a (R_θ) rule with zero non-relational premises, and if ϕ_1 is not of the form $@_a\bot$, then ϕ_1 has the same form as the minimum formula, which is a premise of an introduction rule and hence a quasi-subformula of ϕ_n. If ϕ_1 is not discharged by an instance of $(\bot 1)$, then it is discharged by an instance of $(\rightarrow I)$ with a conclusion that belongs to β or to some branch of order less than p. If ϕ_1 is discharged by an instance of $(\bot 1)$, then we have three cases. We are done if $n = 1$. If $n \neq 1$ and ϕ_1 is the minimum formula of β, then ϕ_1 is a premise of an introduction rule and hence a quasi-subformula of ϕ_n. If $n \neq 1$, but ϕ_1 is not the minimum formula of β, then ϕ_1 is either the major premise of an instance of $(\rightarrow E)$ or a non-relational premise of an instance of a (R_θ) rule. The first case is clear and in the second case ϕ_1 has the same form as the minimum formula which is a premise of an introduction rule and hence quasi-subformula of ϕ_n.

The first two exceptions in the theorem above are inherited from the standard natural deduction system for classical logic, see Prawitz (1965), whereas the third is related to the possibility of having a (R_θ) rule with zero non-relational premises. Remark: If the formula occurrence θ is not covered by one of the three exceptions, then it is a quasi-subformula of ϕ, or of some formula in Γ, or of a formula of the form $@_a c$ or $@_a \Diamond c$ (since relational premises, relational conclusions, and relationally discharged assumptions are of the form $@_a c$ or $@_a \Diamond c$). Note that the formulation of the theorem involves the notion of a branch in what appears to be an indispensable way.

2.2.6 *Discussion*

The natural deduction systems given in the present section share several features with the Gentzen systems given in the next section, for example the feature that all formulas in derivations are satisfaction statements. This feature is also shared by the hybrid-logical tableau and Gentzen systems given by Patrick Blackburn (2000a) (which are similar to the tableau and Gentzen systems considered in Chapter 3). However, since the system of the present section is in natural deduction style, we provide a proof-theoretic analysis in the form of a normalization theorem and a theorem which says that any normal derivation satisfies a version of the subformula property, namely the quasi-subformula property. On the other hand, in general tableau systems and cut-free or analytic Gentzen systems trivially satisfy the subformula property. A difference between our work and Blackburn (2000a) is that we consider additional derivation rules corresponding to first-order conditions expressed by geometric theories whereas Blackburn (2000a) considers tableau systems extended with axioms being pure hybrid-logical formulas, that is, formulas that contain no ordinary propositional symbols (thus, the only propositional symbols in such formulas are nominals).

The use of geometric theories in the context of proof-theory traces back to Alex Simpson's PhD thesis Simpson (1994) where it was pointed out that formulas in basic geometric theories correspond to simple natural deduction rules for intuitionistic modal logic. First-order conditions expressed by geometric theories cover a very wide class of logics. This is for example witnessed by the fact that any *Geach axiom schema*, that is, modal-logical axiom schema of the form

$$\Diamond^k \Box^m \phi \rightarrow \Box^l \Diamond^n \phi$$

where $\Box^j$ (respectively $\Diamond^j$) is an abbreviation for a sequence of j occurrences of $\Box$ (respectively $\Diamond$), corresponds to a formula of the form required in a basic geometric theory. To be precise, such a Geach axiom schema corresponds to the first-order formula

$$\forall a \forall b \forall c((R^k(a,b) \wedge R^l(a,c)) \rightarrow \exists d(R^m(b,d) \wedge R^n(c,d)))$$

where $R^0(a,b)$ means $a = b$ and $R^{j+1}(a,b)$ means $\exists e(R(a,e) \wedge R^j(e,b))$. The displayed formula is then equivalent to a formula of the form required in a basic geometric theory, cf. Simpson (1994) and Basin et al. (1997). An example of a Geach axiom schema is the axiom schema obtained by taking each of the numbers k, m, l, and n to be one. The corresponding first-order condition is called *directedness* (or *Church-Rosser*), see Section 2.2.1 for the natural deduction rule corresponding to this condition. It is notable that this property is not definable in terms of pure formulas involving just nominals and satisfaction operators, cf. Areces and ten Cate (2007, p. 843).

In the natural deduction system considered in Simpson (1994), a distinction is made between the language of ordinary modal logic and a metalanguage involving atomic first-order formulas of the form $R(a,c)$ together with formulas of the form $a : \phi$, where ϕ is a formula of ordinary modal logic. One contribution of the present chapter is to demonstrate that basic geometric theories correspond to natural deduction rules for hybrid logic where no such distinction between an object language and a metalanguage is made. It should be mentioned that a natural deduction system for classical modal logic which is similar to the system of Simpson (1994) has been given in Basin et al. (1997). However, one difference is that only Horn clause theories are considered in the latter work. A slightly modified version of the system in Basin et al. (1997) is described in Section 9.1.

The feature of our natural deduction and Gentzen systems that all formulas in derivations are satisfaction statements is at a general level in line with the fundamental idea of Melvin Fitting's prefixed tableau systems (1983) and Dov Gabbay's labelled deductive systems (1996) which is to prefix formulas in derivations by metalinguistic indexes, or labels, with the aim of regulating the proof process. Note that the work of Simpson (1994) fits naturally into this framework. It should also be mentioned that labelled deductive systems are the basis for the natural deduction systems for substructural logics given in Broda et al. (1999). The crucial difference between the work of Fitting (1983), Gabbay (1996), Simpson (1994), and Broda et al. (1999) and our work is that the indexes, or labels, used in the mentioned work belong to a metalanguage whereas in our systems they are part of the object language, namely the language of hybrid logic.[3] Thus, in the the terminology of Blackburn (2000a), the metalanguage has in our systems been internalized in the object language. We shall return to this issue in more detail in Chapters 9 and 10.

Jerry Seligman's paper (1997) should also be mentioned here: This paper gives a natural deduction system for a logic of situations similar to hybrid logic; the system in question is, however, quite different from ours, see Chapter 4 for a comparison.

---

[3] Labelled systems have the labelling machinery at the metalevel, whereas hybrid-logical systems have machinery with similar effect at the object level. A third option is chosen in Fitting (1972b) where a curious modal-logical axiom system is given in which labelling machinery is incorporated directly into the object language itself. In that system sequences of formulas of ordinary modal logic, delimited by a distinguished symbol $*$, are used as names for possible worlds. To be more specific, a sequence $*\Diamond\phi_1,\ldots,\Diamond\phi_n,\Diamond\phi_{n+1}*$ is used as the name of a world accessible from the world named by $*\Diamond\phi_1,\ldots,\Diamond\phi_n*$ and in which the formula ϕ_{n+1} is true, if there is one. It is allowed to form object language formulas by prefixing ordinary modal-logical formulas with such sequences. Intuitively, a prefixed formula $*\Diamond\phi_1,\ldots,\Diamond\phi_n * \psi$ says that the formula ψ is true at the world named by the prefix. Prefixed formulas can be combined using the usual connectives of classical logic.

2.3 The Basics of Gentzen Systems

In this section we shall sketch the basics of Gentzen systems. See Prawitz (1965) and Troelstra and Schwichtenberg (1996) for further details.

Derivations in Gentzen systems have the form of finite trees where the nodes are labelled with sequents $\Gamma \vdash \Delta$ such that any sequent in a derivation is the conclusion of a rule-instance which has the immediate successors of the sequent in question as the premises. The root of a derivation is called the *end-sequent* of the derivation. The sets of formulas Γ and Δ in a sequent $\Gamma \vdash \Delta$ are finite[4]. The usual intuitive reading of such a sequent is that the truth of all the formulas in Γ implies the truth of at least one formula in Δ. By convention Γ, ϕ and ϕ, Γ are abbreviations for $\Gamma \cup \{\phi\}$, and similarly, Γ, Δ is an abbreviation for $\Gamma \cup \Delta$, thus, a comma on the left hand side of a sequent is intuitively read as a conjunction whereas a comma on the right hand side of a sequent is intuitively read as a disjunction.

We shall make use of the following conventions. A derivation π is called a *derivation of* a sequent $\Gamma \vdash \Delta$ if the end-sequent of π is $\Gamma \vdash \Delta$. If there exists a derivation of the sequent $\Gamma \vdash \Delta$, then we shall simply say that the sequent is *derivable*. The formulas shown explicitly in the conclusion of a rule are called *principal formulas* and the formulas shown explicitly in the premises of a rule are called *side-formulas*. All other formulas in a rule are called *parametric formulas*.

Like natural deduction systems, Gentzen systems are characterised by having two different kinds of rules for each connective, but whereas natural deduction rules either introduce or eliminate a connective, Gentzen rules either introduce a connective on the left hand side of a sequent (that is, the connective occurs in a principal formula on the left hand side of the rule, but not in any side-formula) or introduce the connective on the right hand side of a sequent (the connective occurs in a principal formula on the right hand side of the rule, but not in any side-formula). Rules that introduce a connective on the left hand side of a sequent traditionally have names in the form $(\ldots L \ldots)$, and similarly, rules that introduce a connective on the right hand side of a sequent traditionally have names in the form $(\ldots R \ldots)$. For an instructive and important example, see the standard Gentzen rules for propositional logic in Figure 2.7.

One Gentzen rule which does not introduce a connective is the famous rule called the *cut* rule.

$$\frac{\Gamma \vdash \Delta, \phi \qquad \phi, \Gamma \vdash \Delta}{\Gamma \vdash \Delta} \ (Cut)$$

The formula ϕ is called the *cut-formula*. If a Gentzen system includes the cut rule, then the cut rule can most often be proved to be redundant, that is, the cut rule is admissible in the system obtained by leaving out the cut rule. However, in some

[4] Instead of using sets of formulas, it is possible to use multisets or lists of formulas. In some cases this is more convenient for combinatorial manipulation, but the cost is that rules have to be added to make the multisets or lists behave as sets: In the case of multisets, contraction rules have to be added (allowing formulas to be copied) and in the case of lists, exchange rules also have to be added (allowing formulas to be permuted).

$$\frac{}{\phi,\Gamma \vdash \Delta,\phi}\ (Axiom) \qquad \frac{}{\bot,\Gamma \vdash \Delta}\ (\bot)$$

$$\frac{\phi,\psi,\Gamma \vdash \Delta}{\phi \wedge \psi,\Gamma \vdash \Delta}\ (\wedge L) \qquad \frac{\Gamma \vdash \Delta,\phi \quad \Gamma \vdash \Delta,\psi}{\Gamma \vdash \Delta,\phi \wedge \psi}\ (\wedge R)$$

$$\frac{\Gamma \vdash \Delta,\phi \quad \psi,\Gamma \vdash \Delta}{\phi \to \psi,\Gamma \vdash \Delta}\ (\to L) \qquad \frac{\phi,\Gamma \vdash \Delta,\psi}{\Gamma \vdash \Delta,\phi \to \psi}\ (\to R)$$

Fig. 2.7 Gentzen rules for propositional logic

cases the cut rule is not completely redundant, but a restricted version is needed where the cut-formula is a subformula of some formula already occuring in the end-sequent. A cut rule with this restriction is called an *analytic* cut rule. A derivation is called cut-free if no instances of the cut rule occur in it, and in most Gentzen systems, cut-free derivations satisfy the subformula property which says that any formula in a derivation is a subformula of the end-sequent. Clearly, this property does not hold if the cut rule is allowed (but note that the property is not violated by analytic cuts). Thus, in most Gentzen systems the absence of cuts (to be precise, the absence of non-analytic cuts) in a derivation guarantees that the derivation satisfies the subformula property.[5]

A Gentzen system can often be viewed as a metacalculus for the derivability relation in a natural deduction system, cf. Prawitz (1965, p. 90). See also Scott (1981), §9, for a very instructive discussion. According to this view, a sequent $\Gamma \vdash \phi$ states that the formula ϕ is derivable from the set of formulas Γ in a natural deduction system under consideration (in the interest of simplicity, we here only consider the case with one formula on the right hand side of the sequent). Thus, according to this view, the Gentzen system provides rules for deriving statements of the form $\Gamma \vdash \phi$. Of course, it is a requirement of the Gentzen system that a sequent $\Gamma \vdash \phi$ is derivable if and only if it is the case that ϕ is derivable from Γ in the natural duduction system. Note that this view implies that Gentzen rules are read from top to bottom, namely as derivation rules.

A different view of a Gentzen system is that Gentzen rules are read from bottom to top and the rules step by step attempt to define a counter-model to a sequent, that is, a model together with an assignment which makes all the antecedent formulas true and all the succedent formulas false. Note that according to the second view, the rules of a Gentzen system are directly understood in terms of model-theory. We shall come back to this view in the next chapter.

[5] Note that this is different from natural deduction systems where the subformula property is guaranteed if a derivation has a certain form, namely if it is in normal form.

2.4 Gentzen Systems for Propositional Hybrid Logic

In this section we will give a Gentzen system corresponding to our natural deduction system $\mathbf{N}_{\mathcal{H}(\mathcal{O})} + \mathbf{T}$. We use the normalization theorem for the natural deduction system to prove a result which says that normal derivations in the natural deduction system correspond to cut-free derivations in the Gentzen system (note that there is danger of confusion here as Gentzen discovered natural deduction style as well as what here and elsewhere is called Gentzen style, see Gentzen (1969)). This implies the completeness of the Gentzen system without cuts.

The rules for the Gentzen system are given in Figures 2.8 and 2.9. As with the natural deduction rules given in Section 2.2, all formulas in the Gentzen rules are satisfaction statements. Our Gentzen system for $\mathcal{H}(\mathcal{O})$, which will be denoted $\mathbf{G}_{\mathcal{H}(\mathcal{O})}$, is obtained from the rules given in Figures 2.8 and 2.9 by leaving out the rules for the binders that are not in $\mathcal{O}$. Below is an example of a derivation in $\mathbf{G}_{\mathcal{H}}$.

$$\dfrac{\dfrac{\dfrac{\dfrac{\dfrac{\dfrac{\dfrac{}{@_a\phi \vdash @_a\psi, @_a\phi}\,(Axiom) \qquad \dfrac{}{@_a\psi, @_a\phi \vdash @_a\psi}\,(Axiom)}{@_a(\phi \to \psi), @_a\phi \vdash @_a\psi}\,(\to L)}{@_a(\phi \to \psi), @_a\phi \vdash @_b @_a\psi}\,(@R)}{@_a(\phi \to \psi), @_b @_a\phi \vdash @_b @_a\psi}\,(@L)}{@_a(\phi \to \psi) \vdash @_b(@_a\phi \to @_a\psi)}\,(\to R)}{@_b @_a(\phi \to \psi) \vdash @_b(@_a\phi \to @_a\psi)}\,(@L)}{\vdash @_b(@_a(\phi \to \psi) \to (@_a\phi \to @_a\psi))}\,(\to R)$$

The end-formula of the derivation is the modal axiom K for the satisfaction operator $@_a$ prefixed by a satisfaction operator. Compare this with the natural deduction derivation of the same formula given in Section 2.2.

The Gentzen system $\mathbf{G}_{\mathcal{H}(\mathcal{O})}$ can be extended with additional derivation rules corresponding to geometric first-order conditions on the accessibility relations. This is analogous to the extension of the natural deduction system $\mathbf{N}_{\mathcal{H}(\mathcal{O})}$ which was described in Section 2.2.1. Recall that a basic geometric theory is a geometric theory in which each formula has the form

$$(*) \qquad \forall \overline{a}((S_1 \wedge \cdots \wedge S_n) \to \exists \overline{c} \bigvee_{j=1}^{m} (S_{j1} \wedge \cdots \wedge S_{jn_j}))$$

where $n, m \geq 0$ and $n_1, \ldots, n_m \geq 1$. Exactly as in the case with the natural deduction system, we assume for simplicity that the variables in the list $\overline{a}$ are pairwise distinct, that the variables in $\overline{c}$ are pairwise distinct, and that no variable occurs in both $\overline{c}$ and $\overline{a}$. With a formula θ of the form $(*)$ displayed above, we associate the Gentzen rule (R_θ) given in Figure 2.10 where s_k is of the form $HT(S_k)$ and s_{jk} is of the form $HT(S_{jk})$ (HT is the translation from first-order logic to hybrid logic given in Section 1.2.1). Given a basic geometric theory $\mathbf{T}$, the Gentzen system ob-

$$\frac{}{\phi,\Gamma \vdash \Delta,\phi}\ (Axiom) \qquad \frac{}{@_a\bot,\Gamma \vdash \Delta}\ (\bot)$$

$$\frac{@_a\phi, @_a\psi, \Gamma \vdash \Delta}{@_a(\phi \wedge \psi), \Gamma \vdash \Delta}\ (\wedge L) \qquad \frac{\Gamma \vdash \Delta, @_a\phi \quad \Gamma \vdash \Delta, @_a\psi}{\Gamma \vdash \Delta, @_a(\phi \wedge \psi)}\ (\wedge R)$$

$$\frac{\Gamma \vdash \Delta, @_a\phi \quad @_a\psi, \Gamma \vdash \Delta}{@_a(\phi \rightarrow \psi), \Gamma \vdash \Delta}\ (\rightarrow L) \qquad \frac{@_a\phi, \Gamma \vdash \Delta, @_a\psi}{\Gamma \vdash \Delta, @_a(\phi \rightarrow \psi)}\ (\rightarrow R)$$

$$\frac{\Gamma \vdash \Delta, @_a\Diamond e \quad @_e\phi, \Gamma \vdash \Delta}{@_a\Box\phi, \Gamma \vdash \Delta}\ (\Box L) \qquad \frac{@_a\Diamond c, \Gamma \vdash \Delta, @_c\phi}{\Gamma \vdash \Delta, @_a\Box\phi}\ (\Box R)^*$$

$$\frac{@_a\phi, \Gamma \vdash \Delta}{@_c@_a\phi, \Gamma \vdash \Delta}\ (@L) \qquad \frac{\Gamma \vdash \Delta, @_a\phi}{\Gamma \vdash \Delta, @_c@_a\phi}\ (@R)$$

$$\frac{@_b\phi[e/a], \Gamma \vdash \Delta}{@_b\forall a\phi, \Gamma \vdash \Delta}\ (\forall L) \qquad \frac{\Gamma \vdash \Delta, @_b\phi[c/a]}{\Gamma \vdash \Delta, @_b\forall a\phi}\ (\forall R)^*$$

$$\frac{\Gamma \vdash \Delta, @_ae \quad @_e\phi[e/b], \Gamma \vdash \Delta}{@_a{\downarrow} b\phi, \Gamma \vdash \Delta}\ ({\downarrow} L) \qquad \frac{@_ac, \Gamma \vdash \Delta, @_c\phi[c/b]}{\Gamma \vdash \Delta, @_a{\downarrow} b\phi}\ ({\downarrow} R)^*$$

* c does not occur free in the conclusion.

Fig. 2.8 Gentzen rules for connectives

$$\frac{@_aa, \Gamma \vdash \Delta}{\Gamma \vdash \Delta}\ (Ref) \qquad \frac{\Gamma \vdash \Delta, @_ac \quad \Gamma \vdash \Delta, @_a\phi}{\Gamma \vdash \Delta, @_c\phi}\ (Nom1)^*$$

$$\frac{\Gamma \vdash \Delta, @_ac \quad \Gamma \vdash \Delta, @_a\Diamond b \quad @_c\Diamond b, \Gamma \vdash \Delta}{\Gamma \vdash \Delta}\ (Nom2)$$

* ϕ is a propositional symbol (ordinary or a nominal).

Fig. 2.9 Gentzen rules for nominals

tained by extending $\mathbf{G}_{\mathscr{H}(\mathscr{O})}$ with the set of rules $\{(R_\theta) \mid \theta \in \mathbf{T}\}$ will be denoted by $\mathbf{G}_{\mathscr{H}(\mathscr{O})} + \mathbf{T}$.

2.4.1 Soundness and Completeness

We now use the normalization theorem for our natural deduction system $\mathbf{N}_{\mathscr{H}(\mathscr{O})} + \mathbf{T}$ to prove a lemma which implies the completeness of $\mathbf{G}_{\mathscr{H}(\mathscr{O})} + \mathbf{T}$.

$$\frac{\Gamma \vdash \Delta, s_1 \quad \dots \quad \Gamma \vdash \Delta, s_n \quad s_{11}, \dots, s_{1n_1}, \Gamma \vdash \Delta \quad \dots \quad s_{m1}, \dots, s_{mn_m}, \Gamma \vdash \Delta}{\Gamma \vdash \Delta} \ (R_\theta)^*$$

$*$ None of the nominals in $\overline{c}$ occur free in Γ or Δ.

Fig. 2.10 Gentzen rules for geometric theories

Lemma 2.13. *Let Γ be a set of satisfaction statements, let π be a normal derivation of ψ from Γ in $\mathbf{N}_{\mathscr{H}(\mathscr{O})} + \mathbf{T}$, and moreover, let $\{@_{a_1}\neg\phi_1, \dots, @_{a_n}\neg\phi_n\} \subseteq \Gamma$ where $n \geq 0$, let $\Gamma^* = \Gamma - \{@_{a_1}\neg\phi_1, \dots, @_{a_n}\neg\phi_n\}$, and let $\Delta = \{@_{a_1}\phi_1, \dots, @_{a_n}\phi_n\}$. Then there exists a derivation of the sequent $\Gamma^* \vdash \Delta, \psi$ in $\mathbf{G}_{\mathscr{H}(\mathscr{O})} + \mathbf{T}$.*

Proof. We first prove that the lemma holds for the Gentzen system $\mathbf{G}'_{\mathscr{H}(\mathscr{O})} + \mathbf{T}$ which is obtained from the system $\mathbf{G}_{\mathscr{H}(\mathscr{O})} + \mathbf{T}$ by replacing the axiom $(\bot)$ by the rule

$$\frac{\Gamma \vdash \Delta, @_a\bot}{\Gamma \vdash \Delta, \Delta'}$$

Observe that a derivation τ in the system $\mathbf{G}_{\mathscr{H}(\mathscr{O})} + \mathbf{T}$ ($\mathbf{G}'_{\mathscr{H}(\mathscr{O})} + \mathbf{T}$) of a sequent $\Gamma \vdash \Delta$ can be transformed into a derivation in $\mathbf{G}_{\mathscr{H}(\mathscr{O})} + \mathbf{T}$ ($\mathbf{G}'_{\mathscr{H}(\mathscr{O})} + \mathbf{T}$) of any sequent $\Gamma \cup \Gamma' \vdash \Delta \cup \Delta'$ simply by adding Γ' and Δ' to the sets of formulas in the sequents of τ and by renaming of nominals.

The proof that the lemma holds for $\mathbf{G}'_{\mathscr{H}(\mathscr{O})} + \mathbf{T}$ is by induction on the number of rule instances in π. We only cover the case where ψ is the conclusion of an elimination rule; the other cases are straightforward. Let $\beta = \psi_1, \dots, \psi_n$ be a branch in π such that $\psi_n = \psi$. Since ψ_n is the conclusion of an elimination rule, each formula occurrence in β except ψ_n is the major premise of an elimination rule, cf. Lemma 2.12. Thus, the formula occurrence ϕ_1 cannot be a discharged assumption. So ϕ_1 is either an undischarged assumption, or a relational conclusion, or the conclusion af a (R_θ) rule with zero non-relational premises. The cases where ϕ_1 is a relational conclusion or the conclusion af a (R_θ) rule with zero non-relational premises are straightforward. If ϕ_1 is an undischarged assumption, then we split up in subcases depending on the form of ϕ_1. Note that $\phi_1 \in \Gamma$. We only cover the subcase where ϕ_1 is of the form $@_a\Box\phi$; the other subcases are similar. So we have a derivation of $@_a\Diamond e$ from Γ for some nominal e, and moreover, we have a derivation of ψ from $@_e\phi, \Gamma$. By induction we get derivations in $\mathbf{G}'_{\mathscr{H}(\mathscr{O})} + \mathbf{T}$ of the sequents $\Gamma^* \vdash \Delta, @_a\Diamond e$ and $@_e\phi, \Gamma^* \vdash \Delta, \psi$. It is then easy to build a derivation in $\mathbf{G}'_{\mathscr{H}(\mathscr{O})} + \mathbf{T}$ of $\Gamma^* \vdash \Delta, \psi$ using the rule $(\Box L)$ (note that $@_a\Box\phi, \Gamma^* = \Gamma^*$ since $@_a\Box\phi = \phi_1$ and $\phi_1 \in \Gamma$).

It is straightforward to demonstrate that a derivation τ in $\mathbf{G}_{\mathscr{H}(\mathscr{O})} + \mathbf{T}$ of a sequent $\Gamma \vdash \Delta, @_a\bot$ can be transformed into a derivation in $\mathbf{G}_{\mathscr{H}(\mathscr{O})} + \mathbf{T}$ of the sequent $\Gamma \vdash \Delta$ by removing $@_a\bot$ from the right-hand-side sets of formulas in the sequents of τ and by replacing instances of (*Axiom*) by instances of $(\bot)$. It follows that the lemma holds for $\mathbf{G}_{\mathscr{H}(\mathscr{O})} + \mathbf{T}$.

Intuitively, the lemma above says that normal natural deduction derivations of the system $\mathbf{N}_{\mathscr{H}(\mathscr{O})} + \mathbf{T}$ can be mimicked by cut-free Gentzen derivations of $\mathbf{G}_{\mathscr{H}(\mathscr{O})} + \mathbf{T}$.

Thus, the lemma is in line with the view of a Gentzen system according to which the Gentzen system is a metacalculus for the derivability relation in a natural deduction system. Now soundness and completeness.

Theorem 2.5. *(Soundness and completeness) Let Γ and Δ be sets of satisfaction statements. The two statements below are equivalent.*

1. *$\Gamma \vdash \Delta$ is derivable in $\mathbf{G}_{\mathscr{H}(\mathscr{O})} + \mathbf{T}$.*
2. *For any $\mathbf{T}$-model $\mathfrak{M}$ and any assignment g, if, for any formula $\theta \in \Gamma$, $\mathfrak{M}, g \models \theta$, then for some formula $\psi \in \Delta$, $\mathfrak{M}, g \models \psi$.*

Proof. Soundness is by induction on the structure of the derivation of $\Gamma \vdash \Delta$. Completeness follows from Lemma 2.13.

It is straightforward to show that the system $\mathbf{G}_{\mathscr{H}(\mathscr{O})} + \mathbf{T}$ is still complete if (*Axiom*) is equipped with the side-condition that ϕ has one of the forms $@_a p$, $@_a c$, or $@_a \Diamond c$.

2.4.2 The Form of Derivations

We shall now prove a theorem which says that derivations in the Gentzen system $\mathbf{G}_{\mathscr{H}(\mathscr{O})} + \mathbf{T}$ satisfy a version of the subformula property. The theorem is analogous to Theorem 2.4 for the natural deduction system $\mathbf{N}_{\mathscr{H}(\mathscr{O})} + \mathbf{T}$. We first make the convention that if a side-formula in a premise of a rule has the form $@_a c$ or $@_a \Diamond c$, then it is called a *relational side-formula*.

Theorem 2.6. *(Quasi-subformula property) Let π be a derivation of a sequent $\Gamma \vdash \Delta$ in $\mathbf{G}_{\mathscr{H}(\mathscr{O})} + \mathbf{T}$. Any formula occurrence in π is a quasi-subformula of some formula in Γ or Δ, or is a quasi-subformula of some relational side-formula (quasi-subformulas are defined in Definition 2.7).*

Proof. Induction on the structure of π.

It follows from the theorem above that every formula occurring in the derivation π is a quasi-subformula of a formula in Γ or Δ or is a quasi-subformula of a formula of the form $@_a c$ or $@_a \Diamond c$.

2.4.3 Discussion

The Gentzen system given in the present section is somewhat similar to the Gentzen system given in Blackburn (2000a). As mentioned earlier, these systems share the feature that all formulas in derivations are satisfaction statements. However, the Gentzen system in Blackburn (2000a) is a reformulated tableau system whereas the system in the present section stems from a natural deduction system in the sense that it is designed with the aim of being able to mimic normal natural deduction derivations by cut-free Gentzen derivations. In the next chapter, to be more precise in Section 3.4, we shall consider a Gentzen system which is much closer to the one given in Blackburn (2000a). See Section 3.4 for further discussion of similarities and differences in these systems.

2.5 Axiom Systems for Propositional Hybrid Logic

In this section we shall give a sound and complete Hilbert-style axiom system for the hybrid logic $\mathscr{H}(\mathscr{O})$. Completeness of the axiom system is proved by reduction to completeness of the natural deduction system $\mathbf{N}_{\mathscr{H}(\mathscr{O})}$ given in Section 2.2. See also Section 6.3 where an axiom system for first-order hybrid logic is given and Section 8.3 where an axiom system for intuitionistic hybrid logic is given. The axiom system is comprised of all instances of theorems of propositional logic (that is, tautologies) together with the axioms and rules in Figure 2.11. Our axiom system for the hybrid logic $\mathscr{H}(\mathscr{O})$, which will be denoted $\mathbf{A}_{\mathscr{H}(\mathscr{O})}$, is obtained by leaving out axioms and rules for the binders that are not in $\mathscr{O}$. Some of the axioms and rules, for example the axioms (*Scope*), (*Ref*), and (*Intro*), and the rule ($N_@$), are well-known from the literature on axiom systems for classical hybrid logic, see Blackburn and Tzakova (1999).

$$
\begin{array}{ll}
(\mathit{Distr}_{\rightarrow}) & @_a(\phi \rightarrow \psi) \leftrightarrow (@_a\phi \rightarrow @_a\psi) \\
(\bot) & @_a\bot \rightarrow \bot \\
(\mathit{Scope}) & @_a@_b\phi \leftrightarrow @_b\phi \\
(\mathit{Ref}) & @_a a \\
(\mathit{Intro}) & (a \wedge \phi) \rightarrow @_a\phi \\
(\Box E) & (\Box\phi \wedge \Diamond e) \rightarrow @_e\phi \\
(\forall E) & \forall b\phi \rightarrow \phi[e/b] \\
(\downarrow E) & (\downarrow b\phi \wedge e) \rightarrow @_e\phi[e/b]
\end{array}
$$

$$
\frac{\phi \rightarrow \psi \quad \phi}{\psi}\,(MP) \qquad \frac{\phi}{@_a\phi}\,(N_@) \qquad \frac{@_a\phi}{\phi}\,(\mathit{Name})^*
$$

$$
\frac{(\psi \wedge \Diamond c) \rightarrow @_c\phi}{\psi \rightarrow \Box\phi}\,(\Box I)^{\star} \qquad \frac{\psi \rightarrow \phi[c/b]}{\psi \rightarrow \forall b\phi}\,(\forall I)^{\dagger} \qquad \frac{(\psi \wedge c) \rightarrow @_c\phi[c/b]}{\psi \rightarrow \downarrow b\phi}\,(\downarrow I)^{\ddagger}
$$

$*$ a does not occur free in ϕ.
$\star$ c does not occur free in ϕ or ψ.
$\dagger$ c does not occur free in $\forall b\phi$ or ψ.
$\ddagger$ c does not occur free in $\downarrow b\phi$ or ψ.

Fig. 2.11 Hilbert-style axioms and rules

Clearly, the rules ($\Box I$), ($\forall I$), and ($\downarrow I$) in Figure 2.11 correspond to the natural deduction introduction rules with the same names given in Figure 2.2 of Section 2.2, and similarly, the axioms ($\Box E$), ($\forall E$), and ($\downarrow E$) correspond to the natural deduction elimination rules with the same names.

It is instructive to compare our Hilbert-style axiomatic machinery for the modal operator $\Box$ to the Hilbert-style axiomatic machinery for the first-order universal quantifier in free logic which is a variant of ordinary first-order logic where quantifiers only range over a subset of the universe (but where variables might refer to any member of the universe as in ordinary first-order logic). One original motivation for developing free logic was to avoid the assumption made in ordinary first-order logic that quantifier domains are non-empty as this assumption by a number of philosophers was found undesirable because of the associated "existential commitment". In free logic, the axiomatic machinery for the universal quantifier is constituted by the rule and the axiom

$$\frac{(\psi \wedge E(z)) \rightarrow \phi[z/x]}{\psi \rightarrow \forall x\phi}\ (\mathit{Free}\ \forall I) \qquad\qquad (\mathit{Free}\ \forall E) \quad (\forall x\phi \wedge E(t)) \rightarrow \phi[t/x]$$

where the rule is equipped with the usual side-condition that the variable z does not occur free in ψ or in $\forall x\phi$. Here $E(z)$ is the existence predicate which is defined as $\exists y(y = z)$ where y is a variable distinct from z. The idea in the rule $(\mathit{Free}\ \forall I)$ is that the "guard" formula $E(z)$ in the antecedent ensures that the antecedent is false in the case where the variable z refers to an individual outside the range of the quantifier. This is analogous to the idea in our Hilbert-style rule $(\Box I)$ for hybrid logic which is that the guard formula $\Diamond c$ in the antecedent ensures that the antecedent is false in the case where the nominal c refers a world that is not accessible. Of course, a similar remark applies in connection with the rule $(\mathit{Free}\ \forall E)$ and our Hilbert-style rule $(\Box E)$. The above rule $(\mathit{Free}\ \forall I)$ and axiom $(\mathit{Free}\ \forall E)$ for free logic are actually used in Section 6.3 in connection with first-order hybrid logic. See Bencivenga (2002) for more information on free logic.

The axiom system $\mathbf{A}_{\mathscr{H}(\mathscr{O})}$ can be extended with additional rules corresponding to first-order conditions on the accessibility relation expressed by geometric theories. In analogy with the extension of the natural deduction system $\mathbf{N}_{\mathscr{H}(\mathscr{O})}$ described in Section 2.2.1, the Hilbert-style rule (R_θ) given in Figure 2.6 is associated with a formula θ in a basic geometric theory. Compare to the corresponding natural deduction rule given in Figure 2.6 of Section 2.2.1.

$$\frac{(\psi \rightarrow (s_1 \wedge \ldots \wedge s_n)) \wedge (\wedge_{j=1}^{m}((\psi \wedge s_{j1} \wedge \ldots \wedge s_{jn_j}) \rightarrow \phi))}{\psi \rightarrow \phi}\ (R_\theta)^*$$

$*$ None of the nominals in $\overline{c}$ occur in ϕ or ψ (recall that nominals are identified with first-order variables and note that $\overline{c}$ are the first-order variables existentially quantified over in the formula θ)

Fig. 2.12 Hilbert-style rules for geometric theories

2.5.1 Soundness and Completeness

The axiom system is sound and complete with respect to the semantics given earlier. In what follows, we shall say that a formula is derivable in the axiom system $\mathbf{A}_{\mathcal{H}(\mathcal{O})}$ if there exists a derivation of a formula in question in the axiom system. We first consider soundness.

Theorem 2.7. *(Soundness) If a formula ϕ is derivable in the axiom system $\mathbf{A}_{\mathcal{H}(\mathcal{O})}$, then ϕ is valid.*

Proof. Induction on the structure of the derivation.

We need a lemma to prove completeness.

Lemma 2.14. *Let a finite set $\Gamma = \{\phi_1, \ldots, \phi_n\}$ of satisfaction statements be given. If a satisfaction statement ψ is derivable from Γ in the natural deduction system $\mathbf{N}_{\mathcal{H}(\mathcal{O})}$, cf. Figure 2.2, then the formula*

$$(\phi_1 \wedge \ldots \wedge \phi_n) \rightarrow \psi$$

is derivable in the axiom system $\mathbf{A}_{\mathcal{H}(\mathcal{O})}$.

Proof. Induction on the derivation of ψ.

Now the completeness theorem.

Theorem 2.8. *(Completeness) If a formula ϕ is valid, then ϕ is derivable in the axiom system $\mathbf{A}_{\mathcal{H}(\mathcal{O})}$.*

Proof. If the formula ϕ is valid, then $@_a\phi$ is valid where a is an arbitrary nominal that does not occur in ϕ. Thus, the satisfaction statement $@_a\phi$ is derivable in the natural deduction system $\mathbf{N}_{\mathcal{H}(\mathcal{O})}$, cf. Figure 2.2, as it is complete, so by Lemma 2.14 the formula $\top \rightarrow @_a\phi$ is derivable in the axiom system $\mathbf{A}_{\mathcal{H}(\mathcal{O})}$. It follows by propositional reasoning and the rule (*Name*) that ϕ is derivable in $\mathbf{A}_{\mathcal{H}(\mathcal{O})}$.

It is straightforward to modify the soundness and completeness results above to encompass the rules corresponding to geometric conditions on the accessibility relation: Let $\mathbf{T}$ be any basic geometric theory, cf. Section 2.2.1. A frame is called a $\mathbf{T}$-*frame* if and only if for every hybrid-logical model $\mathfrak{M}$ based on the frame in question and every formula $\theta \in \mathbf{T}$, it is the case that $\mathfrak{M}^* \models \theta$ (recall that $\mathfrak{M}^*$ is the first-order model corresponding to the hybrid-logical model $\mathfrak{M}$). In the soundness and completeness results, validity is then relativised to the class of $\mathbf{T}$-frames and the axiom system $\mathbf{A}_{\mathcal{H}(\mathcal{O})}$ is extended with the set of rules $\{(R_\theta) \mid \theta \in \mathbf{T}\}$ where (R_θ) is the Hilbert-style rule in Figure 2.12 which is associated with a formula θ in the basic geometric theory.

2.5.2 Discussion

A number of axiom systems for hybrid logic can be found in the literature, see for example Blackburn (1993), Blackburn and Tzakova (1999), and Blackburn and ten Cate (2006), but they are all different from the one we give here. The difference can be explained as follows: Axiom systems for ordinary modal logic are usually obtained by extending axiom systems for propositional logic with axioms and rules for modal operators (for example, an axiom system for the ordinary modal logic K can be obtained by extending an axiom system for propositional logic with the axiom $\Box(\phi \rightarrow \psi) \rightarrow (\Box\phi \rightarrow \Box\psi)$ and the necessitation rule, that is, from ϕ derive $\Box\phi$). Beside modal operators, these axioms and rules only involve connectives from propositional logic. Now, other axiom systems for hybrid logic found in the literature "factor through" such axiom systems for ordinary modal logic in the sense that they are obtained by extending such axiom systems with further axioms and rules.

This is different from our axiom systems for hybrid logic where all modal axioms and rules involve satisfaction operators and nominals (beside ordinary propositional connectives). Thus, our axiom system is not an extension of an axiom system for ordinary modal logic. We find this approach of significance since we take it that satisfaction operators and nominals on their own are natural extensions of propositional logic (possibly with the addition of binders). In fact, logics similar to such hybrid logics without the usual modal operators have already been considered in the literature, one example is in the context of situation theory, see Seligman (1997). An example of an intuitionistic hybrid logic that does not involve the usual modal operators can be found in Reed (2007). Of course, such logics can be motivated by an interest in the corresponding models, that is, multi-state domains, where nominals refer to different states and satisfaction operators effect jumps between the states. To sum up, our approach to axiomatisation of modalities is to factor through hybrid logic without modalities rather than factoring through ordinary modal logic. Analogous remarks apply to the axiom systems for first-order hybrid logic and intuitionistic hybrid logic which we shall give in respectively Sections 6.3 and 8.3.

Chapter 3
Tableaus and Decision Procedures for Hybrid Logic

Based on tableau systems, we in this chapter prove decidability results for hybrid logic using tableau systems. The chapter is structured as follows. In the first section of the chapter we sketch the basics of tableau systems. In the second section we give a tableau-based decision procedure for a very expressive hybrid logic including the universal modality. In the third section we show how the decision procedure of the second section can be modified such that simpler tableau-based decision procedures (that is, without loop-checks) are obtained for a weaker hybrid logic where the universal modality is not included. In the fourth section we reformulate the tableau systems of the second and the third sections as Gentzen systems and we discuss how to reformulate the decision procedures. In the fifth section we discuss the results. The results of the second, fourth, and fifth sections of this chapter are taken from Bolander and Braüner (2006). The material in the third section is new (but the tableau systems considered in the third section are obtained by directly modifying the tableau system given in the second section, inspired by a tableau-based decision procedure given in Bolander and Blackburn (2007)).

3.1 The Basics of Tableau Systems

Before giving our hybrid-logical tableau systems, we shall sketch the basics of tableau systems and fix terminology.

A number of persons have played a role in the invention of tableau systems, a leading figure being Jaako Hintikka, see Hintikka (1955). A milestone in the later development of tableau systems is Fitting (1983). See D'Agostino (1999) for further details. Hintikka made the following remarks on the idea behind tableau systems, namely to mimic the recursive truth-conditions in the semantics, whereby a formula is broken down into its components.

> ...the typical situation is one in which we are confronted by a complex formula (or sentence) the truth or falsity of which we are trying to establish by inquiring into its components. Here the rules of truth operate from the complex to the simple: they serve to tell us

T. Braüner, *Hybrid Logic and its Proof-Theory*, Applied Logic Series 37,
DOI 10.1007/978-94-007-0002-4_3,

> what, under the supposition that a given complex formula or sentence is true, can be said about the truth-values of its components. (Hintikka 1955, p. 20)

The method of reasoning in tableau systems is called "backwards" reasoning: Starting with a particular formula whose validity you want to prove, a tableau is built step by step using the rules, whereby more and more information about counter-models for the formula is obtained, and if at some stage it can be concluded that there cannot be such models, it has been proved that the formula in question is valid. This is contrary to natural deduction systems which are forward reasoning systems since you start with natural deduction rules and try to build a derivation of the formula you have in mind, cf. Section 2.1.

A *tableau* is a well-founded tree in which each node is labelled with a formula, and the edges represent applications of tableau rules. Where it is appropriate, we shall blur the distinction between a formula and an occurrence of the formula in a tableau. By applying rules to a tableau, the tableau is expanded, that is, new edges and formulas are added to the leaves. A tableau is displayed such that it grows downwards. Technically, premises and conclusions of tableau rules are finite sets of formulas, and a tableau rule has one premise, and one or more conclusions. Most often the premise contains zero, one, or two formulas whereas a conclusion most often contains one or two formulas. A requirement for applying a rule to a branch in a tableau is that all the formulas in the premise are present at the branch, and the result of applying the rule is that for each conclusion of the rule, the end of the branch is extended with a path containing a node for each of the formulas in the conclusion in question. Thus, if for example the rule has two conclusions, then the result of applying the rule is that the end of the branch is extended with two paths, one path for each conclusion. If the rule only has one conclusion, no splitting takes place. A branch in a tableau is called *open* if for no formula χ occurring on the branch, it is the case that $\neg\chi$ also occurs on the branch. A branch is called *closed* if it is not open. A tableau is called *closed* if all branches are closed.

$$\frac{\neg\neg\phi}{\phi}\,(\neg\neg)$$

$$\frac{\phi \wedge \psi}{\phi, \psi}\,(\wedge) \qquad \frac{\neg(\phi \wedge \psi)}{\neg\phi \mid \neg\psi}\,(\neg\wedge)$$

Fig. 3.1 Tableau rules for propositional logic

For an example of a tableau system, see the standard tableau rules for propositional logic in Figure 3.1. Note that the rule $(\wedge)$ in Figure 3.1 has one conclusion, namely $\{\phi, \psi\}$, which has two formulas, thus, the result of applying this rule to a branch is that the branch is extended with one path containing two nodes which are labelled with respectively ϕ and ψ. On the other hand, the rule $(\neg\wedge)$ has two conclusions, namely $\{\neg\phi\}$ and $\{\neg\psi\}$, thus, the result of applying this rule to a branch is that the branch is extended with two paths each containing one node, where one node is labelled with ϕ and one node is labelled with ψ. Below is a sample tableau to illustrate how we present tableaus.

	$\neg\neg((p \wedge \neg(p \wedge \neg q)) \wedge \neg q)$	1.
	$(p \wedge \neg(p \wedge \neg q)) \wedge \neg q$	2. by $(\neg\neg)$ rule on 1
	$p \wedge \neg(p \wedge \neg q)$	3. by $(\wedge)$ rule on 2
	$\neg q$	4. by $(\wedge)$ rule on 2
	p	5. by $(\wedge)$ rule on 3
	$\neg(p \wedge \neg q)$	6. by $(\wedge)$ rule on 3
7. $\neg p$	$\neg\neg q$	8. by $(\neg\wedge)$ rule on 6
	q	9. by $(\neg\neg)$ rule on 8

The enumeration of the formulas and the notation in the right-hand-side column is not a formal part of the tableau, but has been added to describe how the tableau was constructed. Note that there are two branches and they are both closed (the left-hand-side branch contains p as well as $\neg p$ whereas the right-hand-side branch contains q as well as $\neg q$).

Tableau rules are read from top to bottom, and given an appropriate notion of a model, the intuition behind tableau rules is that the rules step by step attempt to define a model for the root formula of a tableau. This intuition presupposes that tableau rules are sound in the sense that the rules preserve the existence of models, to be more precise, if the premise of a rule has a model (all formulas in the premise are true), then this model is a model for at least one conclusion of the rule (all formulas in the conclusion in question are true). It follows that if the root formula of a tableau has a model, then there is at least one branch in the tableau such that the model for the root formula is a model for all the formulas on the branch, and hence, information about the model for the root formula can be read off from the branch. On the other hand, such a branch obviously has to be open since no formulas χ and $\neg\chi$ can both be true in the same model, so if the tableau does not have any open branches, that is, all its branches are closed, then it can be concluded that the root formula does not have a model. Thus, if a tableau with only closed branches can be constructed having a formula $\neg\phi$ as the root formula, then it has been proved that the formula ϕ is valid. For example, the sample tableau above proves that the formula $\neg((p \wedge \neg(p \wedge \neg q)) \wedge \neg q)$ is valid, which is not a surprise since it is equivalent to $(p \wedge (p \rightarrow q)) \rightarrow q$.

Tableau rules are similar to Gentzen rules, and the rules of a Gentzen system can also be viewed as rules that attempt to define a model, namely a counter-model for the end-sequent, see Section 2.3, but note that according to this view, Gentzen rules are read from bottom to top (whereas tableau rules are read in the opposite direction).

3.2 A tableau System Including the Universal Modality

A central issue in this chapter is the very expressive hybrid logic obtained by extending the hybrid logic $\mathscr{H}$ in Section 1.2 with the universal modality E (which is dual to the modality A considered in Section 1.3). Formally, the notion of a model is kept as it is in Section 1.2, but the definition of the relation $\mathfrak{M}, g, w \models \phi$ is extended with the clause

$$\mathfrak{M}, g, w \models E\phi \text{ iff for some } v \in W, \mathfrak{M}, g, v \models \phi$$

where $\mathfrak{M} = (W, R, \{V_w\}_{w \in W})$ is a model, g is an assignment, and w is an element of W. The hybrid logic $\mathscr{H}$ extended with the universal modality will be denoted $\mathscr{H}(E)$. In the present chapter we define the dual operator A of E by the convention that $A\phi$ is an abbreviation for $\neg E \neg \phi$, thus, E is primitive and A is defined (note that it is opposite in Section 1.3). Moreover, in this chapter we take the connectives $\neg$ and $\Diamond$ to be primitive and $\rightarrow$, $\bot$, and $\Box$ to be defined. It is well-known that the hybrid logic $\mathscr{H}(E)$ is decidable, see Areces et al. (2001), but decision procedures for this logic are usually not based on tableau or Gentzen systems. In the present section, we shall give a decision procedure for $\mathscr{H}(E)$ based on a tableau system. An essential feature of our decision procedure is that it makes use of a technique called loop-checks.

3.2.1 *Tableau Rules for Hybrid Logic*

The rules for the hybrid-logical tableau system are given in Figures 3.2 and 3.3. The tableau system will be denoted $\mathbf{T}_{\mathscr{H}(E)}$. All formulas in the rules are satisfaction statements or negated satisfaction statements, hence, each node in a tableau is labelled with a satisfaction statement or the negation of a satisfaction statement. Note that since we have taken the connectives $\rightarrow$, $\bot$, $\Box$, and A to be defined, not primitive, they do not need separate rules. It is straightforward to check that the rules of $\mathbf{T}_{\mathscr{H}(E)}$ are sound in the sense that if the premise of a rule has a model (strictly speaking together with an assignment), then this model, possibly with modified references to new nominals, is a model for at least one conclusion of the rule.

We shall make use of the following conventions about the tableau rules. The rules $(\neg)$, $(\neg\neg)$, $(\wedge)$, $(\neg\wedge)$, $(@)$, $(\neg @)$, $(\Diamond)$, and (E) will be called *destructive* rules and the remaining rules will be called *non-destructive*. The reason why we call the mentioned rules destructive is that in the systematic tableau construction algorithm we define later in this section, application of destructive rules is restricted such that a destructive rule is applied at most once to a formula (a destructive rule has exactly one formula in the premise).[1] The destructive rules $(\Diamond)$ and (E) will also be called

[1] This terminology is used in a somewhat different sense than is common: Our destructive rules preserve information in the sense that if a conclusion of a destructive rule has a model, then this model is a model for the premise of the rule as well, that is, no models are included (note that this

$$\frac{@_a\neg\phi}{\neg@_a\phi}\,(\neg) \qquad \frac{\neg@_a\neg\phi}{@_a\phi}\,(\neg\neg)$$

$$\frac{@_a(\phi\wedge\psi)}{@_a\phi,\,@_a\psi}\,(\wedge) \qquad \frac{\neg@_a(\phi\wedge\psi)}{\neg@_a\phi\mid\neg@_a\psi}\,(\neg\wedge)$$

$$\frac{@_c@_a\phi}{@_a\phi}\,(@) \qquad \frac{\neg@_c@_a\phi}{\neg@_a\phi}\,(\neg@)$$

$$\frac{@_a\Diamond\phi}{@_c\phi,\,@_a\Diamond c}\,(\Diamond)^{*\star} \qquad \frac{\neg@_a\Diamond\phi,\,@_a\Diamond e}{\neg@_e\phi}\,(\neg\Diamond)$$

$$\frac{@_aE\phi}{@_c\phi}\,(E)^{*} \qquad \frac{\neg@_aE\phi}{\neg@_e\phi}\,(\neg E)^{\dagger}$$

$*$ The nominal c is new.
$\star$ The formula ϕ is not a nominal.
$\dagger$ The nominal e is on the branch.

Fig. 3.2 Tableau rules for connectives

$$\frac{}{@_aa}\,(Ref)^{*} \qquad \frac{@_ac,\,@_a\phi}{@_c\phi}\,(Nom1)^{\star} \qquad \frac{@_ac,\,@_a\Diamond b}{@_c\Diamond b}\,(Nom2)$$

$*$ The nominal a is on the branch.
$\star$ The formula ϕ is a propositional symbol (ordinary or a nominal).

Fig. 3.3 Tableau rules for nominals

existential since they introduce new nominals. Note that non-destructive rules are only applicable to formulas in the forms $@_ap$, $@_ac$, $@_a\Diamond c$, $\neg@_a\Diamond\phi$, and $\neg@_aE\phi$, and conversely, destructive rules are only applicable to formulas not in these forms (in fact, exactly one destructive rule is applicable to any formula which is not in one of these forms). So, the classification of rules as destructive and non-destructive corresponds to a classification of formulas according to their form.

In the remaining part of the present section we shall give a decision procedure $\mathcal{H}(E)$ which works as follows: Given a formula $@_a\phi$ whose validity we have to decide, a systematic tableau construction algorithm constructs a finite tableau having the formula $\neg@_a\phi$ as the root formula. If the tableau has an open branch, then a model for $\neg@_a\phi$ can be defined.[2] Thus, in this case the formula $@_a\phi$ is not valid.

---

is opposite of soundness which says that no models are excluded). In the usual sense destructive rules are rules that do not preserve information, see Fitting (1972a).

[2] An occurrence of a satisfaction statement $@_a\phi$ or the negation of a satisfaction statement $\neg@_a\phi$ in a tableau can be seen as a formula ϕ together with a pair consisting of the representation of a possible world (the nominal a) and the representation of a truth-value (depending on whether the satisfaction statement is negated or not). Note in this connection that in the possible worlds

On the other hand, in the case where there are no open branches in the tableau, it follows from soundness of the tableau rules that $\neg @_a \phi$ does not have a model, hence $@_a \phi$ is valid.

3.2.2 Some Properties of the Tableau System

In this section we shall prove some properties of the tableau system. Only Theorem 3.1 and Proposition 3.1 are used later in the present section, but we find that the other results of the section are of independent interest. In the next section we give a decision procedure based on a tableau system for the weaker hybrid logic $\mathscr{H}$ that does not include the universal modality, and in this connection we shall make crucial use of Corollary 3.1 and a strengthened version of Theorem 3.3 (as well as Theorem 3.1 and Proposition 3.1 again).

The tableau system $\mathbf{T}_{\mathscr{H}(E)}$ satisfies the following variant of the quasi-subformula property.

Theorem 3.1. *(Quasi-subformula property) If a formula $@_a \phi$ occurs in a tableau where ϕ is not a nominal and ϕ is not of the form $\Diamond b$, then ϕ is a subformula of the root formula. If a formula $\neg @_a \phi$ occurs in a tableau, then ϕ is a subformula of the root formula.*

Proof. A simultaneous induction where each rule is checked.

Below we shall give some further results which shows some interesting features of the tableau system. First two definitions.

Definition 3.1. Let Θ be a branch of a tableau and let N^Θ be the set of nominals occurring in the formulas of Θ. Define a binary relation $\sim_\Theta$ on N^Θ by $a \sim_\Theta b$ if and only if the formula $@_a b$ occurs on Θ. Let $\sim^*_\Theta$ be the reflexive, symmetric, and transitive closure of $\sim_\Theta$.

Definition 3.2. An occurrence of a nominal in a formula is *equational* if the occurrence is a formula (that is, if the occurrence is not part of a satisfaction operator).

---

semantics, the semantic value assigned to a formula is a function from possible worlds to truth-values, and set-theoretically, such a function is a set of pairs of possible worlds and truth-values (called the graph of the function). Hence, the pairs of nominals and representations of truth-values associated with formulas in the tableau system can be considered representations of elements of functions constituting semantic values. Thus, the tableau rules step by step build up semantic values of the formulas involved, similar to the way in which the accessibility relation step by step is built up (there is a difference however; the accessibility relation can be *any* relation, but the semantic value of a formula has to be a *function*, that is, a relation where no element of the domain is related to more than one element of the codomain, and this is exactly what is required of an open branch in a tableau, namely that no satisfaction statement is related to more than one truth-value).

For example, the occurrence of the nominal c in the formula $\phi \wedge c$ is equational but the occurrence of c in $\psi \wedge @_c\chi$ is not. The justification for this terminology is that a nominal in the first-order correspondence language (and thereby also in the semantics) gives rise to an equality statement if and only if the nominal occurrence in question occurs equationally. Note that a nominal occurs equationally in a formula if and only if the nominal is a subformula of the formula.

Theorem 3.2. *Let $@_ab$ be a formula occurrence on a branch Θ of a tableau. If the nominals a and b are different, then each of the nominals is identical to, or related by $\sim_\Theta$ to, a nominal with an equational occurrence in the root formula.*

Proof. Check each rule. Theorem 3.1 is needed in a number of the cases. In the case with the rule $(\Diamond)$, we make use of the restriction that the rule cannot be applied to formulas of the form $@_a\Diamond\phi$ where ϕ is a nominal.

Corollary 3.1. *Let Θ be a branch of a tableau. Any non-singleton equivalence class with respect to the equivalence relation $\sim^*_\Theta$ contains a nominal with an equational occurrence in the root formula.*

Proof. Follows directly from Theorem 3.2.

The corollary above says that non-trivial equational reasoning, that is, reasoning involving non-singleton equivalence classes, only takes place in connection with certain nominals in the root formula, namely those that occur equationally. Note that this implies that pure modal input to the tableau only gives rise to reasoning involving singleton equivalence classes.

Definition 3.3. A formula occurrence in a tableau is an *accessibility* formula occurrence if it is an occurrence of the formula $@_a\Diamond c$ generated by the rule $(\Diamond)$.

Note that if the rule $(\Diamond)$ is applied to a formula occurrence $@_a\Diamond\Diamond b$, resulting in the branch being extended with $@_a\Diamond c$ and $@_c\Diamond b$, then the occurrence of $@_a\Diamond c$ is an accessibility formula occurrence, but the occurrence of $@_c\Diamond b$ is not.

Theorem 3.3. *Let $@_a\Diamond b$ be a formula occurrence on a branch Θ of a tableau. Either there is an accessibility formula occurrence $@_{a'}\Diamond b$ on Θ such that $a \sim^*_\Theta a'$ or the formula $\Diamond b$ is a subformula of the root formula.*

Proof. Check each rule. Theorem 3.1 is needed in some of the cases.

The only way new nominals can be introduced to a tableau is by using one of the rules $(\Diamond)$ or (E) which we called existential rules. This motivates the following definition.

Definition 3.4. Let Θ be a branch of a tableau. If a new nominal c is introduced by applying an existential rule to a satisfaction statement $@_a\phi$, then we write $a <_\Theta c$.

The definition above gives us a binary relation $<_\Theta$ on the set N^Θ.

Proposition 3.1. *Let Θ be a branch of a tableau. Assume that if an existential rule is applied to a formula occurrence on Θ, then the existential rule is not applied to any other formula occurrence at Θ having the same form. The graph $(N^\Theta, <_\Theta)$ is the disjoint union of a finite set of well-founded and finitely branching trees.*

Proof. That the graph is well-founded follows from noting that if $a <_\Theta c$, then the first occurrence of a on the branch is before the first occurrence of c. That the graph is the disjoint union of a set of trees follows from well-foundedness together with the observation that if $a <_\Theta c$ and $b <_\Theta c$, then the nominals a and b are identical. That the set of trees is finite follows from the observation that for any nominal c that occurs in the branch, but does not occur in the root formula, there is a nominal a such that $a <_\Theta c$, thus, the nominal c cannot be the root of a tree.

The following argument shows that the trees are finitely branching. Assume conversely that there exists an infinite sequence $a <_\Theta c_1, a <_\Theta c_2, \ldots$ of pairwise distinct edges. For each i, the edge $a <_\Theta c_i$ is generated by applying an existential rule to some formula occurrence χ_i. Consider the sequence $\chi_1, \chi_2, \ldots$ of formula occurrences. These rule applications are distinct since the nominals $c_1, c_2, \ldots$ are distinct, and by assumption, if an existential rule is applied to a formula occurrence, then the existential rule is not applied to any other formula occurrence having the same form, so the formula occurrences in the sequence $\chi_1, \chi_2, \ldots$ are occurrences of infinitely many different formulas. Now, if the edge $a <_\Theta c_i$ is generated by applying the existential rule $(\Diamond)$ to χ_i, then χ_i is of the form $@_a \Diamond \phi_i$ where ϕ_i is not a nominal, and hence, $\Diamond \phi_i$ is a subformula of the root formula by Theorem 3.1, and if $a <_\Theta c_i$ is generated by applying the other existential rule (E) to χ_i, then χ_i is of the form $@_a E \phi_i$, and hence, $E \phi_i$ is a subformula of the root formula, again by Theorem 3.1. But there are only finitely many subformulas of the root formula, which contradicts that infinitely many different formulas occur in the sequence $\chi_1, \chi_2, \ldots$.

Note that in the above results we have not made any assumptions on which rules are applied on the branch Θ, but if we assume that Θ is closed under the rules (Ref) and $(Nom1)$, then $\sim^*_\Theta$ coincides with $\sim_\Theta$.

3.2.3 Systematic Tableau Construction

In this section we give a systematic tableau construction algorithm for $\mathbf{T}_{\mathcal{H}(E)}$. Before giving the algorithm, we need an important definition.

Definition 3.5. Let b and a be nominals occurring on a branch Θ of a tableau in $\mathbf{T}_{\mathcal{H}(E)}$. The nominal a is *included* in the nominal b with respect to Θ if the following is the case: For any subformula ϕ of the root formula, if the formula $@_a \phi$ occurs on Θ, then $@_b \phi$ also occurs on Θ, and similarly, if $\neg @_a \phi$ occurs on Θ, then $\neg @_b \phi$ also occurs on Θ. If a is included in b with respect to Θ, and the first occurrence of b on Θ is before the first occurrence of a, then we write $a \subseteq_\Theta b$.

We are now ready to give the systematic tableau construction algorithm.

Definition 3.6. (Tableau construction) Given a formula $@_a\phi$ of $\mathscr{H}(E)$ whose validity has to be decided, we define by induction a sequence $\mathscr{T}_0, \mathscr{T}_1, \mathscr{T}_2, \ldots$ of finite tableaus in $\mathbf{T}_{\mathscr{H}(E)}$, each of which is embedded in its successor. Let $\mathscr{T}_0$ be the finite tableau constituted by the single formula $\neg @_a\phi$. If possible, apply an arbitrary rule to $\mathscr{T}_n$ with the following three restrictions:

1. If a formula to be added to a branch by applying a rule already occurs on the branch, then the addition of the formula is simply omitted.
2. After the application of a destructive rule to a formula occurrence ϕ on a branch, it is recorded that the rule was applied to ϕ with respect to the branch and the rule will not again be applied to ϕ with respect to the branch or any extension of it.
3. The existential rule $(\Diamond)$ is not applied to a formula occurrence $@_a\Diamond\phi$ on a branch Θ if there exists a nominal b such that $a \subseteq_\Theta b$ (and analogously for the existential rule (E)).

Let $\mathscr{T}_{n+1}$ be the resulting tableau.

Note that due to the first restriction, a formula cannot occur more than once on a branch. Also note that no information is recorded about applications of non-destructive rules. The conditions on applications of the existential rules $(\Diamond)$ and (E) in the third restriction are the loop-check conditions. The intuition behind loop-checks is that an existential rule is not applied in a world if the information in that world can be found already in an ancestor world. Hence, the introduction of a new world by the existential rule is blocked.

Theorem 3.4. *The systematic tableau construction algorithm for $\mathbf{T}_{\mathscr{H}(E)}$ terminates in the sense that there exists an n such that $\mathscr{T}_n = \mathscr{T}_{n+1}$.*

Proof. Assume conversely that the algorithm does not terminate. Then the resulting tableau is infinite, and hence, has an infinite branch Θ. The graph $(N^\Theta, <_\Theta)$ is the disjoint union of a finite set of finitely branching trees cf. Proposition 3.1, so it has an infinite branch $a_1 <_\Theta a_2 <_\Theta a_3, \ldots$ (otherwise $\mathscr{N}^\Theta$ would be finite, and hence, by Theorem 3.1 there would only be finitely many formulas occurring on the branch Θ, contradicting that it is infinite). Now, for each i, let Θ_i be the initial segment of Θ up to, but not including, the first formula containing an occurrence of the nominal a_{i+1}. Thus, an existential rule was applied to a formula occurrence on the branch Θ_i resulting in the generation of a_{i+1}. Let Γ_i be the set of formulas which contains any subformula ϕ of the root formula such that $@_{a_i}\phi$ occurs on the branch Θ_i, and similarly, let Δ_i be the set of formulas which contains any subformula ϕ of the root formula such that $\neg @_{a_i}\phi$ occurs on the branch Θ_i. Since there are only finitely many sets of subformulas of the root formula, there exists j and k such that $j < k$ and $\Gamma_j = \Gamma_k$ as well as $\Delta_j = \Delta_k$. Clearly, the first occurrence of a_j on Θ_k is before the first occurrence of a_k. Moreover, for any subformula ϕ of the root formula, if $@_{a_k}\phi$ occurs on Θ_k, then $\phi \in \Gamma_k$, and hence, $\phi \in \Gamma_j$, but then $@_{a_j}\phi$ occurs on Θ_j which is an initial segment of Θ_k. A similar argument shows that if $\neg @_{a_k}\phi$ occurs on Θ_k, then $\neg @_{a_j}\phi$ also occurs on Θ_k. Hence, a_k is included in a_j with respect to Θ_k. We conclude that $a_k \subseteq_{\Theta_k} a_j$. But this contradicts that an existential rule was applied to a formula occurrence on the branch Θ_k resulting in the addition of the first formula containing an occurrence of the nominal a_{k+1}. Thus, the algorithm terminates.

We have thus given a systematic tableau construction algorithm which step by step builds up a tableau and which terminates with a tableau having the property that no rules are applicable to it except for applications of rules blocked by the three restrictions in Definition 3.6. It is important to note that except for these three restrictions, the tableau construction algorithm does not make any restrictions on the order in which rules are applied. In this sense the algorithm is non-deterministic.

3.2.4 The Model Existence Theorem and Decidability

In this section we give a model existence theorem and we give the decision procedure. The model existence theorem implies that the tableau system $\mathbf{T}_{\mathcal{H}(E)}$ is complete. Throughout the section, we shall assume that Θ is a given branch of a tableau generated by the systematic tableau construction algorithm, Definition 3.6. Where no confusion can occur, we shall often omit reference to the branch Θ. First some machinery.

Definition 3.7. Let W be the subset of N^{Θ} containing any nominal a having the property that there is no nominal b such that $a \subseteq_{\Theta} b$. Let $\approx$ be the restriction of $\sim_{\Theta}$ to W.

Note that W contains all nominals of the root formula since the root formula is the first formula of the branch Θ. Observe that Θ is closed under the rules (Ref) and $(Nom1)$, so the relation $\sim_{\Theta}$ and hence also the relation $\approx$ are equivalence relations. Given a nominal a in W, we let $[a]_{\approx}$ denote the equivalence class of a with respect to $\approx$ and we let $W/\!\approx$ denote the set of equivalence classes.

Definition 3.8. Let R be the binary relation on W defined by aRc if and only if there exists a nominal $c' \approx c$ such that one of the following two conditions is satisfied.

1. The formula $@_a \Diamond c'$ occurs on Θ.
2. There exists a nominal d in N^{Θ} such that the formula $@_a \Diamond d$ occurs on Θ and $d \subseteq_{\Theta} c'$.

Note that the nominal d referred to in the second item in the definition is not an element of W. It follows from Θ being closed under the rule $(Nom2)$ that R is compatible with $\approx$ in the first argument and it is trivial that R is compatible with $\approx$ in the second argument. We let $\overline{R}$ be the binary relation on $W/\!\approx$ defined by $[a]_{\approx}\overline{R}[c]_{\approx}$ if and only if aRc.

Definition 3.9. For any element a of W, let V_a be the function that to each ordinary propositional symbol assigns an element of $\{0,1\}$ such that $V_a(p) = 1$ if $@_a p$ occurs on Θ and $V_a(p) = 0$ otherwise.

It follows from Θ being closed under the rule $(Nom1)$ that V_a is compatible with $\approx$ in the index a, so we let $\overline{V}_{[a]_\approx}$ be defined by $\overline{V}_{[a]_\approx}(p) = V_a(p)$. We are now ready to define a model.

Definition 3.10. Let $\mathfrak{M}$ be the model $(W/{\approx}, \overline{R}, \{\overline{V}_{[a]_\approx}\}_{[a]_\approx \in W/\approx})$ and let the assignment g for $\mathfrak{M}$ be defined by $g(a) = [a]_\approx$.

The model above is in some respects similar to the model defined in Blackburn (2000a). One crucial difference, however, is that the model above is necessarily finite since the tableau branch Θ is finite.

Theorem 3.5. *(Model existence theorem) Assume that the branch Θ is open. For any satisfaction statement $@_a\phi$ which only contains nominals from W, the following two statements hold.*

- *If $@_a\phi$ occurs on Θ, then it is the case that $\mathfrak{M}, g, [a]_\approx \models \phi$.*
- *If $\neg @_a\phi$ occurs on Θ, then it is not the case that $\mathfrak{M}, g, [a]_\approx \models \phi$.*

Proof. Induction on the structure of ϕ. We proceed case by case.

The first case we consider is where ϕ is a nominal, say b. If $@_ab$ occurs on Θ, then $[a]_\approx = [b]_\approx$, and hence, it is the case that $\mathfrak{M}, g, [a]_\approx \models b$.

If $\neg @_ab$ occurs on Θ, then $@_ab$ does not occur at Θ (otherwise Θ would not be open), so $[a]_\approx \neq [b]_\approx$, and hence, it is not the case that $\mathfrak{M}, g, [a]_\approx \models b$.

The case where ϕ is an ordinary propositional symbol is similar to the above case where ϕ is a nominal.

The cases where ϕ are of the forms $\psi \wedge \theta$, $\neg\psi$, and $@_b\psi$ are straightforward applications of the induction hypothesis.

Now the case where ϕ is of the form $\Diamond\psi$. Assume that $@_a\Diamond\psi$ occurs on Θ. We then have to prove that $\mathfrak{M}, g, [a]_\approx \models \Diamond\psi$, that is, for some equivalence class $[c]_\approx$ such that $[a]_\approx\overline{R}[c]_\approx$, it is the case that $\mathfrak{M}, g, [c]_\approx \models \psi$. We have two cases, according to whether the formula ψ is a nominal or not. We first consider the case where ψ is a nominal, say b. So we just have to prove that $[a]_\approx\overline{R}[b]_\approx$ which trivially follows from the definition of the relation $\overline{R}$. We now consider the case where ψ is not a nominal. By the rule $(\Diamond)$ some formulas $@_a\Diamond c$ and $@_c\psi$ also occur on Θ where the nominal c is new (note that $a \in W$, so the application of the rule is not blocked by a loop-check condition). If $c \in W$, then clearly $[a]_\approx\overline{R}[c]_\approx$ and $\mathfrak{M}, g, [c]_\approx \models \psi$ by induction. If $c \notin W$, then by definition of W there exists a nominal d such that $c \subseteq_\Theta d$. Without loss of generality we assume that there does not exist a nominal e such that $d \subseteq_\Theta e$. But this implies that $d \in W$. Moreover, by Theorem 3.1, the formula ψ is a subformula of the root formula, so $@_d\psi$ occurs on Θ. By induction, $\mathfrak{M}, g, [d]_\approx \models \psi$, and clearly, $[a]_\approx\overline{R}[d]_\approx$.

Assume that $\neg @_a\Diamond\psi$ occurs on Θ. We then have to prove that $\mathfrak{M}, g, [a]_\approx \models \Diamond\psi$ does not hold, that is, for any equivalence class $[c]_\approx$ such that $[a]_\approx\overline{R}[c]_\approx$, it is not the case that $\mathfrak{M}, g, [c]_\approx \models \psi$. From $[a]_\approx\overline{R}[c]_\approx$ it follows that there exists a nominal $c' \approx c$ satisfying one of the two conditions in the definition of the relation R. In the first condition in this definition, the formula $@_a\Diamond c'$ occurs on Θ. Thus, by the rule $(\neg\Diamond)$ the formula $\neg @_{c'}\psi$ occurs on Θ. By induction we conclude that $\mathfrak{M}, g, [c']_\approx \models \psi$

does not hold and trivially, $[c']_{\approx} = [c]_{\approx}$. In the second condition in the definition there exists a nominal d in N^{Θ} such that the formula $@_a\Diamond d$ occurs on Θ and $d \subseteq_{\Theta} c'$. By the rule $(\neg\Diamond)$ the formula $\neg @_d \psi$ occurs on Θ. But by Theorem 3.1, the formula ψ is a subformula of the root formula, and $d \subseteq_{\Theta} c'$, so $\neg @_{c'} \psi$ occurs on Θ. By induction we conclude that $\mathfrak{M}, g, [c']_{\approx} \models \psi$ does not hold and trivially, $[c']_{\approx} = [c]_{\approx}$.

Finally the case where ϕ is of the form $E\psi$. Assume that $@_a E\psi$ occurs on Θ. We then have to prove that $\mathfrak{M}, g, [a]_{\approx} \models E\psi$, that is, for some equivalence class $[c]_{\approx}$, it is the case that $\mathfrak{M}, g, [c]_{\approx} \models \psi$. By the rule (E) also a formula $@_c \psi$ occurs on Θ where the nominal c is new (note that $a \in W$, so the application of the rule is not blocked by a loop-check condition). If $c \in W$, then $\mathfrak{M}, g, [c]_{\approx} \models \psi$ by induction. If $c \notin W$, then by definition of W there exists a nominal d such that $c \subseteq_{\Theta} d$. Without loss of generality we assume that there does not exist a nominal e such that $d \subseteq_{\Theta} e$. But this implies that $d \in W$. Moreover, by Theorem 3.1, the formula ψ is a subformula of the root formula, so $@_d \psi$ occurs on Θ. It follows by induction that $\mathfrak{M}, g, [d]_{\approx} \models \psi$.

Assume that the formula $\neg @_a E\psi$ occurs on Θ. We then have to prove that $\mathfrak{M}, g, [a]_{\approx} \models E\psi$ does not hold, that is, for any equivalence class $[c]_{\approx}$, it is not the case that $\mathfrak{M}, g, [c]_{\approx} \models \psi$. By the rule $(\neg E)$, the formula $\neg @_c \psi$ occurs on Θ, so by induction we conclude that $\mathfrak{M}, g, [c]_{\approx} \models \psi$ does not hold.

We are now finally able to give the decision procedure.

Definition 3.11. (Decision procedure) Given a formula $@_a\phi$ of $\mathscr{H}(E)$ whose validity we have to decide, let $\mathscr{T}_n$ be a terminal tableau generated by the tableau construction algorithm, Definition 3.6. If there are no open branches in the tableau $\mathscr{T}_n$, then the root formula $\neg @_a\phi$ of $\mathscr{T}_n$ does not have a model since the tableau rules are sound, hence, the formula $@_a\phi$ is valid. If the tableau $\mathscr{T}_n$ has an open branch, then it follows from the model existence theorem, Theorem 3.5, that the formula $@_a\phi$ is not valid.

As a spin-off from the decision procedure we get the finite model property.

Theorem 3.6. *(Finite model property) If a formula of $\mathscr{H}(E)$ is satisfiable, then it is satisfiable by a finite model.*

Proof. A straightforward application of the decision procedure, Definition 3.11, together with the observation that the model defined in Theorem 3.10 is finite.

We shall finish this section by making some remarks on complexity issues. Consider a branch Θ of a tableau with root formula ψ. If there are n distinct subformulas of ψ, then there are 2^n distinct sets of subformulas of ψ. It follows from inspection of the termination proof, Theorem 3.4, that the height of a tree in the graph $(N^{\Theta}, <_{\Theta})$, cf. Proposition 3.1, is $O(2^n)$. By inspection of Proposition 3.1, the outdegrees of nodes in the graph $(N^{\Theta}, <_{\Theta})$ are bounded by n (to be more precise, the outdegrees are bounded by the number of distinct subformulas of ψ having the form $\Diamond\phi$ or $E\phi$). It follows that the size of N^{Θ} is $O(2^{2^n})$. Combining this with the quasi-subformula property, Theorem 3.1, we can calculate that the length of the branch Θ is $O(2^{2^n})$.

It follows that our algorithm solves the satisfiability problem for $\mathscr{H}(E)$ formulas in nondeterministic double exponential time (2-NEXPTIME) in the size of formulas. However, the satisfiability problem for $\mathscr{H}(E)$ formulas is in fact solvable in exponential time (EXPTIME), cf. Areces et al. (2001), so the algorithm is not optimal from a complexity theoretic point of view. Our aim has been to give a simple and straightforward algorithm, but we believe that by sacrificing some of the simplicity, the algorithm can be optimized by applying the techniques which in Donini and Massacci (2000) are applied to give an optimal EXPTIME tableau-based algorithm for a description logic variant of the modal logic K extended with background theories. The techniques of the paper involve caching of unsatisfiability results for already explored tableau branches. However, according to the handbook chapter Horrocks et al. (2007, p. 220), the optimal EXPTIME algorithm for K with background theories given in Donini and Massacci (2000) has never been implemented, whereas the handbook chapter describes a simple 2-NEXPTIME tableau-based algorithm for the same logic which, again according to the handbook chapter, has proven to work surprisingly well in practice.

3.2.5 Tableau Examples

As a first example, consider the formula $@_a\neg\Diamond(a\wedge\neg\Diamond a)$ which is valid (this is straightforward to see by considering the equivalent formula $@_a\Box(a\rightarrow\Diamond a)$). Given this formula as input, a possible tableau generated by the tableau construction algorithm is the tableau below (recall that the algorithm is non-deterministic).

$\neg@_a\neg\Diamond(a\wedge\neg\Diamond a)$	1.
$@_a\Diamond(a\wedge\neg\Diamond a)$	2. by $(\neg\neg)$ rule on 1
$@_c(a\wedge\neg\Diamond a)$	3. by $(\Diamond)$ rule on 2
$@_a\Diamond c$	4. by $(\Diamond)$ rule on 2
$@_c a$	5. by $(\wedge)$ rule on 3
$@_c\neg\Diamond a$	6. by $(\wedge)$ rule on 3
$\neg@_c\Diamond a$	7. by $(\neg)$ rule on 6
$@_c c$	8. by (Ref) rule
$@_a c$	9. by $(Nom1)$ rule on 5 and 8
$@_a a$	10. by (Ref) rule
$@_c\Diamond c$	11. by $(Nom2)$ rule on 9 and 4
$\neg@_c a$	12. by $(\neg\Diamond)$ rule on 7 and 11

The tableau above only has one branch and that branch is closed since it contains the formula $@_c a$ as well as $\neg@_c a$ (in lines 5 and 12). It follows from the tableau rules being sound that the formula $@_a\neg\Diamond(a\wedge\neg\Diamond a)$ is valid, cf. the decision procedure, Definition 3.11.

As a second example, consider the formula $@_a\neg\Diamond(a\wedge r)$ which is not valid. Given this formula as input, a possible tableau generated by the tableau construction algorithm is the tableau below.

$\neg @_a \neg \Diamond (a \wedge r)$	1.
$@_a \Diamond (a \wedge r)$	2. by $(\neg\neg)$ rule on 1
$@_c (a \wedge r)$	3. by $(\Diamond)$ rule on 2
$@_a \Diamond c$	4. by $(\Diamond)$ rule on 2
$@_c a$	5. by $(\wedge)$ rule on 3
$@_c r$	6. by $(\wedge)$ rule on 3
$@_a r$	7. by $(Nom1)$ rule on 5 and 6
$@_c c$	8. by (Ref) rule
$@_a c$	9. by $(Nom1)$ rule on 5 and 8
$@_c \Diamond c$	10. by $(Nom2)$ rule on 9 and 4
$@_a a$	11. by (Ref) rule

Note that the tableau above only has one branch and that branch is open. It follows from the model existence theorem, Theorem 3.5, that $@_a \neg \Diamond (a \wedge r)$ is not valid, and Definition 3.10 gives a counter-model, namely a model having one world, the set $\{a, c\}$, which is an equivalence class as $@_c a, @_c c, @_a c, @_a a$ are on the branch. The propositional symbol r is true at the world as $@_c r, @_a r$ are on the branch and the world is related to itself by the accessibility relation as $@_a \Diamond c, @_c \Diamond c$ are on the branch.

Now, loop-checks were not needed to ensure termination in the two tableau examples above. Below we shall consider a third and a fourth tableau example where loop-checks are actually needed, namely an example involving the universal modal operator E and an example involving the standard modal operator $\Diamond$. In the example involving E there is no non-trivial equational reasoning, that is, there is no reasoning with formulas like $@_a c$ where the nominals a and c are distinct, however, the example involving $\Diamond$ does make use of such reasoning, which makes it the most complicated of the two examples.

In the third example (the example involving E) we consider $@_b \neg (r \wedge \neg E \neg E r)$ which is not valid. A possible tableau generated by the tableau construction algorithm is the tableau below.

$\neg @_b \neg (r \wedge \neg E \neg E r)$	1.
$@_b (r \wedge \neg E \neg E r)$	2. by $(\neg\neg)$ rule on 1
$@_b r$	3. by $(\wedge)$ rule on 2
$@_b \neg E \neg E r$	4. by $(\wedge)$ rule on 2
$\neg @_b E \neg E r$	5. by $(\neg)$ rule on 4
$@_b b$	6. by (Ref) rule
$\neg @_b \neg E r$	7. by $(\neg E)$ rule on 5
$@_b E r$	8. by $(\neg\neg)$ rule on 7
$@_a r$	9. by (E) rule on 8
$@_a a$	10. by (Ref) rule
$\neg @_a \neg E r$	11. by $(\neg E)$ rule on 5
$@_a E r$	12. by $(\neg\neg)$ rule on 11

The counter-model to $@_b \neg (r \wedge \neg E \neg E r)$ given by Definition 3.10 has two worlds, the equivalence classes $\{b\}$ and $\{a\}$, where the propositional symbol r is true at both. In the tableau above, note that the nominal a is included in the nominal b with respect the branch, cf. Definition 3.5, the reason being that the two sets of subformulas of the root formula which are respectively prefixed by $@_a$ and $\neg @_a$, are the sets $\{r, Er\}$ and $\{\neg Er\}$, whereas the two sets of subformulas of the root formula which are respectively prefixed by $@_b$ and $\neg @_b$, are the sets $\{r \wedge \neg E \neg Er, r, \neg E \neg Er, Er\}$ and $\{\neg(r \wedge \neg E \neg Er), E \neg Er, \neg Er\}$. Of course, the important observations are that

$$\{r, Er\} \subseteq \{r \wedge \neg E \neg Er, r, \neg E \neg Er, Er\}$$

and

$$\{\neg Er\} \subseteq \{\neg(r \wedge \neg E \neg Er), E \neg Er, \neg Er\}.$$

Thus, application of the rule (E) to the occurrence of $@_a Er$ in line 12 is blocked by the loop-check condition in the tableau construction algorithm, that is, the third restriction in Definition 3.6. If the loop-check condition is removed, the tableau construction algorithm can continue in "cycles" as follows.

– – – – – – – – – – – – – – – – –

$@_c r$	11. (E) rule on 10
$\neg @_c \neg Er$	12. $(\neg E)$ rule on 5
$@_c Er$	13. $(\neg\neg)$ rule on 12

– – – – – – – – – – – – – – – – –

$@_d r$	14. (E) rule on 12
$\neg @_d \neg Er$	15. $(\neg E)$ rule on 5
$@_d Er$	16. $(\neg\neg)$ rule on 15

– – – – – – – – – – – – – – – – –

⋮

Of course, the dashed lines that separate the cycles are not a formal part of the tableau. Note that the second cycle, lines 14–16, is identical to the first cycle, lines 11–13, except that the nominal d occurs in the second cycle where the nominal c occurs in the first cycle. What happens is that when a new nominal has been generated, say the nominal c above, the formula $\neg @_b E \neg Er$ in line 5 produces a formula $\neg @_c \neg Er$, which in turn produces $@_c Er$, and this formula generates yet another new nominal, and so on.

In the fourth example (the example involving $\Diamond$) we consider the formula $@_b \neg((b \wedge r) \wedge (\Diamond b \wedge \neg \Diamond \neg \Diamond (b \wedge r)))$ which is not valid. A possible tableau generated by the tableau construction algorithm is the tableau below.

$\neg @_b \neg ((b \wedge r) \wedge (\Diamond b \wedge \neg \Diamond \neg \Diamond (b \wedge r)))$	1.
$@_b ((b \wedge r) \wedge (\Diamond b \wedge \neg \Diamond \neg \Diamond (b \wedge r)))$	2. by $(\neg\neg)$ rule on 1
$@_b (b \wedge r)$	3. by $(\wedge)$ rule on 2
$@_b (\Diamond b \wedge \neg \Diamond \neg \Diamond (b \wedge r))$	4. by $(\wedge)$ rule on 2
$@_b b$	5. by $(\wedge)$ rule on 3
$@_b r$	6. by $(\wedge)$ rule on 3
$@_b \Diamond b$	7. by $(\wedge)$ rule on 4
$@_b \neg \Diamond \neg \Diamond (b \wedge r)$	8. by $(\wedge)$ rule on 4
$\neg @_b \Diamond \neg \Diamond (b \wedge r)$	9. by $(\neg)$ rule on 8
$\neg @_b \neg \Diamond (b \wedge r)$	10. by $(\neg\Diamond)$ rule on 9 and 7
$@_b \Diamond (b \wedge r)$	11. by $(\neg\neg)$ rule on 10
$@_a (b \wedge r)$	12. by $(\Diamond)$ rule on 11
$@_b \Diamond a$	13. by $(\Diamond)$ rule on 11
$@_a b$	14. by $(\wedge)$ rule on 12
$@_a r$	15. by $(\wedge)$ rule on 12
$@_a a$	16. by (Ref) rule
$@_b a$	17. by $(Nom1)$ rule on 14 and 16
$\neg @_a \neg \Diamond (b \wedge r)$	18. by $(\neg\Diamond)$ rule on 9 and 13
$@_a \Diamond (b \wedge r)$	19. by $(\neg\neg)$ rule on 18

The counter-model given by Definition 3.10 has one world, the equivalence class $\{b, a\}$, such that the propositional symbol r is true at the world and such that the world is related to itself by the accessibility relation. In the tableau above, note that the nominal a is included in the nominal b with respect the branch, hence, application of the rule $(\Diamond)$ to the occurrence of $@_a \Diamond (b \wedge r)$ in line 19 is blocked by the loop-check condition. Like in the previous example, if the loop-check condition is removed, the tableau construction algorithm can continue in cycles.

$@_c(b \wedge r)$	20. by $(\Diamond)$ rule on 19
$@_a\Diamond c$	21. by $(\Diamond)$ rule on 19
$@_c b$	22. by $(\wedge)$ rule on 20
$@_c r$	23. by $(\wedge)$ rule on 20
$@_b\Diamond c$	24. by $(Nom2)$ rule on 14 and 21
$\neg @_c \neg\Diamond(b \wedge r)$	25. by $(\neg\Diamond)$ rule on 9 and 24
$@_c\Diamond(b \wedge r)$	26. by $(\neg\neg)$ rule on 25

$@_d(b \wedge r)$	27. by $(\Diamond)$ rule on 26
$@_c\Diamond d$	28. by $(\Diamond)$ rule on 26
$@_d b$	29. by $(\wedge)$ rule on 27
$@_d r$	30. by $(\wedge)$ rule on 27
$@_b\Diamond d$	31. by $(Nom2)$ rule on 22 and 26
$\neg @_d \neg\Diamond(b \wedge r)$	32. by $(\neg\Diamond)$ rule on 9 and 31
$@_d\Diamond(b \wedge r)$	33. by $(\neg\neg)$ rule on 32

⋮

Here the second cycle, lines 27–33, is identical to the first cycle, lines 20–26, except that the nominals d and c occur in the second cycle where the nominals c and a occur in the first cycle.

Note the use of the rule $(Nom2)$ in the tableau example above. Without this rule loop-checks are not needed to ensure termination of the tableau example. This is actually the case for any input to the tableau construction algorithm which only involves the standard modal operator $\Diamond$. This is a consequence of a result in the following section, namely Theorem 3.8, which concerns a tableau system without the rule $(Nom2)$. It is in this connection an interesting observation that the rule $(Nom2)$ can only be used in the presence of non-trivial equational reasoning, the reason being that an application of $(Nom2)$ can only generate a new formula if the nominals a and c in the premise $@_a c$ are distinct (see Figure 3.3). This implies that any tableau example where loop-checks are actually needed to ensure termination, and where only the modal operator $\Diamond$ is involved, must make use of non-trivial equational reasoning. This in turn implies that in any tableau example where loop-checks are actually needed to ensure termination, and where only the modal operator $\Diamond$ is involved, the root formula must contain equationally occurring nominals, cf. Corollary 3.1, like the nominal b in the example above.

If we only consider formulas involving the standard modal operator $\Diamond$, not the universal modal operator E, then it is tempting to ask whether we cannot simply omit the rule $(Nom2)$ and thereby obtain a tableau system that does not require loop-checks. The answer is that the tableau system will not be complete without this rule, that is, without this rule there are valid formulas which do not result in closed tableaus when given as input to the algorithm. This can be seen by considering the first tableau example of this section. In that example the valid formula $@_a\neg\Diamond(a \wedge \neg\Diamond a)$ is given as input, but without the rule $(Nom2)$, the algorithm will terminate already at line 10, resulting in a tableau that is not closed, and by inspecting the example, is straightforward to see that any other tableau having the same root formula will not be closed either, unless $(Nom2)$ is used. Thus, if $(Nom2)$ is omitted, something else must be added to regain completeness, and this will be the topic of the following section.

3.3 A Tableau System Not Including the Universal Modality

If the universal modality is omitted in the decision procedure for $\mathscr{H}(E)$ given in the previous section, then of course a decision procedure for the weaker hybrid logic $\mathscr{H}$ is obtained. However, it turns out that if the universal modality is omitted, then it is possible to give a tableau system such that loop-checks are not needed to ensure termination of the decision procedure. The first tableau-based decision procedure for $\mathscr{H}$, that does not involve loop-checks, was a tableau system given in Thomas Bolander and Patrick Blackburn's paper Bolander and Blackburn (2007). In the present section we shall consider a similar tableau system not involving loop-checks. The system in the present section is obtained by directly modifying the tableau system, and the associated definitions and results, already introduced in the previous section. One crucial modification is the replacement of the rule $(Nom2)$ by two new rules which are variants of a rule in Bolander and Blackburn (2007) however, most of the definitions and results given in the previous section can be reused. A difference between the system in Bolander and Blackburn (2007) and the systems under consideration here is that the present systems make use of unrestricted equational reasoning, which is not the case with the system in Bolander and Blackburn (2007). Bolander and Blackburn's 2007 decision procedure has been implemented in the functional programming language Haskell as described in Hoffmann and Areces (2007).

Now, the rules for the tableau system in the previous section, which we denoted $\mathbf{T}_{\mathscr{H}(E)}$, were given in Figures 3.2 and 3.3. The rules for the tableau system for $\mathscr{H}$ not involving loop-checks are obtained from the rules of Figures 3.2 and 3.3 by omitting the rules (E) and $(\neg E)$ for the universal modality and by replacing the rule $(Nom2)$ by the two rules given in Figure 3.4. The system thus obtained will be denoted $\mathbf{T}_{\mathscr{H}}$. In the previous section we introduced some conventions for the rules of Figures 3.2 and 3.3: Some rules were called destructive, some were called non-destructive, and some were called existential. These conventions are unchanged, but we add the convention that the rule (Id) and $(\neg Id)$ are non-destructive (as destructive rules we only want rules with exactly one formula in the premise).

It is straightforward to check that all results for $\mathbf{T}_{\mathscr{H}(E)}$ in Section 3.2.2 also hold for the tableau system $\mathbf{T}_{\mathscr{H}}$, however, we shall need the following strengthened version of Theorem 3.3.

Theorem 3.7. *Let $@_a\Diamond b$ be a formula occurrence on a branch Θ of a tableau. Either $@_a\Diamond b$ is an accessibility formula occurrence on Θ or the formula $\Diamond b$ is a subformula of the root formula.*

$$\frac{@_a c, @_a \phi}{@_c \phi} \; (Id)^* \qquad \frac{@_a c, \neg @_a \phi}{\neg @_c \phi} \; (\neg Id)^*$$

$*$ The nominal c and the formula ϕ are subformulas of the root formula.

Fig. 3.4 New rules for the tableau system without the universal modality

Proof. Check each rule. Theorem 3.1 is needed in some of the cases.

The systematic tableau construction algorithm for $\mathbf{T}_{\mathscr{H}}$ is defined as follows. Note that the definition below is identical to Definition 3.6, the tableau construction algorithm for $\mathbf{T}_{\mathscr{H}(E)}$, except that the third restriction (the loop-check) of Definition 3.6 is omitted.

Definition 3.12. (Tableau construction) Given a formula $@_a\phi$ of $\mathscr{H}$ whose validity has to be decided, we define by induction a sequence $\mathscr{T}_0$, $\mathscr{T}_1$, $\mathscr{T}_2$, ... of finite tableaus in $\mathbf{T}_{\mathscr{H}}$, each of which is embedded in its successor. Let $\mathscr{T}_0$ be the finite tableau constituted by the single formula $\neg @_a\phi$. If possible, apply an arbitrary rule to $\mathscr{T}_n$ with the following two restrictions:

1. If a formula to be added to a branch by applying a rule already occurs on the branch, then the addition of the formula is simply omitted.
2. After the application of a destructive rule to a formula occurrence ϕ on a branch, it is recorded that the rule was applied to ϕ with respect to the branch and the rule will not again be applied to ϕ with respect to the branch or any extension of it.

Let $\mathscr{T}_{n+1}$ be the resulting tableau.

In the proof of the theorem below we make use of the convention from Section 2.2.2 that the degree of a formula is the number of occurrences of non-nullary connectives in it.

Theorem 3.8. *The tableau construction algorithm for* $\mathbf{T}_{\mathscr{H}}$ *terminates in the sense that there exists an n such that* $\mathscr{T}_n = \mathscr{T}_{n+1}$.

Proof. Assume conversely that the algorithm does not terminate. Then the resulting tableau is infinite, and hence, has an infinite branch Θ. Analogous to the proof of Theorem 3.4, it follows from Proposition 3.1 and Theorem 3.1 that the graph $(N^\Theta, <_\Theta)$ has an infinite branch $a_1 <_\Theta a_2 <_\Theta a_3, \ldots$. Now, for any i, consider the set of formula occurrences on Θ either having the form $@_{a_i}\phi$ where ϕ is not of the form $\Diamond b$ or having the form $\neg @_{a_i}\phi$. Let d_i be the maximal degree of such formula occurrences and let d_i be 0 if there are no such formula occurrences (by Theorem 3.1 the degrees of such formula occurrences are bounded by the degree of the root formula plus two). By inspection of the rules, it is straightforward to see that $d_i > d_{i+1}$ for any i such that $d_{i+1} > 0$, where in the case with the rule $(\neg\Diamond)$ we use Theorem 3.7 and in the cases with the rules $(@)$ and $(\neg @)$ we use Theorem 3.1. Hence, there exists a j such that any formula $@_{a_j}\phi$ occurring on Θ has the property that ϕ

is of the form $\Diamond b$ or ϕ has degree 0, and any formula $\neg @_{a_j}\phi$ occurring on Θ has the property that ϕ has degree 0. This contradicts $a_j <_\Theta a_{j+1}$ being the case. Thus, the algorithm terminates.

Informally, the proof above is based on the observation that formulas $@_a\phi$ and $\neg @_a\phi$ in the branch Θ get smaller when the path from the nominal a to a root in the graph $(N^\Theta, <_\Theta)$ gets longer (except for formulas of the form $@_a\Diamond b$). A similar observation is also the basis of the standard termination proof for prefixed tableau systems for the modal logic K, cf. Fitting (1983).

In what follows, we shall assume that Θ is a given branch of a tableau generated by the systematic tableau construction algorithm Definition 3.12. Before we come to the model existence theorem, we introduce some important machinery. Given a nominal a in N^Θ, we let $[a]_\sim$ denote the equivalence class of a with respect to the binary relation $\sim$ and we let $N^\Theta/\!\sim$ denote the set of equivalence classes. Note that it follows from Corollary 3.1 that any non-singleton equivalence class $[a]_\sim$ contains a nominal with an equational occurrence in the root formula.

Definition 3.13. Given some fixed total order on N^Θ, we define a function u from N^Θ to N^Θ as follows: If $[a]_\sim$ is non-singleton, then we let $u(a)$ be the smallest nominal in $[a]_\sim$ with an equational occurrence in the root formula, and if $[a]_\sim$ is singleton, then we let $u(a)$ be a. The nominal $u(a)$ is called the *urfather* of a.

The idea of letting a function pick out an equivalent nominal occurring equationally in the root formula, if such a nominal exists, stems from Bolander and Blackburn (2007) where a similar function is defined. The definition above leads to the following proposition.

Proposition 3.2. *(Urfather closure property) Assume that the branch Θ is open. Let ϕ be a subformula of the root formula. If $@_a\phi$ occurs on Θ, then also $@_{u(a)}\phi$ occurs at Θ, and similarly, if $\neg @_a\phi$ occurs on Θ, then also $\neg @_{u(a)}\phi$ occurs on Θ.*

Proof. Follows straightforwardly from applications of the rules (Id) and $(\neg Id)$.

The urfather closure property is a basic idea behind the tableau system given in Bolander and Blackburn (2007). Intuitively, the urfather closure property allows information to be moved freely from any world to identical worlds referred to in the root formula.

Definition 3.14. Let R be the binary relation on N^Θ defined by aRc if and only if there exists a nominal $c' \sim c$ such that $@_{u(a)}\Diamond c'$ occurs on Θ.

We let $\overline{R}$ be the binary relation on $N^\Theta/\!\sim$ defined by $[a]_\sim \overline{R}[c]_\sim$ if and only if aRc.

Definition 3.15. For any element a of N^Θ, let V_a be the function that to each ordinary propositional symbol assigns an element of $\{0,1\}$ such that $V_a(p) = 1$ if $@_a p$ occurs on Θ and $V_a(p) = 0$ otherwise.

We let $\overline{V}_{[a]_\sim}$ be defined by $\overline{V}_{[a]_\sim}(p) = V_a(p)$. We are now ready to define a model.

Definition 3.16. Let $\mathfrak{M}$ be the model $(N^\Theta/\!\sim, \overline{R}, \{\overline{V}_{[a]_\sim}\}_{[a]_\sim \in N^\Theta/\sim})$ and let the assignment g for $\mathfrak{M}$ be defined by $g(a) = [a]_\sim$.

Theorem 3.9. *(Model existence theorem) Assume that the branch Θ is open. For any satisfaction statement $@_a\phi$ where ϕ is a subformula of the root formula, the following two statements hold.*

- *If $@_a\phi$ occurs on Θ, then it is the case that $\mathfrak{M}, g, [a]_\sim \models \phi$.*
- *If $\neg @_a\phi$ occurs on Θ, then it is not the case that $\mathfrak{M}, g, [a]_\sim \models \phi$.*

Proof. induction on the structure of ϕ. We only cover the case where ϕ is of the form $\Diamond\psi$, all the other cases are exactly as in the proof of Theorem 3.5.

Assume that $@_a\Diamond\psi$ occurs on Θ. We then have to prove that $\mathfrak{M}, g, [a]_\sim \models \Diamond\psi$, that is, for some equivalence class $[c]_\sim$ such that $[a]_\sim\overline{R}[c]_\sim$, it is the case that $\mathfrak{M}, g, [c]_\sim \models \psi$. If ψ is a nominal, say b, we just have to prove that $[a]_\sim\overline{R}[b]_\sim$, which trivially follows from the definition of the relation $\overline{R}$ together with the observation that if $@_a\Diamond b$ occurs on Θ, then by Proposition 3.2 also $@_{u(a)}\Diamond b$ occurs on Θ. If ψ is not a nominal, by Proposition 3.2 and the rule $(\Diamond)$ also some formulas $@_{u(a)}\Diamond c$ and $@_c\psi$ occur on Θ. Clearly, $[a]_\sim\overline{R}[c]_\sim$ and $\mathfrak{M}, g, [c]_\sim \models \psi$ by induction.

Assume that $\neg @_a\Diamond\psi$ occurs on Θ. We then have to prove that $\mathfrak{M}, g, [a]_\sim \models \Diamond\psi$ does not hold, that is, for any equivalence class $[c]_\sim$ such that $[a]_\sim\overline{R}[c]_\sim$, it is not the case that $\mathfrak{M}, g, [c]_\sim \models \psi$. From $[a]_\sim\overline{R}[c]_\sim$ it follows that there exists a nominal $c' \sim c$ such that $@_{u(a)}\Diamond c'$ occurs on Θ. By Proposition 3.2, $\neg @_{u(a)}\Diamond\psi$ occurs on Θ. Thus, by the rule $(\neg\Diamond)$ the formula $\neg @_{c'}\psi$ occurs on Θ. By induction we conclude that $\mathfrak{M}, g, [c']_\sim \models \psi$ does not hold and trivially, $[c']_\sim = [c]_\sim$.

Given the machinery introduced above, the decision procedure is defined exactly as in Definition 3.11.

It should be mentioned that the rules (Id) and $(\neg Id)$ can be restricted to the case where ϕ is of the form $\Diamond\psi$ without affecting the results for $\mathbf{T}_{\mathscr{H}}$ we consider in this section, except Proposition 3.2, the urfather closure property, which is restricted in the same way.

As an example, consider the formula $@_a\neg\Diamond(a \wedge \neg\Diamond a)$ which also was considered in the first tableau example for $\mathbf{T}_{\mathscr{H}(E)}$ given in section 3.2.5. Given this formula as input, a possible tableau generated by the tableau construction algorithm for $\mathbf{T}_{\mathscr{H}}$, Definition 3.12, is the tableau below where we have imposed the restriction on the rules (Id) and $(\neg Id)$ mentioned in the previous paragraph.

$\neg @_a\neg\Diamond(a \wedge \neg\Diamond a)$	1.
$@_a\Diamond(a \wedge \neg\Diamond a)$	2. by $(\neg\neg)$ rule on 1
$@_c(a \wedge \neg\Diamond a)$	3. by $(\Diamond)$ rule on 2
$@_a\Diamond c$	4. by $(\Diamond)$ rule on 2
$@_c a$	5. by $(\wedge)$ rule on 3
$@_c\neg\Diamond a$	6. by $(\wedge)$ rule on 3
$\neg @_c\Diamond a$	7. by $(\neg)$ rule on 6
$@_c c$	8. by (Ref) rule
$@_a c$	9. by $(Nom1)$ rule on 5 and 8
$@_a a$	10. by (Ref) rule
$\neg @_a\Diamond a$	11. by $(\neg Id)$ rule on 5 and 7
$\neg @_c a$	12. by $(\neg\Diamond)$ rule on 11 and 4

Note that lines 1–10 in the tableau above are identical to lines 1–10 in the tableau example for $@_a\neg\Diamond(a\wedge\neg\Diamond a)$ given in Section 3.2.5, thus, it is only in the last two lines that the difference between $\mathbf{T}_{\mathscr{H}}$ and $\mathbf{T}_{\mathscr{H}(E)}$ crops up. Note that the branch is closed since it contains the formula $@_c a$ as well as $\neg @_c a$.

3.3.1 A Hybrid-Logical Version of Analytic Cuts

In this section we shall give an alternative version of the tableau system without loop-checks, $\mathbf{T}_{\mathscr{H}}$, which was considered above. Now, at the end of Section 3.2.5 we concluded that if the rule (*Nom*2) is omitted from $\mathbf{T}_{\mathscr{H}(E)}$, something else must be added to regain completeness, and in the version of $\mathbf{T}_{\mathscr{H}}$ considered above, the rules (*Id*) and ($\neg$*Id*) in Figure 3.4 were added. In the alternative version of $\mathbf{T}_{\mathscr{H}}$ we give in this section, we shall investigate how much standard proof-theoretic machinery in the form of cuts is needed to replace the (*Nom*2) rule (with the implicit requirement that loop-checks are avoided). The decision procedure given in the present section has been implemented in the logic programming language PROLOG as described in the Roskilde University student project report Wenningsted-Torgard (2008).

The alternative version is obtained by replacing the rules in Figure 3.4 by the rule in Figure 3.5, and moreover, by changing the definition of an open branch in a tableau such that a branch is called *open* if for no satisfaction statements $@_a\chi$ and $@_a b$ occurring on the branch, it is the case that $\neg @_b\chi$ also occurs on the branch. To avoid excessive proliferation of terminology, we use the notation $\mathbf{T}_{\mathscr{H}}$ also for the alternative system. As in the case of the rules (*Id*) and ($\neg$*Id*), the rule (*Quasi-analytic cut*) is classified as non-destructive. Note that in the alternative version of $\mathbf{T}_{\mathscr{H}}$, the rules (*Id*) and ($\neg$*Id*) are derivable.[3] Of course, the rule (*Quasi-analytic cut*) is a hybrid-logical version of the standard analytic cut rule, formulated as appropriate for tableau systems, see Section 2.3.

$$\frac{}{@_a\phi \mid \neg @_a\phi}\ (\textit{Quasi-analytic cut})^{*\star}$$

$*$ The nominal a and the formula ϕ are subformulas of the root formula.
$\star$ None of the formulas $@_a\phi$ and $\neg @_a\phi$ are on the branch.

Fig. 3.5 A hybrid-logical version of the analytic cut rule

Note that in one branch of a tableau, there can only be finitely many applications of the rule (*Quasi-analytic cut*) if the cut-formulas are different, the reason being that there are only finitely many subformulas of the root formula. It is straightforward to check that all the results for the first version of $\mathbf{T}_{\mathscr{H}}$ also hold for the alternative version considered in this section. A difference between the two versions of $\mathbf{T}_{\mathscr{H}}$ is that in the first version, the urfather closure property is essentially built-in

[3] This was pointed out to the author by Jens Ulrik Hansen.

as two derivation rules, namely (*Id*) and ($\neg$*Id*), whereas in the alternative version, the urfather closure property follows from the presence of other rules, in particular (*Quasi-analytic cut*), together with a more general closure condition on branches. At a more intuitive level, a difference between the two versions of $\mathbf{T}_{\mathcal{H}}$ is that the rules (*Id*) and ($\neg$*Id*) have a semantical motivation (the intuition being that they allow information to be moved freely from any world to identical worlds referred to in the root formula) whereas the rule (*Quasi-analytic cut*) has a proof-theoretical motivation (it is a hybrid-logical version of standard proof-theoretic machinery, namely the analytic cut rule).

A couple of more general remarks should be made in connection with analytic cut rules. A defence of analytic cuts can be found in D'Agostino and Mondadori (1994) where it is pointed out that ordinary cut-free tableau and Gentzen systems have a number of anomalies that can be avoided in proof systems allowing analytic cuts. According to that paper, cut-free systems are anomalous from three different points of view.

1. From a proof-theoretical point of view, it is an anomaly that cut-free systems cannot represent lemmas in proofs.
2. From a semantical point of view, it is an anomaly that cut-free systems cannot express the bivalence of classical logic.
3. From a computational point of view, it is an anomaly that for some classes of propositional formulas, decision procedures based on cut-free systems are incomparably slower than the truth-table method (in the more precise technical sense that there is no polynomial time computable function that maps truth-table proofs of such formulas to proofs of the same formulas in cut-free tableau or Gentzen systems).

In relation to the computational anomaly, see also Boolos (1984) where examples of first-order formulas are given whose derivations in cut-free systems are much larger than their derivations in natural deduction systems, which implicitly allow unrestricted cuts (in one case more than 10^{38} characters compared to less than 3280 characters). However, at present it is not clear to which extent the discussion outlined above is directly relevant to the proof-theory of hybrid logic.

Above it was mentioned that the rules (*Id*) and ($\neg$*Id*) of Figure 3.4 can be restricted to the case where ϕ is of the form $\Diamond\psi$. This also applies to the rule (*Quasi-analytic cut*) in the alternative version of $\mathbf{T}_{\mathcal{H}}$ considered in this section, however, in the light of the remarks in the previous paragraph, it is not clear whether this restriction on (*Quasi-analytic cut*) is desirable.

As an example we consider the formula $@_a\neg\Diamond(a\wedge\neg\Diamond a)$ which was also considered in connection with the first version of $\mathbf{T}_{\mathcal{H}}$ (and in connection with $\mathbf{T}_{\mathcal{H}(E)}$ in Section 3.2.5). A possible tableau generated by the tableau construction algorithm for the alternative version of $\mathbf{T}_{\mathcal{H}}$ is the tableau below where we have imposed the restriction on (*Quasi-analytic cut*) mentioned in the previous paragraph.

$\neg @_a \neg \Diamond (a \wedge \neg \Diamond a)$	1.
$@_a \Diamond (a \wedge \neg \Diamond a)$	2. by $(\neg\neg)$ rule on 1
$@_c (a \wedge \neg \Diamond a)$	3. by $(\Diamond)$ rule on 2
$@_a \Diamond c$	4. by $(\Diamond)$ rule on 2
$@_c a$	5. by $(\wedge)$ rule on 3
$@_c \neg \Diamond a$	6. by $(\wedge)$ rule on 3
$\neg @_c \Diamond a$	7. by $(\neg)$ rule on 6
$@_c c$	8. by (Ref) rule
$@_a c$	9. by $(Nom1)$ rule on 5 and 8
$@_a a$	10. by (Ref) rule

Left branch:

11. $@_a \Diamond a$

Right branch:

$\neg @_a \Diamond a$	12. by (*Quasi-analytic cut*) rule
$\neg @_c a$	13. by $(\neg\Diamond)$ rule on 12 and 4

Note that lines 1–10 in the tableau above are identical to lines 1–10 in the tableau in connection with the first version of $\mathbf{T}_{\mathscr{H}}$ (and the tableau in connection with $\mathbf{T}_{\mathscr{H}(E)}$ in Section 3.2.5). Also, note that both branches are closed, and in the case of the left-hand-side branch, we make use of the more general closure condition for the alternative version of $\mathbf{T}_{\mathscr{H}}$ (the branch contains $@_a \Diamond a$ and $@_a c$ as well as $\neg @_c \Diamond a$).

Remark: Let $@_a \phi$ be a formula whose validity has to be decided. Thus, according to the decision procedure, a tableau is constructed with $\neg @_a \phi$ as the root formula. Clearly, $\neg @_a \phi$ is equivalent to the formula

$$\neg @_a \phi \wedge \bigwedge_{i=1}^{n} \bigwedge_{j=1}^{m} (@_{c_i} \psi_j \vee \neg @_{c_i} \psi_j)$$

where $c_1, \ldots, c_n$ and $\psi_1, \ldots, \psi_m$ are respectively the nominals and the formulas that are subformulas of $\neg @_a \phi$. If the tableau having $\neg @_a \phi$ as the root is closed, then a closed tableau having the displayed formula (strictly speaking prefixed by a "dummy" satisfaction operator to fit the format of the tableau system) as root can be constructed without applying (*Quasi-analytic cut*). Conversely, if the tableau having $\neg @_a \phi$ as the root has an open branch Θ, then a tableau having the displayed formula (again, strictly speaking prefixed by a dummy satisfaction operator) as root can be constructed without applying (*Quasi-analytic cut*) such that the tableau has an open branch containing all the formulas of Θ. Thus, applications of (*Quasi-analytic cut*) are dispensable in the sense that they can be simulated by preprocessing the input formula to the tableau construction algorithm, where in the preprocessing a conjunct is added for each possible application of (*Quasi-analytic cut*).[4]

[4] This was pointed out to the author by Thomas Bolander.

3.4 The Tableau Systems Reformulated as Gentzen Systems

In this section we reformulate the tableau systems $\mathbf{T}_{\mathscr{H}(E)}$ and $\mathbf{T}_{\mathscr{H}}$ given in the two previous sections, Sections 3.2 and 3.3, as Gentzen systems and we discuss how to reformulate the decision procedures that are based on these tableau systems. See Section 2.3 for a sketch of the basics of Gentzen systems.

We first reformulate the tableau system $\mathbf{T}_{\mathscr{H}(E)}$. The Gentzen system corresponding to $\mathbf{T}_{\mathscr{H}(E)}$ will be denoted $\mathbf{G}_{\mathscr{H}(E)}$ (in the interest of simplicity we use notation similar to the notation used in connection with the Gentzen system given in Section 2.4). The rules for the Gentzen system $\mathbf{G}_{\mathscr{H}(E)}$ are given in Figures 3.6 and 3.7. All formulas in the rules are satisfaction statements. Note how the Gentzen rules, except the rule (*Axiom*), correspond one-to-one to the tableau rules of Figures 3.2 and 3.3, and conversely, the tableau rules correspond one-to-one to the Gentzen rules. All the Gentzen rules are sound in the sense that if the premise sequents of a Gentzen rule are valid, then the conclusion sequent of the rule is valid as well, but this should not be a surprise since the tableau rules are sound.

$$\frac{}{\phi, \Gamma \vdash \Delta, \phi}\ (Axiom)$$

$$\frac{@_a\phi, @_a\psi, \Gamma \vdash \Delta}{@_a(\phi \wedge \psi), \Gamma \vdash \Delta}\ (\wedge L) \qquad \frac{\Gamma \vdash \Delta, @_a\phi \quad \Gamma \vdash \Delta, @_a\psi}{\Gamma \vdash \Delta, @_a(\phi \wedge \psi)}\ (\wedge R)$$

$$\frac{\Gamma \vdash \Delta, @_a\phi}{@_a\neg\phi, \Gamma \vdash \Delta}\ (\neg L) \qquad \frac{@_a\phi, \Gamma \vdash \Delta}{\Gamma \vdash \Delta, @_a\neg\phi}\ (\neg R)$$

$$\frac{@_a\phi, \Gamma \vdash \Delta}{@_c@_a\phi, \Gamma \vdash \Delta}\ (@L) \qquad \frac{\Gamma \vdash \Delta, @_a\phi}{\Gamma \vdash \Delta, @_c@_a\phi}\ (@R)$$

$$\frac{@_a\Diamond c, @_c\phi, \Gamma \vdash \Delta}{@_a\Diamond\phi, \Gamma \vdash \Delta}\ (\Diamond L)^{**} \qquad \frac{@_a\Diamond e, \Gamma \vdash \Delta, @_a\Diamond\phi, @_e\phi}{@_a\Diamond e, \Gamma \vdash \Delta, @_a\Diamond\phi}\ (\Diamond R)$$

$$\frac{@_c\phi, \Gamma \vdash \Delta}{@_aE\phi, \Gamma \vdash \Delta}\ (EL)^{*} \qquad \frac{\Gamma \vdash \Delta, @_aE\phi, @_e\phi}{\Gamma \vdash \Delta, @_aE\phi}\ (ER)^{\dagger}$$

∗ The nominal c is new.
⋆ The formula ϕ is not a nominal.
† The nominal e occurs in the conclusion.

Fig. 3.6 Gentzen rules for connectives

Note that the Gentzen system given in the present section is different from the Gentzen system given in Section 2.4 which was designed with the aim of being able to mimic normal natural deduction derivations by cut-free Gentzen derivations, see Lemma 2.13 of that section as well as the discussion in Section 2.4.3. One consequence of this is that different rules for equality (in connection with nominals)

$$\frac{@_a a, \Gamma \vdash \Delta}{\Gamma \vdash \Delta}\ (Ref)^* \qquad \frac{@_c \phi, @_a c, @_a \phi, \Gamma \vdash \Delta}{@_a c, @_a \phi, \Gamma \vdash \Delta}\ (Nom1)^\star \qquad \frac{@_c \Diamond b, @_a c, @_a \Diamond b, \Gamma \vdash \Delta}{@_a c, @_a \Diamond b, \Gamma \vdash \Delta}\ (Nom2)$$

$*$ The nominal a occurs in, or below, the conclusion.
$\star$ The formula ϕ is a propositional symbol (ordinary or a nominal).

Fig. 3.7 Gentzen rules for nominals

are used. Another difference is that in the Gentzen system of the present section the connective $\neg$ is primitive and $\rightarrow$ defined, reflecting the symmetric character of classical Gentzen sequents, whereas in Section 2.4 it is opposite, that is, $\rightarrow$ is primitive and $\neg$ defined, reflecting the asymmetric character of natural deduction derivations.

We shall make use of the following conventions about the Gentzen rules, analogous to the corresponding conventions about the tableau rules. The rules $(\wedge L)$, $(\wedge R)$, $(\neg L)$, $(\neg R)$, $(@L)$, $(@R)$, $(\Diamond L)$, and (EL) will be called *destructive* rules and the remaining rules, except *(Axiom)*, will be called *non-destructive*. Note that the rules $(\Diamond R)$ and (ER) are non-destructive. The destructive rules $(\Diamond L)$ and (EL) will also be called *existential* rules since they introduce new nominals (rules are here read from bottom to top). It is important to note that all non-destructive rules have in-built contraction, thus, no formulas are discarded when a non-destructive rule is applied (read from bottom to top).[5] On the other hand, when a destructive rule is applied, its principal formula might be discarded, depending on which sets Γ and Δ of parametric formulas are chosen.

The decision procedure works by searching backwards from a sequent for possible derivations of it. The search procedure finds a derivation if a derivation exists or at some stage it terminates with the information that no derivations exist. To be more precise, if no derivation exists, then at the terminal stage a counter-model to the sequent can be defined, where a counter-model to a sequent is a model (strictly speaking together with an assignment) which makes all the antecedent formulas true and all the succedent formulas false. Note that this is in line with the view of a Gentzen system according to which Gentzen rules are rules that step by step attempt to define a counter-model to a sequent, see Section 2.3.

Incomplete derivations in the search algorithm are formalized using the notion of a *pseudo-derivation* which is a well-founded tree where the nodes are labelled with sequents such that any non-leaf sequent in a pseudo-derivation is the conclusion of a rule-instance which has the immediate successors of the sequent in question as the premises. A derivation is trivially a pseudo-derivation, and note also that a finite pseudo-derivation where any leaf sequent is an axiom, is a derivation. Note that the rule *(Axiom)* of the Gentzen system $\mathbf{G}_{\mathscr{H}(E)}$ does not correspond to any tableau rule, rather it corresponds to a finite tableau branch not being open. In fact,

[5] It follows that if the conclusion sequent of a rule is valid, then the premise sequents of the rule are valid as well (this is opposite of soundness)

a pseudo-derivation in the Gentzen system corresponds to a tableau and a derivation corresponds to a finite tableau with no open branches.

Having reformulated the tableau system $\mathbf{T}_{\mathscr{H}(E)}$ as the Gentzen system $\mathbf{G}_{\mathscr{H}(E)}$, all the definitions and results in Sections 3.2.2, 3.2.3, and 3.2.4 for $\mathbf{T}_{\mathscr{H}(E)}$ are reformulated in terms of Gentzen systems. In particular, Definition 3.5 is reformulated to the definition below.

Definition 3.17. Let b and a be nominals occurring on a branch Θ of a pseudo-derivation in $\mathbf{G}_{\mathscr{H}(E)}$. The nominal a is *included* in the nominal b with respect to Θ if the following is the case: For any subformula ϕ of a formula in the end-sequent, if the formula $@_a\phi$ occurs as an antecedent (succedent) formula in some sequent in Θ, then $@_b\phi$ also occurs as an antecedent (succedent) formula in some sequent in Θ. If a is included in b with respect to Θ, and the lowest sequent with an occurrence of b is lower in the branch than the lowest sequent with an occurrence of a, then we write $a \subseteq_\Theta b$.

The tableau construction algorithm Definition 3.6 is reformulated to the definition below.

Definition 3.18. (Search algorithm) Given a sequent of $\mathscr{H}(E)$ whose validity has to be decided, we define by induction a sequence $\pi_0, \pi_1, \pi_2, \ldots$ of finite pseudo-derivations in $\mathbf{G}_{\mathscr{H}(E)}$, each of which is embedded in its successor. Let π_0 be the pseudo-derivation constituted by the single sequent whose validity has to be decided. If possible, apply a rule to an arbitrary leaf sequent of the pseudo-derivation π_n with the following three restrictions:

1. A non-destructive rule is not applied to the leaf sequent in a branch if a premise sequent of the application is identical to the conclusion sequent of the application.
2. A destructive rule is not applied to the leaf sequent in a branch if the principal formula of the application already occurs as the principal formula of a lower application of the rule.
3. The existential rule $(\Diamond L)$ is not applied to the leaf sequent $@_a\Diamond\phi, \Gamma \vdash \Delta$ in a branch Θ if there exists a nominal b such that $a \subseteq_\Theta b$ (and analogously for the existential rule (EL)).

Let π_{n+1} be the resulting pseudo-derivation.

Note that the definition above defines an algorithm that constructs a finite pseudo-derivation whereas Definition 3.6 defines an algorithm that constructs a finite tableau. Note also that the search algorithm does not involve backtracking.[6]

As a first example, consider the Gentzen sequent $@_a\Diamond(a \wedge \neg\Diamond a) \vdash$ which is equivalent to the valid formula $@_a\neg\Diamond(a \wedge \neg\Diamond a)$ that was considered in the first tableau example of Section 3.2.5 (and also in Section 3.3). Given this sequent as input, a possible pseudo-derivation generated by the search algorithm is the pseudo-derivation below which actually is a derivation (the algorithm is non-deterministic as in the tableau-case).

[6] Essentially, backtracking is not needed since the premise sequents of a rule are valid if the conclusion sequent is valid, hence, no information is lost when a rule is applied.

$$
\begin{array}{cl}
\hline
@_c\Diamond c, @_a c, @_c c, @_c a, @_a\Diamond c \vdash @_c\Diamond a, @_c a & (Axiom) \\
\hline
\boxed{@_c\Diamond c}, @_a c, @_c c, @_c a, @_a\Diamond c \vdash \boxed{@_c\Diamond a} & (\Diamond R) \\
\hline
@_c\Diamond c, @_a c, @_c c, @_c a, \boxed{@_c\neg\Diamond a}, @_a\Diamond c \vdash & (\neg L) \\
\hline
\boxed{@_a c}, @_c c, @_c a, @_c\neg\Diamond a, \boxed{@_a\Diamond c} \vdash & (Nom2) \\
\hline
\boxed{@_c c}, \boxed{@_c a}, @_c\neg\Diamond a, @_a\Diamond c \vdash & (Nom1) \\
\hline
@_c a, @_c\neg\Diamond a, @_a\Diamond c \vdash & (Ref) \\
\hline
@_a\Diamond c, \boxed{@_c(a \wedge \neg\Diamond a)} \vdash & (\wedge L) \\
\hline
\boxed{@_a\Diamond(a \wedge \neg\Diamond a)} \vdash & (\Diamond L)
\end{array}
$$

The principal formulas of each rule application have been indicated by putting frames around them (this is obviously not a formal part of the pseudo-derivation).

As a second example, consider the Gentzen sequent $@_a\Diamond(a\wedge r)\vdash$ which is equivalent to the formula $@_a\neg\Diamond(a\wedge r)$ that was considered in the second tableau example of Section 3.2.5. A possible pseudo-derivation generated by the search algorithm is the pseudo-derivation below.

$$
\begin{array}{cl}
@_a a, @_c\Diamond c, @_a c, @_c c, @_a r, @_c a, @_c r, @_a\Diamond c \vdash & \\
\hline
@_c\Diamond c, @_a c, @_c c, @_a r, @_c a, @_c r, @_a\Diamond c \vdash & (Ref) \\
\hline
\boxed{@_a c}, @_c c, @_a r, @_c a, @_c r, \boxed{@_a\Diamond c} \vdash & (Nom2) \\
\hline
\boxed{@_c c}, @_a r, \boxed{@_c a}, @_c r, @_a\Diamond c \vdash & (Nom1) \\
\hline
@_a r, @_c a, @_c r, @_a\Diamond c \vdash & (Ref) \\
\hline
\boxed{@_c a}, \boxed{@_c r}, @_a\Diamond c \vdash & (Nom1) \\
\hline
@_a\Diamond c, \boxed{@_c(a\wedge r)} \vdash & (\wedge L) \\
\hline
\boxed{@_a\Diamond(a\wedge r)} \vdash & (\Diamond L)
\end{array}
$$

The pseudo-derivation is not a derivation as it only has one branch and the leaf-sequent of that branch is not an axiom. Compare this pseudo-derivation with the second example tableau given in Section 3.2.5. Since the pseudo-derivation is not a derivation, it follows from a Gentzen version of the model existence theorem, Theorem 3.5, that the sequent $@_a\Diamond(a\wedge r)\vdash$ is not valid, which of course is not a surprise since in Section 3.2.5 we established that the formula $@_a\neg\Diamond(a\wedge r)$ has a counter-model. The model in question is a counter-model to the sequent since it makes all the antecedent formulas true (there is only one antecedent formula, namely $@_a\Diamond(a\wedge r)$) and it makes all the succedent formulas false (there are none).

There is a significant difference between the tableau system $\mathbf{T}_{\mathscr{H}(E)}$ and the Gentzen system $\mathbf{G}_{\mathscr{H}(E)}$: When a rule is applied to a formula occurrence on a tableau branch resulting in one or two extensions of the branch, the formula occurrence in question is also a formula occurrence of the new branches. Such structure sharing does not take place in Gentzen derivations. Thus, in connection with tableau systems, we can directly refer to particular formula occurrences on a branch and talk about such formula occurrences being identical, but in connection with Gentzen systems, we cannot refer to particular formula occurrences across applications of rules. Accordingly, when in the first two restrictions of the tableau-based algorithm, Definition 3.6, we talk about particular formula occurrences, in the first two restrictions of the Gentzen-based algorithm, Definition 3.18, we talk about the form of formulas. It follows that the book-keeping machinery needed to record that a destructive tableau rule has been applied to a formula occurrence with respect to a certain branch is not used in the case with Gentzen rules (but for each application of a destructive Gentzen rule, we do, strictly speaking, need to record the form of the principal formula).

We finally reformulate the tableau system $\mathbf{T}_{\mathscr{H}}$ to a Gentzen system. We gave two versions of the tableau system $\mathbf{T}_{\mathscr{H}}$, cf. Section 3.3.1, and we here consider the alternative version of $\mathbf{T}_{\mathscr{H}}$ which involves the proof-theoretically motivated rule (*Quasi-analytic cut*) (it is a hybrid-logical version of the standard analytic cut rule, see Section 2.3). The Gentzen system corresponding to $\mathbf{T}_{\mathscr{H}}$ will be denoted $\mathbf{G}_{\mathscr{H}}$ (we use the same notation as in connection with the Gentzen system given in Section 2.4). Rather than describing how $\mathbf{G}_{\mathscr{H}}$ is obtained directly from $\mathbf{T}_{\mathscr{H}}$, we describe how $\mathbf{G}_{\mathscr{H}}$ is obtained from the Gentzen system $\mathbf{G}_{\mathscr{H}(E)}$. The rules for $\mathbf{G}_{\mathscr{H}}$ are obtained from the rules of Figures 3.6 and 3.7 for $\mathbf{G}_{\mathscr{H}(E)}$ by omitting the rules (EL) and (ER) for the universal modality and by replacing the rules (*Axiom*) and (*Nom*2) by the rules given in Figure 3.8. Note that the version of (*Axiom*) given in Figure 3.8 in the presence of the rule (*Ref*) is a generalization of the version of (*Axiom*) given in Figure 3.7. Replacing the Gentzen rule (*Axiom*) in Figure 3.6 by the Gentzen rule with the same name in Figure 3.8 corresponds to changing the definition of an open branch such that a branch in a tableau of $\mathbf{T}_{\mathscr{H}}$ is open if for no satisfaction statements $@_a\chi$ and $@_ab$ occurring on the branch, $\neg @_b\chi$ also occurs on the branch, rather than if for no satisfaction statement $@_a\chi$ occurring on the branch, $\neg @_a\chi$ also occurs on the branch. Having reformulated the tableau system $\mathbf{T}_{\mathscr{H}}$ as the Gentzen system $\mathbf{G}_{\mathscr{H}}$, all the definitions and results in Section 3.3 for $\mathbf{T}_{\mathscr{H}}$ are reformulated in terms of Gentzen systems. Details are left to the reader, we only remark that the search algorithm for $\mathbf{G}_{\mathscr{H}}$ requires that a destructive rule always is applied in a way such that the principal formula is dublicated, thus, since non-destructive rules have in-built contraction, no formulas are discarded by any rule applications (which is different from the search algorithm Definition 3.18 for $\mathbf{G}_{\mathscr{H}(E)}$ which does not require that destructive rules dublicate principal formulas).

$$\frac{}{@_ab, @_a\phi, \Gamma \vdash \Delta, @_b\phi}\ (Axiom) \qquad \frac{\Gamma \vdash \Delta, @_a\phi \qquad @_a\phi, \Gamma \vdash \Delta}{\Gamma \vdash \Delta}\ (Quasi\text{-}analytic\ cut)^*$$

$*$ The nominal a and the formula ϕ are subformulas of formulas in the end-sequent.

Fig. 3.8 New rules for the Gentzen system without the universal modality

3.5 Discussion

The tableau system $\mathbf{T}_{\mathscr{H}(E)}$ given in Section 3.2 is a slightly simplified version of a system given in Bolander and Braüner (2006), that is, the system given in Bolander and Braüner (2006) includes the rule

$$\frac{@_c\Diamond a, @_a b}{@_c\Diamond b}\ (Bridge)$$

which has turned out to be superflous. The Gentzen system $\mathbf{G}_{\mathscr{H}(E)}$ considered in the previous section is also a slightly simplified version of a system considered in Bolander and Braüner (2006), that is, analogous to the tableau case, the Gentzen system of Bolander and Braüner (2006) includes a Gentzen version of $(Bridge)$ which is superflous. The tableau system of Bolander and Braüner (2006) is a modified, and also extended, version of a tableau system originally given in Blackburn (2000a), to be precise, the rules are identical to the rules given in Blackburn (2000a) except that in the latter system the rules for the universal modality are not included, and moreover, in the latter system the tableau rule $(Nom1)$ is not restricted to propositional symbols, and consequently, the tableau rule $(Nom2)$ is omitted. It turns out that the more general version of $(Nom1)$ given in Blackburn (2000a) is not needed, and restricting it as we have done here simplifies certain technical considerations (in particular, with this restriction the classification of tableau rules as destructive and non-destructive corresponds directly to a classification of formulas). An analogous remark applies to the Gentzen system $\mathbf{G}_{\mathscr{H}(E)}$ in comparison to the Gentzen system considered in Blackburn (2000a).

The tableau-based decision procedure given in Bolander and Braüner (2006) was published already in Bolander and Braüner (2005). We are not aware of any Gentzen-based decision procedures for hybrid logic published before the publication of Bolander and Braüner (2006) and we are only aware of one tableau-based decision procedure for hybrid logic published before the publication of Bolander and Braüner (2005), namely the prefixed tableau system given in Miroslava Tzakova (1999). However, it turns out that Tzakova's termination proof is flawed.

Now, the Gentzen system $\mathbf{G}_{\mathscr{H}(E)}$ considered in the previous section is obtained by reformulating the tableau system $\mathbf{T}_{\mathscr{H}(E)}$. By reformulating the tableau system $\mathbf{T}_{\mathscr{H}(E)}$ as a Gentzen system, and sketching a decision procedure based on the Gentzen system as done in the previous section, it has been made clear that the loop-check technique does not depend on particular features of the Gentzen or tableau systems, but can be applied in connection with different kinds of proof systems. This is also corroborated by the fact that in Bolander and Braüner (2006) loop-checks are applied in connection with yet another two kinds of proof systems for the hybrid logic $\mathscr{H}(E)$, namely a prefixed tableau system along the lines of the system given in Tzakova (1999) and a tableau system involving a rule for nominal substitution. Thus, loop-checks are applicable in connection with a spectrum of different proof systems.

In ordinary modal logic, loop-checks are used in connection with standard Fitting-style prefixed tableau systems for transitive logics such as K4, see Goré (1999) and Massacci (2000). The loop-check technique can be tracked to Fitting (1983), although a similar idea was involved in a graphical formalism for deciding validity of modal-logical formulas in the earlier book Hughes and Cresswell (1968). Now, a simple prefixed tableau system can be formulated for the modal logic K such that a systematic tableau construction always terminates, cf. Fitting (1983). The systematic tableau construction algorithm for K does not involve loop-checks. However, if the tableau system for K is extended with the standard prefixed tableau rule for transitivity

$$\frac{(a, \Box\phi), R(a,b)}{(b, \Box\phi)}$$

(the notation should be self-explanatory) whereby a tableau system for K4 is obtained, then a systematic tableau construction may not terminate. Intuitively, the problem is that the rule allows information to be moved forward from a world to any accessible world. The standard way to fix this problem is to incorporate loop-check conditions on the applications of existential rules. The intuitive reason why this technique works in the context of hybrid logic, is that the problem here is also that information can be moved between worlds, namely in connection with applications of the rules $(Nom1)$ and $(Nom2)$ in the tableau system described in Section 3.2. Intuitively, these rules allow atomic information to be moved freely between worlds that are identical.

There is a close connection between the hybrid logic $\mathscr{H}(E)$ and description logics, see Blackburn and Tzakova (1998) and Carlos Areces' PhD thesis, (2000). The hybrid logic $\mathscr{H}(E)$ is the mono-modal hybrid logic $\mathscr{H}$ extended with the universal modality, but all the results in the present chapter also hold if a multi-modal version of the hybrid logic $\mathscr{H}$ is extended with the universal modality, that is, if the single modal operator $\Diamond$ in the hybrid logic is replaced by an arbitrary, finite number of modal operators $\Diamond_1, \ldots, \Diamond_m$. Such a multi-modal hybrid logic with the universal modality can be seen as a natural generalization of a description logic. Now, the description logic called $\mathscr{ALC}$ is a notational variant of ordinary multi-modal logic, that is, propositional logic extended with a finite number of modal operators $\Diamond_1, \ldots, \Diamond_m$. The *concept expressions* of $\mathscr{ALC}$ simply correspond to formulas of multi-modal logic and vice versa. Given a description logic, for example $\mathscr{ALC}$, a knowledge base is a set of metalinguistic statements expressing relationships between concepts and individuals. There are two kinds of metalinguistic statements; they are called *TBox-statements* and *ABox-statements* respectively. A TBox-statement $\phi \sqsubseteq \psi$ expresses that the concept ϕ is subsumed by the concept ψ, that is, that any individual that belongs to the extension of ϕ also belongs to the extension of ψ. An ABox-statement $\phi(a)$ expresses that the individual a belongs to the extension of the concept ϕ and an ABox-statement $R_i(a,c)$ expresses that the individual a is R_i-related to the individual c. This can all be expressed in terms of the multi-modal hybrid logic with the universal modality: The TBox-statement above is expressed by the formula $A(\phi \rightarrow \psi)$, where A is the universal modality, and the ABox-statements are expressed by the formulas $@_a\phi$ and $@_a\Diamond_i c$. Note that no binders are needed. Of course, a nominal is here considered a name of an individual. Thus, the hybrid logic here can be seen as a generalized version of this description logic where no distinction between an object language and a metalanguage is made.

Nominals are often used in description logics, and certain tableau-based decision procedures for such logics also make use of loop-checks. An example is the decision procedure given in Horrocks and Sattler (2005) which is based on a prefixed tableau system that uses metalinguistic prefixes and accessibility formulas. The logic given in that paper, and other similar logics, do not involve satisfaction operators or the universal modality, but it is well-known that if a description logic has transitive and inverse roles together with role hierarchies, which is the case with the logic in Horrocks and Sattler (2005), then general concept inclusion axioms can be internalised into concepts, as described in Horrocks et al. (1999, pp. 164–165). This technique can also be used to define an "approximation" of the universal modality: Given roles $R_1,\ldots,R_n$ occurring in a description-logical formula ϕ and a new role U, a set of role axioms

$$\{\mathsf{Trans}(U), U \sqsubseteq \mathsf{Inv}(U), R_1 \sqsubseteq U, ..., R_n \sqsubseteq U\}$$

is defined ensuring that the role U is a transitive and symmetric role containing all the other roles. In the terminology of modal logic, a model satisfying the axioms has the property that the submodel generated by a world w is identical to the equivalence class of w with respect to the equivalence relation obtained by taking the reflexive closure of U. Consequently, a formula $\psi \vee \exists U.\psi$ is true at a world w if and only if ψ is true somewhere at the submodel generated by w, and furthermore, all worlds in the submodel generated by w generate the same submodel. It follows that a formula ϕ is satisfiable with respect to arbitrary models if and only if the formula ϕ' obtained by replacing any universal modality $E\psi$ in ϕ by $\psi \vee \exists U.\psi$ is satisfiable with respect to models satisfying the axioms. In case nominals are involved, further axioms have to be added such that $\psi \vee \exists U.\psi$ is true at a world w if and only if ψ is true somewhere at the submodel generated by the set of worlds consisting of w together with the denotations of all nominals in ψ. In this sense, the universal modality can be approximated if further machinery is present, namely axioms involving transitive and inverse roles as well as role hierarchies.

However, we think that the universal modality and satisfaction operators are so important and widely used that it justifies independent and direct tableau-based and Gentzen-based decision procedures, as given in the present chapter. Also, it seems unnecessarily complicated to obtain a decision procedure encompassing the universal modality (which is first-order definable) by a reduction to a decision procedure involving axioms for a new role (which implicitly amounts to imposing a second-order condition on models, namely the condition that there exists a relation satisfying the axioms).

Chapter 4
Comparison to Seligman's Natural Deduction System

In this chapter we compare and contrast the natural deduction system given in Section 2.2 to a modified version of a hybrid-logical natural deduction system given by Jerry Seligman. The chapter is structured as follows. In the first section of the chapter we describe the natural deduction systems under consideration, in particular, we define our version of Seligman's system. In the second and third sections, we give translations of derivations backwards and forwards between the systems, and in the fourth section we devise a set of reduction rules for our version of Seligman's system by translation of the reduction rules for the system given in Section 2.2. In the final section we discuss the results. All the results of this chapter are taken from Braüner (2004b).

4.1 The Natural Deduction Systems Under Consideration

In this section we describe the natural deduction systems under consideration. We consider two different systems, both for the hybrid logic $\mathcal{H}(\mathcal{O})$. The first natural deduction system is the system $\mathbf{N}_{\mathcal{H}(\mathcal{O})}$ given in Section 2.2. The rules for $\mathbf{N}_{\mathcal{H}(\mathcal{O})}$ can be found in Figures 2.2 and 2.3 of that section. The second system, which we denote $\mathbf{N'}_{\mathcal{H}(\mathcal{O})}$, is obtained from the rules given in Figures 4.1 and 4.2 by leaving out the rules for the binders that are not in the set $\mathcal{O}$. Note that the rule $(Term)$ in Figure 4.2 discharges $n+1$ parcels.

The most notable concrete difference between the systems $\mathbf{N}_{\mathcal{H}(\mathcal{O})}$ and $\mathbf{N'}_{\mathcal{H}(\mathcal{O})}$ is that all formulas occurring in the system $\mathbf{N}_{\mathcal{H}(\mathcal{O})}$ are satisfaction statements, whereas any formula can occur in the system $\mathbf{N'}_{\mathcal{H}(\mathcal{O})}$. From a semantical point of view, the restriction that all formulas in $\mathbf{N}_{\mathcal{H}(\mathcal{O})}$ have to be satisfaction statements ensures that any formula is given an explicit, that is, named, world of evaluation. This is in line with a general idea of hybrid logic, namely to build in semantic notions in the object language. At a more abstract level, it is notable that the ways of reasoning in the two systems are very different from each other. For example, in some important cases reasoning in $\mathbf{N}_{\mathcal{H}(\mathcal{O})}$ is closer to semantic intuition than reasoning in $\mathbf{N'}_{\mathcal{H}(\mathcal{O})}$. In

T. Braüner, *Hybrid Logic and its Proof-Theory*, Applied Logic Series 37,
DOI 10.1007/978-94-007-0002-4_4, © Springer Science+Business Media B.V. 2011

$$\frac{\phi \quad \psi}{\phi \wedge \psi}\,(\wedge I) \qquad \frac{\phi \wedge \psi}{\phi}\,(\wedge E1) \qquad \frac{\phi \wedge \psi}{\psi}\,(\wedge E2)$$

$$\frac{\begin{array}{c}[\phi]\\ \vdots\\ \psi\end{array}}{\phi \rightarrow \psi}\,(\rightarrow I) \qquad \frac{\phi \rightarrow \psi \quad \phi}{\psi}\,(\rightarrow E)$$

$$\frac{\begin{array}{c}[\neg\phi]\\ \vdots\\ \bot\end{array}}{\phi}\,(\bot)^{*}$$

$$\frac{a \quad \phi}{@_a\phi}\,(@I) \qquad \frac{a \quad @_a\phi}{\phi}\,(@E)$$

$$\frac{\begin{array}{c}[\Diamond c]\\ \vdots\\ @_c\phi\end{array}}{\Box\phi}\,(\Box I)^{\star} \qquad \frac{\Box\phi \quad \Diamond e}{@_e\phi}\,(\Box E)$$

$$\frac{\begin{array}{c}[c]\\ \vdots\\ @_c\phi[c/b]\end{array}}{\downarrow b\phi}\,(\downarrow I)^{\dagger} \qquad \frac{\downarrow b\phi \quad e}{@_e\phi[e/b]}\,(\downarrow E)$$

$$\frac{\phi[c/b]}{\forall b\phi}\,(\forall I)^{\ddagger} \qquad \frac{\forall b\phi}{\phi[e/b]}\,(\forall E)$$

$*$ ϕ is a propositional letter.
$\star$ c does not occur free in $\Box\phi$ or in any undischarged assumptions other than the specified occurrences of $\Diamond c$.
$\dagger$ c does not occur free in $\downarrow b\phi$ or in any undischarged assumptions other than the specified occurrences of c.
$\ddagger$ c does not occur free in $\forall b\phi$ or in any undischarged assumptions.

Fig. 4.1 Seligman-style natural deduction rules for connectives

Section 4.2 we shall come back to this issue in connection with a translation from $\mathbf{N}'_{\mathcal{H}(\mathcal{O})}$ to $\mathbf{N}_{\mathcal{H}(\mathcal{O})}$.

The natural deduction system $\mathbf{N}_{\mathcal{H}(\mathcal{O})}$ corresponds to the class of all frames, but it can be extended with additional derivation rules corresponding to first-order conditions on the accessibility relations expressed by geometric theories, cf. Section 2.2.1. This can also be done with the system $\mathbf{N}'_{\mathcal{H}(\mathcal{O})}$ but we shall not consider this issue here. The natural deduction systems $\mathbf{N}_{\mathcal{H}(\mathcal{O})}$ and $\mathbf{N}'_{\mathcal{H}(\mathcal{O})}$ are both sound and com-

$$\dfrac{\phi_1 \;\ldots\; \phi_n \qquad \begin{array}{c}[\phi_1]\ldots[\phi_n][a] \\ \vdots \\ \psi\end{array}}{\psi}\;(Term)^* \qquad\qquad \dfrac{\begin{array}{c}[a] \\ \vdots \\ \psi\end{array}}{\psi}\;(Name)^\star$$

$*$ $\phi_1, \ldots, \phi_n$, and ψ are all satisfaction statements and there are no undischarged assumptions in the derivation of ψ besides the specified occurrences of $\phi_1, \ldots, \phi_n$, and a.
$\star$ a does not occur free in ψ or in any undischarged assumptions other than the specified occurrences of a.

Fig. 4.2 Seligman-style natural deduction rules for nominals

plete. Soundness and completeness of $\mathbf{N}_{\mathcal{H}(\mathcal{O})}$ was proved in Section 2.2.3. Soundness and completeness of $\mathbf{N}'_{\mathcal{H}(\mathcal{O})}$ is dealt with in Section 4.3 of the present chapter.

The natural deduction system $\mathbf{N}'_{\mathcal{H}(\mathcal{O})}$ is a modified version of a natural deduction system given in Jerry Seligman (1997). Below we shall describe how and why we have modified the system of Seligman (1997). The system of Seligman (1997) was originally meant to be a natural deduction system for a logic of situations similar to hybrid logic (but this difference is not directly of importance here). See also the Gentzen system for hybrid logic given in Seligman (2001). Analogous to the natural deduction system $\mathbf{N}'_{\mathcal{H}(\mathcal{O})}$, the Gentzen system of Seligman (2001) allows arbitrary formulas to occur in derivation rules. Such a Gentzen system can also be found in Kushida and Okada (2007). The latter Gentzen system makes use of proof-theoretical machinery of ordinary (non-hybrid) modal logic, that is, it makes use of a standard Gentzen rule for the ordinary modal logic K, which makes it quite different from the Gentzen system of Seligman (2001) and the natural deduction system $\mathbf{N}'_{\mathcal{H}(\mathcal{O})}$ as well.

4.1.1 Seligman's Original System

The natural deduction system $\mathbf{N}'_{\mathcal{H}(\mathcal{O})}$ and the original system given in Seligman (1997) differ in several ways. Some differences are insignificant here, but there is one difference which is significant: In Seligman (1997), another version of the rule $(Term)$ of Figure 4.2 is used, namely the rule

$$\dfrac{\begin{array}{c}[a] \\ \vdots \\ \psi\end{array}}{\psi}\;(Term)$$

which is equipped with the side-condition that ψ and all undischarged assumptions other than the specified occurrences of a are satisfaction statements.

We should explain why we have chosen a different version of the rule $(Term)$. According to Sections 2.2.4 and 2.2.5, the natural deduction system $\mathbf{N}_{\mathcal{H}(\mathcal{O})}$ can be equipped with reduction rules such that a normalization theorem holds and such that normal derivations satisfy a version of the subformula property. On the other hand, Seligman (1997) does not consider reduction rules for the natural deduction system given there, and reduction rules for the system are not considered elsewhere. One aim here is to discuss reduction rules for the system given in Seligman (1997), but it turns out that in its original form, that is, with the original version of the rule $(Term)$, this system does not have the property called *closure under substitution*, which is a prerequisite for rewriting derivations using reduction rules. Closure under substitution requires that for any two derivations

$$\begin{array}{c} \vdots \\ \phi \end{array} \qquad\qquad \begin{array}{c} (\phi) \\ \vdots \\ \psi \end{array}$$

the result of substituting the left-hand-side derivation for the undischarged assumption (ϕ) in the right hand-side-derivation, that is, the following

$$\begin{array}{c} \vdots \\ \phi \\ \vdots \\ \psi \end{array}$$

is a correct derivation (after renaming of nominals, if necessary). For example, if the derivations

$$\dfrac{b \qquad \theta}{@_b\theta}\,(@I) \qquad\qquad \begin{array}{cc} [a] & (@_b\theta) \\ & \vdots \\ & \dfrac{\psi}{\psi}\,(Term) \end{array}$$

are considered, then the following

$$\begin{array}{cc} & \dfrac{b \qquad \theta}{@_b\theta}\,(@I) \\ [a] & \\ & \vdots \\ & \dfrac{\psi}{\psi}\,(Term) \end{array}$$

has to be a correct derivation, but this is not the case since the side-condition on the rule $(Term)$ is now violated as b is not a satisfaction statement. We have solved the problem by modifying the original rule $(Term)$ such that substitutions are made explicit[1] as shown in Figure 4.2. It is straightforward to check that closure under substitution is satisfied by the system $\mathbf{N}'_{\mathcal{H}(\mathcal{O})}$. Note that the side-condition of our modified version of $(Term)$ requires that *all* assumptions are discharged. It should be emphasized that reduction rules are not considered in Seligman (1997) and therefore the problem that closure under substitution is not satisfied does not crop up.

4.2 Translation from Seligman-Style Derivations

In this and the following section we translate derivations backwards and forwards between the natural deduction systems $\mathbf{N}_{\mathcal{H}(\mathcal{O})}$ and $\mathbf{N}'_{\mathcal{H}(\mathcal{O})}$. Our aim with the translations is to compare the systems and clarify the differences.

In the present section we translate derivations of $\mathbf{N}'_{\mathcal{H}(\mathcal{O})}$ to derivations of $\mathbf{N}_{\mathcal{H}(\mathcal{O})}$ as follows. For any formula ϕ and any set of formulas Γ, a derivation π of ϕ from Γ in $\mathbf{N}'_{\mathcal{H}(\mathcal{O})}$ is translated to a derivation $\pi^{\bullet}$ of $@_d\phi$ from $@_d\Gamma$ in $\mathbf{N}_{\mathcal{H}(\mathcal{O})}$ where d is a new nominal and $@_d\Gamma$ is the set of formulas $\{@_d\psi \mid \psi \in \Gamma\}$. (Recall that any formula can occur in the system $\mathbf{N}'_{\mathcal{H}(\mathcal{O})}$ whereas only satisfaction statements can occur in $\mathbf{N}_{\mathcal{H}(\mathcal{O})}$.) In the translation we shall make use of the rule

$$\frac{@_a c \qquad @_a\phi}{@_c\phi}\ (Nom)$$

which is admissible in $\mathbf{N}_{\mathcal{H}(\mathcal{O})}$ according to Proposition 2.1 of Section 2.2.2. The translation is defined by induction on the structure of derivations. We only give a selection of cases.

- A derivation on the form

$$@_a\phi$$

  is translated to the derivation

$$@_d@_a\phi$$

- A derivation on the form

$$\frac{\begin{array}{c}[\phi] \\ \vdots\ \tau \\ \psi\end{array}}{\phi \rightarrow \psi}\ (\rightarrow I)$$

  is translated to the derivation

[1] A historical remark is relevant here: An analogous problem appears in connection with intuitionistic linear logic. This problem was originally solved by Benton et al. (1992, 1993), and by the author of the present book. See the account given in Braüner (1996). The same problem appears in connection with a natural deduction system for the modal logic S4, see Bierman and de Paiva (2000).

$$\dfrac{\begin{array}{c}[@_d\phi]\\ \vdots\,\tau^{\bullet}\\ @_d\psi\end{array}}{@_d(\phi\rightarrow\psi)}\,(\rightarrow I)$$

- A derivation on the form

$$\dfrac{\begin{array}{c}\vdots\,\tau\\ \phi\rightarrow\psi\end{array}\qquad\begin{array}{c}\vdots\,\sigma\\ \phi\end{array}}{\psi}\,(\rightarrow E)$$

  is translated to the derivation

$$\dfrac{\begin{array}{c}\vdots\,\tau^{\bullet}\\ @_d(\phi\rightarrow\psi)\end{array}\qquad\begin{array}{c}\vdots\,\sigma^{\bullet}\\ @_d\phi\end{array}}{@_d\psi}\,(\rightarrow E)$$

- A derivation on the form

$$\dfrac{\begin{array}{c}\vdots\,\tau\\ a\end{array}\qquad\begin{array}{c}\vdots\,\sigma\\ \phi\end{array}}{@_a\phi}\,(@I)$$

  is translated to the derivation

$$\dfrac{\dfrac{\begin{array}{c}\vdots\,\tau^{\bullet}\\ @_d a\end{array}\qquad\begin{array}{c}\vdots\,\sigma^{\bullet}\\ @_d\phi\end{array}}{@_a\phi}\,(Nom)}{@_d@_a\phi}\,(@I)$$

- A derivation on the form

$$\dfrac{\begin{array}{c}\vdots\,\tau\\ a\end{array}\qquad\begin{array}{c}\vdots\,\sigma\\ @_a\phi\end{array}}{\phi}\,(@E)$$

  is translated to the derivation

$$\dfrac{\dfrac{\begin{array}{c}\vdots\,\tau^{\bullet}\\ @_d a\end{array}}{@_a d}\,(Sym)\qquad\dfrac{\begin{array}{c}\vdots\,\sigma^{\bullet}\\ @_d@_a\phi\end{array}}{@_a\phi}\,(@E)}{@_d\phi}\,(Nom)$$

- A derivation on the form

$$\dfrac{\begin{matrix}\vdots\,\tau_1\\ \phi_1\end{matrix}\quad\cdots\quad\begin{matrix}\vdots\,\tau_n\\ \phi_n\end{matrix}\qquad\qquad\begin{matrix}[\phi_1]\ldots[\phi_n][a]\\ \vdots\,\pi\\ \psi\end{matrix}}{\psi}\,(Term)$$

  is translated to the derivation

$$\begin{array}{c}
\begin{matrix}\vdots\,\tau_1^\bullet\\ \dfrac{\dfrac{@_d\phi_1}{\phi_1}\,(@E)}{@_a\phi_1}\,(@I)\end{matrix}\quad\cdots\quad\begin{matrix}\vdots\,\tau_n^\bullet\\ \dfrac{\dfrac{@_d\phi_n}{\phi_n}\,(@E)}{@_a\phi_n}\,(@I)\end{matrix}\quad\dfrac{}{@_a a}\,(Ref)\\
\vdots\,\pi^\bullet[a/d]\\
\dfrac{\dfrac{@_a\psi}{\psi}\,(@E)}{@_d\psi}\,(@I)
\end{array}$$

- A derivation on the form

$$\dfrac{\begin{matrix}[a]\\ \vdots\,\tau\\ \psi\end{matrix}}{\psi}\,(Name)$$

  is translated to the derivation

$$\begin{array}{c}
\dfrac{}{@_d d}\,(Ref)\\
\vdots\,\tau^\bullet[d/a]\\
@_d\psi
\end{array}$$

All the other cases are similar to the cases given above. The rule (Sym) used in the case for the rule $(@E)$ is straightforwardly derivable using $(Nom1)$ and (Ref). It is notable that the reasoning in the system $\mathbf{N}_{\mathcal{H}(\mathcal{O})}$ which takes place in the last two cases given above is analogous to the reasoning in the corresponding two cases of the soundness proof for the system $\mathbf{N'}_{\mathcal{H}(\mathcal{O})}$, Theorem 4.1. Thus, in these cases reasoning in $\mathbf{N}_{\mathcal{H}(\mathcal{O})}$ is closer to semantic intuition than reasoning in $\mathbf{N'}_{\mathcal{H}(\mathcal{O})}$. See also the remark following Theorem 4.1.

4.3 Translation to Seligman-Style Derivations

In the present section we translate derivations of $\mathbf{N}_{\mathcal{H}(\mathcal{O})}$ to derivations of $\mathbf{N'}_{\mathcal{H}(\mathcal{O})}$. In the next section we shall investigate how this translation behaves in connection with substitution of derivations for parcels of undischarged assumptions, and since a derivation is substituted for each undischarged assumption in a specified parcel, we need to be able to keep track of the identity of parcels when translating a derivation. To this end we introduce a few further conventions: A set of annotated satisfaction statements will be called a *context* and the metavariables Φ, Ψ, ... will range over contexts. Moreover, a derivation π is a *derivation from* a context Φ if and only if each undischarged assumption in π is an occurrence of an annotated satisfaction statement in Φ. Note that we have previously considered derivations as being derivations from sets of satisfaction statements, that is, we have ignored numbers annotating undischarged assumptions. Keeping the numbers, that is, considering derivations as being derivations from contexts, enables us to keep track of the identity of parcels of undischarged assumptions when translating a derivation.

Having introduced the required conventions, we are now ready to define the translation. The technique used in the translation of some of the rules of $\mathbf{N}_{\mathcal{H}(\mathcal{O})}$ (namely the introduction and elimination rules, except the introduction and elimination rules for satisfaction operators) is similar to a technique used in Seligman (2001) to replace hybrid logical Gentzen rules involving only satisfaction statements with corresponding Gentzen rules involving arbitrary formulas. For any satisfaction statement ϕ and any context Φ, a derivation π of ϕ from Φ in $\mathbf{N}_{\mathcal{H}(\mathcal{O})}$ is translated to a derivation π° of ϕ from Φ in $\mathbf{N}'_{\mathcal{H}(\mathcal{O})}$. Let $\{\theta_1^{r_1}, \ldots, \theta_n^{r_n}\} \subseteq \Phi$ be the set of annotated satisfaction statements that occur as undischarged assumptions in the derivation π. The translation is defined by induction on the structure of derivations. We only give a selection of cases.

- A derivation on the form

$$@_a\phi^r$$

  is translated to the derivation

$$@_a\phi^r$$

- A derivation on the form

$$\dfrac{\begin{matrix}[@_a\phi]\\ \vdots\,\tau\\ @_a\psi\end{matrix}}{@_a(\phi\rightarrow\psi)}\,(\rightarrow I)$$

  is translated to the derivation

$$\dfrac{\theta_1^{r_1}\ldots\theta_n^{r_n} \qquad \dfrac{[a] \qquad \dfrac{\dfrac{[a] \qquad \begin{matrix}[\theta_1]\ldots[\theta_n] \quad \dfrac{[a] \quad [\phi]}{@_a\phi}\,(@I)\\ \vdots\,\tau^\circ\\ @_a\psi\end{matrix}}{\psi}\,(@E)}{\phi\rightarrow\psi}\,(\rightarrow I)}{@_a(\phi\rightarrow\psi)}\,(@I)}{@_a(\phi\rightarrow\psi)}\,(Term)$$

- A derivation on the form

$$\dfrac{\begin{matrix}\vdots\,\tau\\ @_a(\phi\to\psi)\end{matrix}\qquad \begin{matrix}\vdots\,\sigma\\ @_a\phi\end{matrix}}{@_a\psi}\,(\to E)$$

is translated to the derivation

$$\dfrac{\theta_1^{r_1}\dots\theta_n^{r_n}\qquad \dfrac{[a]\qquad \dfrac{\dfrac{[a]\qquad \begin{matrix}[\theta_1]\dots[\theta_n]\\ \vdots\,\tau^\circ\\ @_a(\phi\to\psi)\end{matrix}}{\phi\to\psi}\,(@E)\qquad \dfrac{[a]\qquad \begin{matrix}[\theta_1]\dots[\theta_n]\\ \vdots\,\sigma^\circ\\ @_a\phi\end{matrix}}{\phi}\,(@E)}{\psi}\,(\to E)}{@_a\psi}\,(@I)}{@_a\psi}\,(Term)$$

- A derivation on the form

$$\dfrac{\begin{matrix}\vdots\,\tau\\ @_a\phi\end{matrix}}{@_c@_a\phi}\,(@I)$$

is translated to the derivation

$$\dfrac{\theta_1^{r_1}\dots\theta_n^{r_n}\qquad \dfrac{[c]\qquad \begin{matrix}[\theta_1]\dots[\theta_n]\\ \vdots\,\tau^\circ\\ @_a\phi\end{matrix}}{@_c@_a\phi}\,(@I)}{@_c@_a\phi}\,(Term)$$

- A derivation on the form

$$\dfrac{\begin{matrix}\vdots\,\tau\\ @_c@_a\phi\end{matrix}}{@_a\phi}\,(@E)$$

is translated to the derivation

$$\dfrac{\theta_1^{r_1}\dots\theta_n^{r_n}\qquad \dfrac{[c]\qquad \begin{matrix}[\theta_1]\dots[\theta_n]\\ \vdots\,\tau^\circ\\ @_c@_a\phi\end{matrix}}{@_a\phi}\,(@E)}{@_a\phi}\,(Term)$$

- A derivation on the form

$$\dfrac{}{@_a a}\,(Ref)$$

is translated to the derivation

$$\dfrac{\dfrac{[a] \quad [a]}{@_a a}\,(@I)}{@_a a}\,(Term)$$

- A derivation on the form

$$\dfrac{\begin{array}{c}\vdots\,\tau \\ @_a c\end{array} \qquad \begin{array}{c}\vdots\,\sigma \\ @_a \phi\end{array}}{@_c \phi}\,(Nom)$$

is translated to the derivation

$$\dfrac{\theta_1^{r_1}\ldots\theta_n^{r_n} \qquad \dfrac{\dfrac{[a] \qquad \begin{array}{c}[\theta_1]\ldots[\theta_n]\\ \vdots\,\tau^\circ \\ @_a c\end{array}}{c}\,(@E) \qquad \dfrac{[a] \qquad \begin{array}{c}[\theta_1]\ldots[\theta_n]\\ \vdots\,\sigma^\circ \\ @_a \phi\end{array}}{\phi}\,(@E)}{@_c \phi}\,(@I)}{@_c \phi}\,(Term)$$

All the other cases are similar to the cases given above. Note that the cases with $(Nom1)$ and $(Nom2)$ are covered by translation of the more general rule (Nom), which is admissible in the system $\mathbf{N}_{\mathcal{H}(\mathcal{O})}$ according to Proposition 2.1 of Section 2.2.2. Also, note that in some of the cases $\{\theta_1^{r_1},\ldots,\theta_n^{r_n}\}$ is the union of the undischarged assumptions of two derivations, for example in the last case given above where $\{\theta_1^{r_1},\ldots,\theta_n^{r_n}\}$ is the union of the undischarged assumptions of the derivations τ and σ, and in such a case an annotated satisfaction statement $\theta_j^{r_j}$ need not occur as an undischarged assumption in both derivations (but it obviously has to occur as an undischarged assumption in at least one of the derivations).

We now consider soundness and completeness of $\mathbf{N}'_{\mathcal{H}(\mathcal{O})}$.

Theorem 4.1. *The first statement below implies the second statement and vice versa.*

1. *ψ is derivable from Γ in $\mathbf{N}'_{\mathcal{H}(\mathcal{O})}$.*
2. *For any model $\mathcal{M}$, any world w, and any assignment g, if, for any formula $\theta \in \Gamma$, $\mathcal{M},g,w \models \theta$, then $\mathcal{M},g,w \models \psi$.*

Proof. The soundness proof is by induction on the structure of the derivation of ψ. We only consider the cases with the rules $(Term)$ and $(Name)$. Let $\mathcal{M}$ be a model, w a world, and g an assignment such that for any formula $\theta \in \Gamma$, $\mathcal{M},g,w \models \theta$. In the case involving the rule $(Term)$ it follows by induction that $\mathcal{M},g,w \models \phi_i$, where $i \in \{1,\ldots,n\}$, cf. Figure 4.2, and hence $\mathcal{M},g,g(a) \models \phi_i$ as ϕ_i is a satisfaction statement. By induction it follows that $\mathcal{M},g,g(a) \models \psi$ and hence $\mathcal{M},g,w \models \psi$

as ψ is a satisfaction statement. In the case involving the rule (*Name*) we let g' be the assignment such that $g' \stackrel{a}{\sim} g$ and $g'(a) = w$ and we let $\Gamma' \subseteq \Gamma$ be the set of undischarged assumptions in the derivation of ψ, cf. Figure 4.2. Then for any formula $\theta \in \Gamma'$, $\mathcal{M}, g', w \models \theta$. By induction it follows that $\mathcal{M}, g', w \models \psi$ and hence $\mathcal{M}, g, w \models \psi$.

The proof of completeness is similar to a completeness proof for a Gentzen system given in Seligman (2001). Assume that for any model $\mathcal{M}$, any world w, and any assignment g, if, for any formula $\theta \in \Gamma$, $\mathcal{M}, g, w \models \theta$, then $\mathcal{M}, g, w \models \psi$. Let d be a new nominal. It follows that for any model $\mathcal{M}$ and any assignment g, if, for any formula $@_d\theta \in @_d\Gamma$, $\mathcal{M}, g \models @_d\theta$, then $\mathcal{M}, g \models @_d\psi$ (recall that $@_d\Gamma$ denotes the set of formulas $\{@_d\theta \mid \theta \in \Gamma\}$). By completeness of the system $\mathbf{N}_{\mathcal{H}(\mathcal{O})}$, Theorem 2.2 of Section 2.2.3, there exists a derivation π of $@_d\psi$ from $@_d\Gamma$ in $\mathbf{N}_{\mathcal{H}(\mathcal{O})}$. By annotating the undischarged assumptions in π with distinct numbers, translating the resulting derivation into $\mathbf{N}'_{\mathcal{H}(\mathcal{O})}$ using the translation $(\cdot)^\circ$, and then removing the numbers again, a derivation of $@_d\psi$ from $@_d\Gamma$ in $\mathbf{N}'_{\mathcal{H}(\mathcal{O})}$ is obtained. By application of the rules $(@I)$, $(@E)$, and (*Name*), it follows that ψ is derivable from Γ in $\mathbf{N}'_{\mathcal{H}(\mathcal{O})}$.

It is instructive to take a closer look at the intuitions behind the two cases considered in the soundness proof above. In the case involving the rule (*Term*), the world of evaluation is shifted from the actual world to a hypothetical world, namely the world where the nominal a is true, then some reasoning is performed, and finally the world of evaluation is shifted back to the actual world. The side-condition that the assumptions $\phi_1, \ldots, \phi_n$ and the conclusion ψ all have to be satisfaction statements, ensures that their truth-values are not affected when the world of evaluation is shifted. On the other hand, in the case involving the rule (*Name*), the world of evaluation is kept constant, but the world to which the nominal a refers is shifted to the actual world, then some reasoning is performed, and finally the reference of a is shifted back to the world to which it originally referred. The side-condition that the nominal a must not occur in any of the undischarged assumptions or in the conclusion ψ ensures that their truth-values are not affected.

4.4 Reduction Rules

In this section we use the translation $(\cdot)^\circ$ from $\mathbf{N}_{\mathcal{H}(\mathcal{O})}$ to $\mathbf{N}'_{\mathcal{H}(\mathcal{O})}$ to devise a set of reduction rules for $\mathbf{N}'_{\mathcal{H}(\mathcal{O})}$ in the following way: We already have reduction rules for the system $\mathbf{N}_{\mathcal{H}(\mathcal{O})}$, namely those given in Section 2.2.4. Moreover, a desirable property of a translation is that it preserves reductions, that is, if a derivation π reduces to a derivation τ, then the translation of π has to reduce to the translation of τ. The reason why preservation of reductions is a desirable property, is that the application of a reduction rule to a derivation is supposed to leave the identity of the proof represented by the derivation unchanged. Rather, the application of a reduction rule just removes a "detour" in the derivation. See the discussion in Prawitz (1971, p. 257). Thus, by considering the image of the reduction rules for $\mathbf{N}_{\mathcal{H}(\mathcal{O})}$ under the translation $(\cdot)^\circ$ from $\mathbf{N}_{\mathcal{H}(\mathcal{O})}$ to $\mathbf{N}'_{\mathcal{H}(\mathcal{O})}$, and requiring that the translation preserves reductions, a set of reduction rules for $\mathbf{N}'_{\mathcal{H}(\mathcal{O})}$ can be read off. In other words, in this section we show that the natural deduction system $\mathbf{N}'_{\mathcal{H}(\mathcal{O})}$ can be equipped with a set of reduction rules that are proof-theoretically well-behaved in the sense that they can simulate reductions in $\mathbf{N}_{\mathcal{H}(\mathcal{O})}$.

Above we gave an outline of the section. In what follows we shall be more specific. First some conventions in connection with the rules of $\mathbf{N}'_{\mathcal{H}(\mathcal{O})}$. If a premise of a rule has the form c or $\Diamond c$, then it is called a *relational premise*. The premise of the form ϕ in the rule $(\rightarrow E)$ is called the *minor premise*. A premise of an elimination rule that is neither minor nor relational is called *major*.

A *maximum formula* in a derivation is a formula occurrence that is both the conclusion of an introduction rule and the major premise of an elimination rule. Maximum formulas can be removed by applying *proper reduction rules*. Of course, this terminology is analogous to the case with $\mathbf{N}_{\mathcal{H}(\mathcal{O})}$, cf. Section 2.2.4. The rules for proper reductions are as follows. We consider each case in turn.

$(\wedge I)$ followed by $(\wedge E1)$ (analogously in the case involving $(\wedge E2)$)

$$\dfrac{\dfrac{\begin{matrix}\vdots\ \pi_1\\ \phi\end{matrix}\qquad\begin{matrix}\vdots\ \pi_2\\ \psi\end{matrix}}{(\phi\wedge\psi)}}{\phi}\qquad\rightsquigarrow\qquad\begin{matrix}\vdots\ \pi_1\\ \phi\end{matrix}$$

$(\rightarrow I)$ followed by $(\rightarrow E)$

$$\dfrac{\dfrac{\begin{matrix}[\phi]\\ \vdots\ \pi_1\\ \psi\end{matrix}}{(\phi\rightarrow\psi)}\qquad\begin{matrix}\vdots\ \pi_2\\ \phi\end{matrix}}{\psi}\qquad\rightsquigarrow\qquad\begin{matrix}\vdots\ \pi_2\\ \phi\\ \vdots\ \pi_1\\ \psi\end{matrix}$$

$(@I)$ followed by $(@E)$

$$\dfrac{\begin{matrix}\vdots\ \pi_1\\ a\end{matrix}\qquad\dfrac{\begin{matrix}\vdots\ \pi_2\\ a\end{matrix}\qquad\begin{matrix}\vdots\ \pi_3\\ \phi\end{matrix}}{@_a\phi}}{\phi}\qquad\rightsquigarrow\qquad\begin{matrix}\vdots\ \pi_3\\ \phi\end{matrix}$$

$(\Box I)$ followed by $(\Box E)$

$$\dfrac{\dfrac{\begin{array}{c}[\Diamond c]\\ \vdots\ \pi_1\\ @_c\phi\end{array}}{\Box\phi} \qquad \begin{array}{c}\vdots\ \pi_2\\ \Diamond e\end{array}}{@_e\phi} \quad \rightsquigarrow \quad \begin{array}{c}\vdots\ \pi_2\\ \Diamond e\\ \vdots\ \pi_1[e/c]\\ @_e\phi\end{array}$$

$(\downarrow I)$ followed by $(\downarrow E)$

$$\dfrac{\dfrac{\begin{array}{c}[c]\\ \vdots\ \pi_1\\ @_c\phi[c/b]\end{array}}{\downarrow b\phi} \qquad \begin{array}{c}\vdots\ \pi_2\\ e\end{array}}{@_e\phi[e/b]} \quad \rightsquigarrow \quad \begin{array}{c}\vdots\ \pi_2\\ e\\ \vdots\ \pi_1[e/c]\\ @_e\phi[e/b]\end{array}$$

$(\forall I)$ followed by $(\forall E)$

$$\dfrac{\dfrac{\begin{array}{c}\vdots\ \pi\\ \phi[c/b]\end{array}}{\forall b\phi}}{\phi[e/b]} \quad \rightsquigarrow \quad \begin{array}{c}\vdots\ \pi[e/c]\\ \phi[e/b]\end{array}$$

In the proof that the translation $(\cdot)^\circ$ preserves reductions, that is, Lemma 4.1 and Theorem 4.2, we shall need six additional reduction rules in connection with the derivation rules for nominals (note that we have already given one reduction rule in connection with the nominal rules, namely the proper reduction above where a maximum formula of the form $@_a\phi$ is removed). These additional reduction rules will be called *auxiliary reduction rules*.

The first auxiliary reduction rule concerns the case where $(@E)$ is followed by $(@I)$ such that the prefixed satisfaction operators are identical.

$$\dfrac{\begin{array}{c}\vdots\ \pi_1\\ a\end{array} \qquad \dfrac{\begin{array}{c}\vdots\ \pi_2\\ a\end{array} \qquad \begin{array}{c}\vdots\ \pi_3\\ @_a\phi\end{array}}{\phi}\,(@E)}{@_a\phi}\,(@I) \quad \rightsquigarrow \quad \begin{array}{c}\vdots\ \pi_3\\ @_a\phi\end{array}$$

The second auxiliary reduction rule concerns the case where an instance of the rule $(Term)$ is followed by an instance of $(@E)$ such that the nominal discharged by $(Term)$ is identical to the nominal in the prefixed satisfaction operator in $(@E)$.

$$\dfrac{\begin{array}{c}\vdots\,\sigma\\ a\end{array} \qquad \dfrac{\begin{array}{c}\vdots\,\tau_1\\ \phi_1\end{array}\ \ldots\ \begin{array}{c}\vdots\,\tau_n\\ \phi_n\end{array} \qquad \begin{array}{c}[\phi_1]\ldots[\phi_n][a]\\ \vdots\,\pi\\ @_a\psi\end{array}}{@_a\psi}\,(\mathit{Term})}{\psi}\,(@E)$$

$$\rightsquigarrow$$

$$\dfrac{\begin{array}{c}\vdots\,\sigma\\ a\end{array} \qquad \begin{array}{c}\begin{array}{c}\vdots\,\tau_1\\ \phi_1\end{array}\ \ldots\ \begin{array}{c}\vdots\,\tau_n\\ \phi_n\end{array}\ \ \begin{array}{c}\vdots\,\sigma\\ a\end{array}\\ \vdots\,\pi\\ @_a\psi\end{array}}{\psi}\,(@E)$$

The third auxiliary reduction rule concerns the case involving two successive instances of the rule (Term).

$$\dfrac{\dfrac{\begin{array}{c}\vdots\,\tau_1\\ \phi_1\end{array}\ \ldots\ \begin{array}{c}\vdots\,\tau_n\\ \phi_n\end{array} \qquad \begin{array}{c}[\phi_1]\ldots[\phi_n][a]\\ \vdots\,\pi\\ \chi\end{array}}{\chi}\,(\mathit{Term}) \quad \begin{array}{c}\vdots\,\sigma_1\\ \theta_1\end{array}\ \ldots\ \begin{array}{c}\vdots\,\sigma_m\\ \theta_m\end{array} \qquad \begin{array}{c}[\chi][\theta_1]\ldots[\theta_m][b]\\ \vdots\,\rho\\ \psi\end{array}}{\psi}\,(\mathit{Term})$$

$$\rightsquigarrow$$

$$\dfrac{\begin{array}{c}\vdots\,\tau_1\\ \phi_1\end{array}\ \ldots\ \begin{array}{c}\vdots\,\tau_n\\ \phi_n\end{array}\ \ \begin{array}{c}\vdots\,\sigma_1\\ \theta_1\end{array}\ \ldots\ \begin{array}{c}\vdots\,\sigma_m\\ \theta_m\end{array} \qquad \begin{array}{c}\dfrac{[\phi_1]\ldots[\phi_n] \qquad \begin{array}{c}[\phi_1]\ldots[\phi_n][a]\\ \vdots\,\pi\\ \chi\end{array}}{\chi}\,(\mathit{Term}) \quad [\theta_1]\ldots[\theta_m][b]\\ \vdots\,\rho\\ \psi\end{array}}{\psi}\,(\mathit{Term})$$

The fourth auxiliary reduction rule concerns the case where an instance of the rule (Term) does not discharge any nominals.

$$\dfrac{\begin{array}{c}\vdots\,\tau_1\\ \phi_1\end{array}\ \ldots\ \begin{array}{c}\vdots\,\tau_n\\ \phi_n\end{array} \qquad \begin{array}{c}[\phi_1]\ldots[\phi_n]\\ \vdots\,\pi\\ \psi\end{array}}{\psi}\,(\mathit{Term}) \quad \rightsquigarrow \quad \begin{array}{c}\begin{array}{c}\vdots\,\tau_1\\ \phi_1\end{array}\ \ldots\ \begin{array}{c}\vdots\,\tau_n\\ \phi_n\end{array}\\ \vdots\,\pi\\ \psi\end{array}$$

The fifth auxiliary reduction rule concerns the case where an instance of the rule $(Term)$ does not discharge any assumptions in a specified parcel.

$$\dfrac{\begin{matrix} & & & & [\phi_1]\dots[\phi_n][a] \\ \vdots\sigma & \vdots\tau_1 & & \vdots\tau_n & \vdots\pi \\ \chi & \phi_1 & \dots & \phi_n & \psi \end{matrix}}{\psi}\,(Term)$$

$$\rightsquigarrow$$

$$\dfrac{\begin{matrix} & & & [\phi_1]\dots[\phi_n][a] \\ \vdots\tau_1 & & \vdots\tau_n & \vdots\pi \\ \phi_1 & \dots & \phi_n & \psi \end{matrix}}{\psi}\,(Term)$$

The sixth auxiliary reduction rule concerns the case where two derivations in an instance of the rule $(Term)$ are identical. Then the corresponding two parcels are amalgamated to one parcel.

$$\dfrac{\begin{matrix} & & & & & [\chi][\chi][\phi_1]\dots[\phi_n][a] \\ \vdots\sigma & \vdots\sigma & \vdots\tau_1 & & \vdots\tau_n & \vdots\pi \\ \chi & \chi & \phi_1 & \dots & \phi_n & \psi \end{matrix}}{\psi}\,(Term)$$

$$\rightsquigarrow$$

$$\dfrac{\begin{matrix} & & & & [\chi][\phi_1]\dots[\phi_n][a] \\ \vdots\sigma & \vdots\tau_1 & & \vdots\tau_n & \vdots\pi \\ \chi & \phi_1 & \dots & \phi_n & \psi \end{matrix}}{\psi}\,(Term)$$

Note that in the reduction rule above two identical derivations are replaced by one derivation.

The fifth and sixth auxiliary reduction rules can be considered bookkeeping rules ensuring that explicit substitutions behave as expected: The fifth rule ensures that if a derivation has to be substituted for an undischarged parcel which is empty, then the explicit substitution can be disregarded. The sixth rule ensures that if a derivation has to be substituted for two different undischarged parcels, then the derivation can instead be substituted for the parcel obtained by amalgamating the two parcels in question. Rules that are analogous to the fifth and sixth auxiliary reduction rules are considered in Bierman and de Paiva (2000) in connection with categorical logic. The fourth and fifth auxiliary reduction rules are similar to what Prawitz calls immediate simplifications, cf. Prawitz (1971, p. 254).

We shall need a lemma which says that the translation commutes up to reduction with substitution of derivations for undischarged assumptions in derivations.

Lemma 4.1. *Let Φ and Ψ be disjoint contexts and let ϕ^r be an annotated satisfaction statement such that $\phi^r \notin \Phi \cup \Psi$. Moreover, let τ and π be derivations in $\mathbf{N}_{\mathcal{H}(\mathcal{O})}$ such that τ is a derivation of ϕ from Φ and π is a derivation from $\{\phi^r\} \cup \Phi \cup \Psi$. Let κ be the derivation obtained by substituting τ° for ϕ^r in π° and let λ be the derivation obtained by substituting τ for ϕ^r in π (note that κ and λ are both derivations from $\Phi \cup \Psi$). Then $\kappa \rightsquigarrow \lambda^\circ$ where only the third and sixth auxiliary reduction rules have been applied.*

Proof. Induction on the structure of the derivation of π. Observe that if τ is not an undischarged assumption, then τ° has the form

$$\dfrac{\theta_1^{r_1} \ldots \theta_n^{r_n} \qquad \begin{matrix} [\theta_1] \ldots [\theta_n][a] \\ \vdots \\ \psi \end{matrix}}{\psi}\ (Term)$$

where $\{\theta_1^{r_1}, \ldots, \theta_n^{r_n}\} \subseteq \Phi$ is the set of annotated satisfaction statements that occur as undischarged assumptions in τ.

We can now give the theorem which says that the translation $(\cdot)^\circ$ from $\mathbf{N}_{\mathcal{H}(\mathcal{O})}$ to $\mathbf{N}'_{\mathcal{H}(\mathcal{O})}$ is well behaved with respect to reduction in the sense that it preserves reductions.

Theorem 4.2. *(Preservation of reductions) Let π be a derivation in the system $\mathbf{N}_{\mathcal{H}(\mathcal{O})}$. If $\pi \rightsquigarrow \tau$, then $\pi^\circ \rightsquigarrow \tau^\circ$.*

Proof. By induction on the structure of the derivation of π. It is straightforward to check each case. If the end-formula of π is the conclusion of an elimination rule that is involved in a reduction, then we use Lemma 4.1 and the observation that the translation commutes with substitution of nominals for nominals. Otherwise we use the induction hypothesis.

4.5 Discussion

It should be noted that we do not know whether a normalization theorem holds for the reduction rules for the natural deduction system $\mathbf{N}'_{\mathcal{H}(\mathcal{O})}$ given here. Moreover, the reduction rules for $\mathbf{N}'_{\mathcal{H}(\mathcal{O})}$ can simulate reductions in the system $\mathbf{N}_{\mathcal{H}(\mathcal{O})}$ cf. Theorem 4.2, but it should be noted that this does not imply that normal derivations in $\mathbf{N}'_{\mathcal{H}(\mathcal{O})}$ satisfy the subformula property or any reasonable version thereof. Consider for example the derivation below.

$$
\dfrac{\dfrac{b \qquad \dfrac{@_a b \quad @_a p \quad @_a q \qquad \dfrac{\dfrac{[a] \quad [@_a b]}{b}\,(@E) \qquad \dfrac{\dfrac{[a] \quad [@_a p]}{p}\,(@E) \qquad \dfrac{[a] \quad [@_a q]}{q}\,(@E)}{p \wedge q}\,(\wedge I)}{@_b(p \wedge q)}\,(@I)}{@_b(p \wedge q)}\,(\mathit{Term})}{p \wedge q}\,(@E)}{p}\,(\wedge E1)
$$

No reduction rules can be applied to this derivation, but it contains an occurrence of a formula, namely $@_b(p \wedge q)$, which is not a subformula of the end-formula or one of the undischarged assumptions. Of course, this also applies to the occurrences of $p \wedge q$. The problem is that the instance of the rule (Term) prevents application of proper reduction rules for the connectives @ and $\wedge$. So it would be desirable to find a more complete set of reduction rules for the natural deduction system $\mathbf{N}'_{\mathcal{H}(\mathcal{O})}$.

Chapter 5
Functional Completeness for a Hybrid Logic

In this chapter we prove a functional completeness result for the hybrid logic of the universal modality. The chapter is structured as follows. In the first section of the chapter we describe the natural deduction system under consideration, in the second section we give an introduction to the notion of functional completeness, and in the third section we give general rule schemas for natural deduction rules. In the fourth section we prove the functional completeness result and in the final section we discuss the result. The result of this chapter are taken from Braüner (2005c). Besides a functional completeness result involving the universal modality, this paper also gives a functional completeness result for a hybrid logic involving what is called the difference modality, but the two results are similar, wherefore the second result has not been included in the present book.

5.1 The Natural Deduction System Under Consideration

In this section we shall introduce the hybrid logic and the natural deduction system that our functional completeness result is about. Recall that the hybrid logic $\mathcal{H}$ defined in Section 1.2 includes one modal operator, namely $\Box$. The hybrid logic we shall consider in this chapter is obtained by replacing the modal operator $\Box$ by the universal modality A. Thus, the language under consideration includes the modal operator A, but no other modal operators. The universal modality has a fixed interpretation, namely the universal relation. This can be compared to the difference modality which also has a fixed interpretation, namely the relation of inequality.[1]

[1] The fact that functional completeness results can be given for hybrid logics involving modalities interpreted using respectively the universal relation and the relation of inequality makes it tempting to draw a parallel to Tarski (1986) where Alfred Tarski argues that given an arbitrary set, there are only four binary relations on the set which should be called "logical", namely the empty relation, the universal relation, the relation of equality, and the relation of inequality. Tarski's argument is based on the observation that these relations are exactly the binary relations on the set which are mapped to themselves by all bijections on the set in question. Tarski motivated his argument by an

T. Braüner, *Hybrid Logic and its Proof-Theory*, Applied Logic Series 37,
DOI 10.1007/978-94-007-0002-4_5,

As indicated above, in the present chapter we shall consider the hybrid logic of the universal modality. Formally, the notion of a model defined in Section 1.2, that is, Definition 1.1, is adjusted by omitting the accessibility relation, and in the definition of the relation $\mathfrak{M}, g, w \models \phi$ the clause for $\Box$ is replaced by the clause

$$\mathfrak{M}, g, w \models A\phi \text{ iff for any } v \in W, \mathfrak{M}, g, v \models \phi$$

where $\mathfrak{M} = (W, \{V_w\}_{w \in W})$ is a model, g is an assignment, and w is an element of W. In the present chapter we let $\mathscr{H}$ denote the hybrid logic thus obtained (note that the hybrid logic denoted $\mathscr{H}$ in Section 1.2 includes the modal operator $\Box$; this is not the case here, but no confusion should be possible). We let the operator E be defined by the convention that $E\phi$ is an abbreviation for $\neg A \neg \phi$. If $\overline{p}$ is a list of pairwise distinct ordinary propositional symbols and $\overline{\phi}$ is a list of formulas of the same length as $\overline{p}$, then $\psi[\overline{\phi}/\overline{p}]$ is the formula ψ where the formulas $\overline{\phi}$ have been simultaneously substituted for all occurrences of the propositional symbols $\overline{p}$.

The quantification over all worlds in the clause for A makes this modality an S5 modality. The quantification over all worlds also explains why the modality is called the "universal" modality.

Now, the natural deduction rules for $\mathscr{H}$ are given in Figures 5.1 and 5.2. All formulas in the rules are satisfaction statements. Compare to the natural deduction rules in Figures 2.2 and 2.3 of Section 2.2. The natural deduction system will be denoted $\mathbf{N}_{\mathscr{H}}$.

The natural deduction system $\mathbf{N}_{\mathscr{H}}$ is sound and complete in the usual model-theoretic sense.

Theorem 5.1. *(Soundness and completeness) Let ψ be a satisfaction statement and let Γ be a set of satisfaction statements. The statements below are equivalent.*

1. *ψ is derivable from Γ in $\mathbf{N}_{\mathscr{H}}$.*
2. *For any model $\mathfrak{M}$ and any assignment g, if, for any formula $\theta \in \Gamma$, $\mathfrak{M}, g \models \theta$, then $\mathfrak{M}, g \models \psi$.*

Proof. Soundness and completeness follows from the soundness and completeness result for the natural deduction system given in Section 2.2. It is straightforward to prove that the system considered here is equivalent to a version of the system given in Section 2.2 including the derivation rule corresponding to the condition $\forall a \forall b R_A(a,b)$ on the accessibility relation R_A for the modal operator A.

---

analogy to Klein's Erlangen Programm, named after the famous matematician Felix Klein (1849–1925). According to the Erlangen Programm, geometrical notions should be classified in terms of which transformations (bijective functions from a space to itself) they preserve. For example, all notions in Euclidean geometry are preserved by similarity transformations which are transformations that decrease or increase the size of a geometrical figure uniformly in all directions, hence, a triangle is transformed into a triangle with the same angles but possibly with proportionlly smaller or larger sides. It follows that notions which are not preserved by all similarity transformations, for example the notion of the distance between two pointe, cannot be formulated in Euclidean geometry. In an analogous way Tarski proposed to distinguish between logical and non-logical notions.

$$\frac{@_c\phi \qquad @_c\psi}{@_c(\phi \wedge \psi)}\ (\wedge I) \qquad \frac{@_c(\phi \wedge \psi)}{@_c\phi}\ (\wedge E1) \qquad \frac{@_c(\phi \wedge \psi)}{@_c\psi}\ (\wedge E2)$$

$$\begin{array}{c}[@_c\phi]\\ \vdots \\ @_c\psi \\ \hline @_c(\phi \to \psi)\end{array}\ (\to I) \qquad \frac{@_c(\phi \to \psi) \qquad @_c\phi}{@_c\psi}\ (\to E)$$

$$\begin{array}{c}[@_a\neg\phi]\\ \vdots \\ @_a\bot \\ \hline @_a\phi\end{array}\ (\bot 1)^* \qquad \frac{@_a\bot}{@_c\bot}\ (\bot 2)$$

$$\frac{@_a\phi}{@_c@_a\phi}\ (@I) \qquad \frac{@_c@_a\phi}{@_a\phi}\ (@E)$$

$$\frac{@_c\phi}{@_aA\phi}\ (AI)^\star \qquad \frac{@_aA\phi}{@_e\phi}\ (AE)$$

$*$ ϕ is a propositional symbol.
$\star$ c does not occur in $@_aA\phi$ or in any undischarged assumptions.

Fig. 5.1 Natural deduction rules for connectives

$$\frac{}{@_aa}\ (Ref) \qquad \frac{@_ac \qquad @_a\phi}{@_c\phi}\ (Nom1)^*$$

$*$ ϕ is a propositional symbol.

Fig. 5.2 Natural deduction rules for nominals

Below is an example of a derivation in $\mathbf{N}_{\mathscr{H}}$.

$$\cfrac{\cfrac{\cfrac{\cfrac{@_aA\neg c^1}{@_c\neg c}\ (AE) \qquad \cfrac{}{@_cc}\ (Ref)}{@_c\bot}\ (\to E)}{@_a\bot}\ (\bot 2)}{@_aEc}\ (\to I)^1$$

Recall that the formula $@_aEc$ is identical to $@_a(A\neg c \to \bot)$. We remark that the rule (AI) is unsound if the side-condition regarding the nominal c is dropped; consider the very simple derivation below.

$$\frac{}{@_c c}\ (Ref)$$

If the side-condition on (AI) is dropped, then this rule can be applied to the derivation whereby a derivation of the formula $@_a Ac$ is obtained, and this formula is obviously not valid.

5.2 Introduction to Functional Completeness

In this section we shall make a few introductory remarks on functional completeness. The notion of functional completeness we shall consider is defined as follows. Let a set of natural deduction rules involving a set a connectives Σ be given. Moreover, let a set of general rule schemas involving a new connective $\sharp$ be given. The general rule schemas can be instantiated to a set of natural deduction introduction and elimination rules for $\sharp$. A functional completeness result then says that any such new connective $\sharp$ is explicitly definable in terms of the original connectives Σ in the following sense: For any formula ϕ built using the connectives $\Sigma \cup \{\sharp\}$, there exists a formula ψ built using only the original connectives Σ such that ψ can be proved to be equivalent to ϕ using the original derivation rules as well as the introduction and elimination rules for the new connective $\sharp$. Thus, such a functional completeness result says that the original natural deduction system involving only the connectives Σ can define explicitly any connective with introduction and elimination rules of a certain form. It is required that the extension with $\sharp$ is conservative, that is, if a formula θ built using only the original connectives Σ is provable using the original derivation rules as well as the rules for $\sharp$, then θ is also provable using only the original derivation rules. We prove conservativity via normalization theorems for the natural deduction systems under consideration. Our general rule schemas are based on rule schemas for propositional logic given in Zucker and Tragesser (1978) and our functional completeness results are in line with the results for propositional logic given in Prawitz (1978). Furthermore, see von Kutschera (1968). See also Heinrich Wansing (1996) where functional completeness results are proved for a number of classical modal and tense logics which are presentable in terms of Belnap's display logic.

Note that the notion of functional completeness defined above is purely proof-theoretic, that is, it makes no reference to model-theoretic notions. A different notion of functional completeness which does refer to model-theoretic notions is exemplified by the well-known result in ordinary classical propositional logic which says that any truth-functional connective is definable in terms of the connectives $\{\neg, \wedge, \top\}$. Another model-theoretic functional completeness result says that any monotonic truth-functional connective is definable in terms of the connectives $\{\wedge, \top, \vee, \bot\}$. Note that functional completeness with respect to a greater class of truth-functions (all truth-functions in comparison to monotone ones) requires a more expressive set of connectives (the set of connectives $\{\neg, \wedge, \top\}$ in comparison to the set $\{\wedge, \top, \vee, \bot\}$). See McCullough (1969) for a model-theoretic functional completeness result for intuitionistic logic and see Wansing (2006) for model-theoretic functional completeness results for various intuitionistic modal logics involving what is called strong negation (also considered in Section 8.4 of the present book).

Clearly, a model-theoretic functional completeness result presupposes a model-theoretic semantics. On the other hand, proof-theoretic functional completeness is usually associated with proof-theoretic semantics, which is a research programme having as a goal to explain the meaning of a logical connective in terms of a set of derivation rules (rather than in terms of model-theoretic truth-conditions). Thus, in proof-theoretic functional completeness the general rule schemas delimits the possible meanings of the new connective $\sharp$ by delimiting the possible forms of the rules for $\sharp$. In model-theoretic semantics the possible meanings of a new connective is delimited by considering a particular set of truth-functions, for example arbitrary functions or monotone ones. In Section 10.4 we shall return to the discussion of model-theoretic semantics versus proof-theoretic semantics. See also Wansing (2000) for an informative comparison between proof-theoretic and model-theoretic semantics (and other semantic paradigms as well).

With the aim of proving functional completeness results for our natural deduction formulation of the logic $\mathscr{H}$, we shall in the next section specify a set of general rule schemas involving the new connective $\sharp$. The connective $\sharp$ takes as arguments an arbitrary, but fixed, number of nominals and an arbitrary, but again fixed, number of formulas. Our rule schemas are obtained by adapting the propositional rule schemas given by Zucker and Tragesser (1978) to hybrid logic. The rule schemas satisfy the inversion principle, cf. Section 2.1, which is a prerequisite for obtaining the normalization theorem on which our conservativity result is based.

5.3 The General Rule Schemas

In this section we shall lay the foundation of the functional completeness results which will be given in the following section. We first consider some relevant earlier work on proof-theoretic functional completeness for ordinary propositional logic and we then give our general hybrid-logical rule schemas. We finish the section by proving normalization and conservativity.

5.3.1 Earlier Work on Functional Completeness

We first introduce some conventions and definitions. The language we consider is the language of ordinary propositional logic extended with a new connective $\sharp$ that takes as arguments an arbitrary, but fixed, number of formulas. Now, rule schemas talk about rules in the sense that a rule schema delimits a set of rules having a specific common form (which is the form of the rule schema). The rules we want to talk about often involve formulas having a specific common form. This is for example the case with the standard natural deduction rules for conjunction

$$\frac{\phi \quad \psi}{\phi \wedge \psi}\,(\wedge I) \qquad\qquad \frac{\phi \wedge \psi}{\phi}\,(\wedge E1) \quad \frac{\phi \wedge \psi}{\psi}\,(\wedge E2)$$

where the conclusion of the introduction rule as well as the premises of the elimination rules all have the form of conjunctions. These rules are taken from Figure 2.1 in Section 2.1. To formalize a specific common form of formulas, we shall introduce *formula schemas*. Thus, using a formula schema we want to be able to delimit a set of formulas having a specific common form. A formula schema is defined exactly as a formula except that a formula schema has metavariables ϕ, ψ, θ, ... for formulas where a formula has propositional symbols p, q, r, ... So for example $p \wedge q$ is a formula whereas $\phi \wedge \psi$ is a formula schema. Technically, the propositional formula schemas are defined by the grammar

$$S \ ::= \ \boldsymbol{\phi} \mid S \wedge S \mid S \rightarrow S \mid \bot \mid \sharp(S_1,\ldots,S_m)$$

where $\boldsymbol{\phi}$ ranges over metavariables for formulas.[2] In what follows, the metavariables **F**, **G**, **H**, ... range over formula schemas.

We shall often substitute formulas for metavariables for formulas in a formula schema, whereby an ordinary formula is obtained, called an *instance* of the formula schema in question, for example, the formula $p \wedge q$ is an instance of the formula schema $\phi \wedge \psi$. This is analogous to a rule-instance being an instance of a rule. To sum up, we use rule schemas to talk about rules, we use rules to talk about rule-instances, and we use formula schemas to talk about formulas.

Now, in the paper Zucker and Tragesser (1978), rule schemas for sets of natural deduction introduction and elimination rules are given where each introduction rule comes together with n elimination rules. Below is a generalized version of Zucker and Tragesser's rule schemas (the rule schemas below are generalized in the sense that they have formula schemas where Zucker and Tragesser's original rule schemas have metavariables for formulas).

$$\frac{\begin{array}{c}[\mathbf{H}_{11}]\ldots[\mathbf{H}_{1k_1}]\\ \vdots \\ \mathbf{G}_1\end{array} \quad \cdots \quad \begin{array}{c}[\mathbf{H}_{n1}]\ldots[\mathbf{H}_{nk_n}]\\ \vdots \\ \mathbf{G}_n\end{array}}{\sharp(\overline{\boldsymbol{\theta}})}\,(\sharp I) \qquad\qquad \frac{\sharp(\overline{\boldsymbol{\theta}}) \quad \mathbf{H}_{i1} \quad \ldots \quad \mathbf{H}_{ik_i}}{\mathbf{G}_i}\,(\sharp E i)$$

where $1 \leq i \leq n$. It is required that the formula schemas $\mathbf{G}_i$ and $\mathbf{H}_{ij}$ contain no other metavariables for formulas than those in the list $\overline{\boldsymbol{\theta}} = \boldsymbol{\theta}_1,\ldots,\boldsymbol{\theta}_m$ (this requirement is a prerequisite for the inversion principle being satisfied).

Prawitz (1978) introduces the following useful classifications: A natural deduction introduction rule of the form displayed above is *explicit* if all connectives in the

[2] Note that the symbol $\boldsymbol{\phi}$ ranges over metavariables, thus, the symbol is a metametavariable. In general we let the metametavariables $\boldsymbol{\phi}$, $\boldsymbol{\psi}$, $\boldsymbol{\theta}$, ... range over the metavariables ϕ, ψ, θ, ... Observe that metametavariables are printed in boldface.

formula schemas $\mathbf{G}_i$ and $\mathbf{H}_{ij}$ are different from the connective $\sharp$ being introduced by the rule; otherwise the rule is *implicit*. Moreover, an introduction rule is *independent* if all the formula schemas $\mathbf{G}_i$ and $\mathbf{H}_{ij}$ are metavariables for formulas; otherwise the rule is *dependent*. Thus, in this terminology Zucker and Tragesser's original rule schemas only admit independent rules whereas our more general rule schemas also admit dependent rules. Clearly, an independent rule is explicit.

Observe that the standard introduction and elimination rules for the propositional connective $\wedge$ fit the displayed rule schemas, and moreover, observe that the introduction rule is independent. These observations also apply to the standard rules for the propositional connective $\rightarrow$ given in Figure 2.1 of Section 2.1. An example of introduction and elimination rules where the introduction rule is explicit but dependent are the rules for classical disjunction

$$\frac{\begin{array}{c}[\neg\phi]\;[\neg\psi]\\ \vdots\\ \bot\end{array}}{\phi\vee\psi}\,(\vee I) \qquad\qquad \frac{\phi\vee\psi \quad \neg\phi \quad \neg\psi}{\bot}\,(\vee E)$$

The above introduction rule for classical disjunction is taken from Prawitz (1978) and the elimination rule can be read off from the rule schemas.

The following functional completeness result can be proved: If the standard natural deduction system for classical propositional logic, see Figure 2.1 in Section 2.1, are extended with explicit introduction and elimination rules for the connective $\sharp$ that fit Zucker and Tragesser's rule schemas, then for any formula ϕ built using the connectives $\{\wedge,\rightarrow,\bot,\sharp\}$, there exists a formula ψ built using only the original connectives $\{\wedge,\rightarrow,\bot\}$ such that the formula $\phi\leftrightarrow\psi$ is provable using the original derivation rules as well as the introduction and elimination rules for the new connective $\sharp$. (It should be mentioned that Zucker and Tragesser's paper, which is very informal, does not explicitly give this or other functional completeness results and the paper is oriented towards intuitionistic rather than classical logic.)

Besides functional completeness, the natural deduction system of Figure 2.1 extended with Zucker and Tragesser's rules for $\sharp$ enjoys another desirable feature, namely that he set of introduction and elimination rules associated with each connective satisfies the inversion principle. We remarked already in Section 2.1 that the standard introduction and elimination rules for the connectives $\wedge$ and $\rightarrow$ satisfy the inversion principle and it is straightforward to check that introduction and elimination rules for $\sharp$ in general satisfy the inversion principle if they fit Zucker and Tragesser's rule schemas. This follows from the observation that a derivation in which an occurrence of the connective $\sharp$ is both introduced by an instance of $(\sharp I)$ and eliminated by an instance of $(\sharp Ei)$ can be rewritten such that the occurrence of $\sharp$ disappears. That is, a derivation of the form

$$\dfrac{\dfrac{\begin{matrix}[\phi_{11}]\cdots[\phi_{1k_1}] & & [\phi_{n1}]\cdots[\phi_{nk_n}]\\ \vdots\,\pi_1 & & \vdots\,\pi_n\\ \psi_1 & \cdots & \psi_n\end{matrix}}{\sharp(\overline{\theta})}\,(\sharp I) \qquad \begin{matrix}\vdots\,\tau_1 & & \vdots\,\tau_{k_i}\\ \phi_{i1} & \cdots & \phi_{ik_i}\end{matrix}}{\psi_i}\,(\sharp E i)$$

can be rewritten to the derivation

$$\begin{matrix}\vdots\,\tau_1 & & \vdots\,\tau_{k_i}\\ \phi_{i1} & \cdots & \phi_{ik_i}\\ & \vdots\,\pi_i & \\ & \psi_i & \end{matrix}$$

In fact, the rewriting displayed above gives rise to a general form of proper reduction rules for the system such that if the introduction rule for $\sharp$ is explicit, then a normalization theorem can be proved, and moreover, it can be proved that normalization implies conservativity (we shall return to hybrid-logical versions of these results in Section 5.3.3).

Now, Zucker and Tragesser also consider the case where a connective has more than one introduction rule of the form displayed above. The standard introduction and elimination rules for intuitionistic disjunction

$$\dfrac{\phi}{\phi\vee\psi}\,(\vee I1) \qquad \dfrac{\psi}{\phi\vee\psi}\,(\vee I2) \qquad \dfrac{\phi\vee\psi \quad \begin{matrix}[\phi]\\ \vdots\\ \xi\end{matrix} \quad \begin{matrix}[\psi]\\ \vdots\\ \xi\end{matrix}}{\xi}\,(\vee E)$$

is one example of this. However, the rule schemas do not fit in the case of more than one introduction rule. Zucker and Tragesser conclude that "in this case, in general, there is (apparently) no suitable set of E-rules", cf. p. 505 in their paper.

It should be mentioned that the rules for intuitionistic disjunction actually do fit a different set of rule schemas given in Prawitz (1978). A functional completeness result like the result for Zucker and Tragesser's rule schemas can be proved for Prawitz' rule schemas, and also, an intuitionistic version of the functional completeness result, where the rules for intuitionistic disjunction have been included, can be proved, cf. Prawitz (1978). However, Prawitz' rule schemas do not satisfy the inversion principle in general. In some cases they do, for example in the case with the rules for intuitionistic disjunction, but in general they do not. Moreover, there is no general and straightforward way in which Prawitz' rule schemas can be equipped with reduction rules. This is itself a deficiency and it moreover follows that we cannot prove conservativity via normalization.

Given these observations, we have based our work on functional completeness for hybrid logics on Zucker and Tragesser's rule schemas rather than Prawitz' rule schemas. The fact that the rules for intuitionistic disjunction do not fit Zucker and

Tragesser's rule schemas is not a loss since these rules are not proof-theoretically well-behaved in the context of classical logic. (The problem is that the classical rule for $\perp$ cannot, in the presence of the rules for intuitionistic disjunction, be restricted to propositional symbols, which is a prerequisite for obtaining a subformula property; an important property of a natural deduction system. We shall not consider this issue further, see Prawitz (1965) as well as Section 5.3.3.) In any case, it should be mentioned that the rules for intuitionistic disjunction are admissible in the classical natural deduction system.

5.3.2 Rule Schemas for Hybrid Logic

The first we do is to extend the language $\mathscr{H}$ with a new connective $\sharp$ that takes as arguments an arbitrary, but fixed, number of nominals and an arbitrary, but again fixed, number of formulas. The extended language will be denoted $\mathscr{H}(\sharp)$. One decisive reason why the connective $\sharp$ takes as argument a number of nominals (besides a number of formulas) is that we want to have satisfaction operators as instances. We then extend the natural deduction system $\mathbf{N}_{\mathscr{H}}$ with introduction and elimination rules for the connective $\sharp$. The rules are instances of rule schemas which are obtained by adapting Zucker and Tragesser's propositional rule schemas (see the previous section) to hybrid logic such that all formulas in the rules are satisfaction statements, that is, our hybrid-logical rules are obtained by decorating all formulas in the propositional rules with satisfaction operators.[3]

In the definition of the rule schemas, we shall make use of the metametavariables $\mathbf{a}, \mathbf{b}, \mathbf{c}, \ldots$ that range over the metavariables $a, b, c, \ldots$ (that in turn range over nominals) and as in the propositional case, we shall make use of the metametavariables $\boldsymbol{\phi}, \boldsymbol{\psi}, \boldsymbol{\theta}, \ldots$ that range over the metavariables $\phi, \psi, \theta, \ldots$ (that in turn range over formulas). Moreover, we shall make use of hybrid-logical formula schemas defined by the grammar

$$S \ ::= \ \boldsymbol{\phi} \mid \mathbf{a} \mid S \wedge S \mid S \rightarrow S \mid \perp \mid @_{\mathbf{a}} S \mid AS \mid \sharp(\mathbf{b}_1, \ldots, \mathbf{b}_r, S_1, \ldots, S_m)$$

where $\boldsymbol{\phi}$ ranges over metavariables for formulas and $\mathbf{a}, \mathbf{b}_1, \ldots, \mathbf{b}_r$ range over metavariables for nominals. As in the propositional case, the metavariables $\mathbf{F}$, $\mathbf{G}$, $\mathbf{H}$, ... range over formula schemas. Note that a formula schema does not contain ordinary propositional symbols.

In what follows, we shall make use of three different types of substitution: (i) In formulas, nominals are substituted for nominals, and moreover, formulas are substi-

[3] In fact, if the language $\mathscr{H}$ is extended with a finite number $\sharp_1, \ldots, \sharp_s$ of new connectives like the connective $\sharp$ and the natural deduction system $\mathbf{N}_{\mathscr{H}}$ is extended with introduction and elimination rules for each of the new connectives, then all results of this chapter still holds provided no introduction rule for a connective $\sharp_i$ exhibits a connective $\sharp_j$ where $j \neq i$. This would for example enable us to consider connectives with a varying number of inputs, like conjunction (but formally, the connectives $\sharp_1, \ldots, \sharp_s$ are independent of each other even though they might be equipped with similar rules, like the rules for ternary conjunction are similar to the rules for binary conjunction).

tuted for ordinary propositional symbols. (ii) In formula schemas, metavariables for nominals are substituted for metavariables for nominals, and moreover, metavariables for formulas are substituted for metavariables for formulas. (iii) In formula schemas, nominals are substituted for metavariables for nominals, and moreover, formulas are substituted for metavariables for formulas. The second and third types of substitution are defined in analogy with the first type which of course is the standard one. The use of metametavariables allows us to distinguish between the case where the same metavariable occurs more than once in a rule and the case where the same formula or nominal is substituted for different metavariables.

We will make use of the following terminology. A formula schema of the form $@_{\mathbf{a}}\mathbf{F}$ is called a *satisfaction formula schema*. An occurrence of a metavariable for nominals in a formula schema is called *equational* if the occurrence in question is a formula schema, that is, if it is generated by the second clause in the grammar for hybrid-logical formula schemas displayed above. For example, the occurrences of $\mathbf{b}$ in the formula schemas $@_{\mathbf{a}}(\boldsymbol{\phi} \wedge \mathbf{b})$ and $@_{\mathbf{a}}\neg\mathbf{b}$ are equational whereas the occurrences of $\mathbf{a}$ are not. The justification for using the term equational is that an equational metavariable for nominals, when it is instantiated to a nominal, gives rise to a statement saying that two worlds are identical, cf. the definition of the relation $\models$.

We now give the technical details of the general rule schemas. The rule schemas are given in Figure 5.3 where $\overline{\mathbf{e}} = \mathbf{e}_1,\ldots,\mathbf{e}_r$, $\overline{\mathbf{d}} = \mathbf{d}_1,\ldots,\mathbf{d}_q$, $\overline{\mathbf{f}} = \mathbf{f}_1,\ldots,\mathbf{f}_q$, and $\overline{\boldsymbol{\theta}} = \boldsymbol{\theta}_1,\ldots,\boldsymbol{\theta}_m$. All formula schemas in the rule schemas are satisfaction formula schemas. Note that in Figure 5.3 we have made use of substitution of metavariables (that range over nominals) for metavariables (that range over nominals). In the interest of simplicity, it is assumed that the metavariables $\mathbf{c},\mathbf{e}_1,\ldots,\mathbf{e}_r,\mathbf{d}_1,\ldots,\mathbf{d}_q,\mathbf{f}_1,\ldots,\mathbf{f}_q$ are pairwise distinct. Also, it is assumed that the satisfaction formula schemas $\mathbf{G}_i$ and $\mathbf{H}_{ij}$ contain no other metavariables for formulas than $\boldsymbol{\theta}_1,\ldots,\boldsymbol{\theta}_m$ and that they contain no other metavariables for nominals than $\mathbf{c},\mathbf{e}_1,\ldots,\mathbf{e}_r,\mathbf{d}_1,\ldots,\mathbf{d}_q$. Such introduction and elimination rules will be called *schematic*. The natural deduction system which is obtained by extending $\mathbf{N}_{\mathscr{H}}$ with schematic introduction and elimination rules will be denoted $\mathbf{N}_{\mathscr{H}(\sharp)}$.

$$\dfrac{\begin{matrix}[\mathbf{H}_{11}]\ldots[\mathbf{H}_{1k_1}] & & [\mathbf{H}_{n1}]\ldots[\mathbf{H}_{nk_n}] \\ \vdots & & \vdots \\ \mathbf{G}_1 & \ldots & \mathbf{G}_n\end{matrix}}{@_{\mathbf{c}}\sharp(\overline{\mathbf{e}},\overline{\boldsymbol{\theta}})}\ (\sharp I)^* \qquad \dfrac{@_{\mathbf{c}}\sharp(\overline{\mathbf{e}},\overline{\boldsymbol{\theta}}) \quad \mathbf{H}_{i1}[\overline{\mathbf{f}}/\overline{\mathbf{d}}] \quad \ldots \quad \mathbf{H}_{ik_i}[\overline{\mathbf{f}}/\overline{\mathbf{d}}]}{\mathbf{G}_i[\overline{\mathbf{f}}/\overline{\mathbf{d}}]}\ (\sharp Ei)$$

$*$ The nominals $\overline{\mathbf{d}}$ are pairwise distinct and do not occur in $@_{\mathbf{c}}\sharp(\overline{\mathbf{e}},\overline{\boldsymbol{\theta}})$ or in any undischarged assumptions other than the specified occurrences of $\mathbf{H}_{i1},\ldots,\mathbf{H}_{ik_i}$.

Fig. 5.3 Hybrid-logical rule schemas for the connective $\sharp$.

We shall classify schematic introduction rules in a number of ways: A schematic introduction rule is *non-equational* if neither the metavariable $\mathbf{c}$ nor any of the metavariables $\overline{\mathbf{d}}$ occur equationally in the satisfaction formula schemas $\mathbf{G}_i$ and $\mathbf{H}_{ij}$. The distinction between explicit and implicit rules given in the previous section is kept as it is whereas the distinction between independent and dependent rules is adjusted to the hybrid-logical case by taking a schematic introduction rule to be independent if all the satisfaction formula schemas $\mathbf{G}_i$ and $\mathbf{H}_{ij}$ are of the form $@_{\mathbf{a}}\boldsymbol{\phi}$. In Braüner (2003) we considered the case of functional completeness for independent introduction rules. Note a significant difference: The conditions mentioned in the classifications above apply to *rules* (this should be compared to the side-condition on the rule ($\sharp I$) which applies to *rule-instances*).

It is straightforward to check that the introduction and elimination rules for the connectives $\wedge$, $\rightarrow$, $@$, and A given in Figure 5.1 are schematic (strictly speaking, we here consider new versions of the rules where the connectives have been renamed). Moreover, all the introduction rules are independent, non-equational, and explicit. The rules for modal operator given in Figure 2.2 of Section 2.2 are examples of schematic introduction and elimination rules where the introduction rule is implicit. Another example of schematic introduction and elimination rules are the rules

$$\dfrac{\begin{matrix}[@_d\phi] \\ \vdots \\ @_d\psi\end{matrix}}{@_c(\phi \prec \psi)}\,(\prec I) \qquad\qquad \dfrac{@_c(\phi \prec \psi) \quad @_f\phi}{@_f\psi}\,(\prec E)$$

for the binary connective $\prec$ where the rule $(\prec I)$ is equipped with the side-condition that the nominal d does not occur in $@_c(\phi \prec \psi)$ or in any undischarged assumptions other than the specified occurrences of $@_d\phi$ (note that we have used infix notation for the new connective instead of the usual prefix notation). Clearly, the introduction rule is independent, non-equational, and explicit. The connective $\prec$ is called *strict implication*.

Like in the propositional case considered in the previous section, it is straightforward to check that schematic introduction and elimination rules satisfy the inversion principle, that is, a derivation in which an occurrence of the connective $\sharp$ is both introduced by an instance of ($\sharp I$) and eliminated by an instance of ($\sharp Ei$) can be rewritten such that the occurrence of $\sharp$ disappears (where we beside substitution of derivations for undischarged assumptions allow substitution of nominals for nominals in derivations). The inversion principle is related to normalization and hence conservativity, which we shall return to in the next section.

5.3.3 Normalization and Conservativity

Before proving normalization and conservativity, we shall give some conventions, a proposition, and a lemma. We need two conventions to prove the proposition. To any formula ϕ of $\mathscr{H}(\sharp)$ a natural number $depth(\phi)$ is assigned as follows.

$$depth(\phi) = \begin{cases} 0 & \text{if } \phi \text{ is atomic} \\ sup\{depth(\psi), depth(\theta)\} & \text{if } \phi \in \{\psi \wedge \theta, \psi \rightarrow \theta\} \\ depth(\psi) & \text{if } \phi \in \{@_a\psi, A\psi\} \\ 1 + sup\{depth(\theta_1), \ldots, depth(\theta_m)\} & \text{if } \phi = \sharp(\overline{e}, \overline{\theta}) \end{cases}$$

(Recall that $\overline{\theta} = \theta_1, \ldots, \theta_m$.) Thus, the function *depth* measures the maximal number of occurrences of the connective $\sharp$ that are nested within each other. The second convention we need simply says that $degree(\phi)$ is the number of occurrences of connectives in the formula ϕ which are different from $\perp$ (this convention is also used in Section 2.2.2 and elsewhere in the present book). Now the proposition.

Proposition 5.1. *Let a set of schematic introduction and elimination rules for the connective $\sharp$ be given where the introduction rule is explicit. The rules*

$$\dfrac{\begin{matrix}[@_a\neg\phi] \\ \vdots \\ @_a\perp\end{matrix}}{@_a\phi}\,(\perp) \qquad\qquad \dfrac{@_a d \quad @_a\phi}{@_d\phi}\,(Nom)$$

are admissible in $\mathbf{N}_{\mathscr{H}(\sharp)}$.

Proof. The proof that the rule $(\perp)$ is admissible is an induction proof along the lines of the proof of Proposition 2.1 in Section 2.2. However, we make use of another order on formulas: To any formula ϕ of $\mathscr{H}(\sharp)$ we assign the pair of natural numbers $(depth(\phi), degree(\phi))$. The set of such pairs of natural numbers is equipped with the lexicographic order. If the introduction rule for $\sharp$ is independent, then it is possible to make use of the simpler order on formulas obtained by assigning the natural number $degree(\phi)$ to ϕ, this is actually the order used in Proposition 2.1. We prove that the rule (Nom) is admissible by proving that the more general rule

$$\dfrac{@_a d \quad \psi[a/b]}{\psi[d/b]}$$

is admissible. This proof is analogous to the proof that $(\perp)$ is admissible.

The side-conditions on the rules $(\perp 1)$ and $(Nom1)$ enable a normalization theorem (Theorem 5.2) to be proved such that normal derivations satisfy a version of the subformula property, and moreover, such that a conservativity theorem (Theorem 5.3) can be proved via the normalization theorem. See also the remarks in Section 2.2.

With the aim of formulating the lemma, we introduce a convention. A formula in $\mathscr{H}(\sharp)$ of the form $\phi \leftrightarrow \psi$ is called a *provable equivalence* in $\mathbf{N}_{\mathscr{H}(\sharp)}$ if for any nominal c, $@_c(\phi \leftrightarrow \psi)$ is derivable in $\mathbf{N}_{\mathscr{H}(\sharp)}$. Note that it does not make sense to ask whether the formulas ϕ and ψ are equivalent in the usual model-theoretic sense, that is, whether $\phi \leftrightarrow \psi$ is valid, since the connective $\sharp$ has not been assigned a model-theoretic interpretation. It is instructive to consider an example of provable equivalence: The formula $(\phi \prec \psi) \leftrightarrow A(\phi \rightarrow \psi)$ of the language $\mathscr{H}(\prec)$ considered in the previous section is a provable equivalence in $\mathbf{N}_{\mathscr{H}(\prec)}$. We shall return to this example in the remarks following Lemma 5.2 in the next section. Now the lemma.

Lemma 5.1. *(Replacement lemma) Let a set of schematic introduction and elimination rules for the connective* $\sharp$ *be given where the introduction rule is explicit. If the formula* $\psi \leftrightarrow \theta$ *is a provable equivalence in* $\mathbf{N}_{\mathscr{H}(\sharp)}$, *then for any formula* ϕ *in* $\mathscr{H}(\sharp)$, $\phi[\psi/q] \leftrightarrow \phi[\theta/q]$ *is a provable equivalence in* $\mathbf{N}_{\mathscr{H}(\sharp)}$.

Proof. The proof is by induction on the formula ϕ where we use the same order on formulas as we did in Proposition 5.1.

Later in the paper we shall also make use of the following convention. A formula schema of the form $\mathbf{F} \leftrightarrow \mathbf{G}$ is called a *provable equivalence* in $\mathbf{N}_{\mathscr{H}(\sharp)}$ if any instance of it, where the formulas substituted for metavariables for formulas belong to $\mathscr{H}(\sharp)$, is a provable equivalence in $\mathbf{N}_{\mathscr{H}(\sharp)}$. Before proving conservativity, we prove normalization.

Theorem 5.2. *(Normalization) Let a set of schematic introduction and elimination rules for the connective* $\sharp$ *be given where the introduction rule is explicit. Any derivation in* $\mathbf{N}_{\mathscr{H}(\sharp)}$ *can be rewritten to a normal derivation in* $\mathbf{N}_{\mathscr{H}(\sharp)}$ *by repeated applications of reductions of the kind induced by the inversion principle, as described in Section 2.1 (and cf. Sections 5.3.1 and 5.3.2).*

Proof. The proof is an induction proof along the lines of the proof of Proposition 2.3 in Section 2.2. However, we make use of another order on formulas, namely the order on formulas used in Proposition 5.1.

Given the normalization theorem above, conservativity can be proved in a straightforward way.

Theorem 5.3. *(Conservativity) Let a set of schematic introduction and elimination rules for the connective* $\sharp$ *be given where the introduction rule is explicit. Let* ϕ *be a satisfaction statement that does not contain the connective* $\sharp$, *and similarly, let* Γ *be a set of satisfaction statements that do not contain* $\sharp$. *If* ϕ *is derivable from* Γ *in* $\mathbf{N}_{\mathscr{H}(\sharp)}$, *then* ϕ *is derivable from* Γ *in* $\mathbf{N}_{\mathscr{H}}$.

Proof. If ϕ is derivable from Γ in $\mathbf{N}_{\mathscr{H}(\sharp)}$, then by Theorem 5.2 there exists a normal derivation π of ϕ from Γ in $\mathbf{N}_{\mathscr{H}(\sharp)}$. It can then be proved that π does not contain the connective $\sharp$, the proof is along the lines of the proof of Proposition 2.4, the quasi-subformula property, in Section 2.2.

5.4 Functional Completeness

In this section we prove a functional completeness theorem for the natural deduction system $\mathbf{N}_{\mathscr{H}}$. The rules under consideration in the theorems are non-equational and explicit.

Theorem 5.4. *(Normal form of formula schemas) Let* $\mathbf{F}$ *be a formula schema of* $\mathscr{H}$*. Then there exists a formula schema* $\mathbf{I}$ *of* $\mathscr{H}$ *of the form*

$$\bigwedge_{i=1}^{l}(\mathbf{J}_i \vee \bigvee_{j=1}^{h_i} @_{\mathbf{b}_{ij}}\mathbf{K}_{ij})$$

such that

1. *all metavariables for formulas and nominals that occur in* $\mathbf{I}$ *also occur in* $\mathbf{F}$*;*
2. *if a metavariable for nominals occurs equationally in* $\mathbf{I}$*, then it also occurs equationally in* $\mathbf{F}$*;*
3. *the formula schemas* $\mathbf{J}_i$ *and* $\mathbf{K}_{ij}$ *do not contain satisfaction operators;*
4. *for any i, the metavariables* $\mathbf{b}_{i1},\ldots,\mathbf{b}_{ih_i}$ *are pairwise distinct; and*
5. *the formula schema* $\mathbf{F} \leftrightarrow \mathbf{I}$ *is a provable equivalence in* $\mathbf{N}_{\mathscr{H}}$*.*

Proof. The proof is similar to a proof of a normal form result given in Areces (2000). Step one of the proof: Using the equivalence

$$(\mathbf{G} \to \mathbf{H}) \leftrightarrow \neg(\mathbf{G} \wedge \neg\mathbf{H})$$

the formula schema $\mathbf{F}$ is rewritten in terms of the connectives $\wedge$, $\top$, $\neg$, $@$, and A. Step two: Using standard propositional equivalences and the equivalences

$$\neg A\mathbf{G} \leftrightarrow E\neg\mathbf{G} \qquad \neg E\mathbf{G} \leftrightarrow A\neg\mathbf{G}$$
$$\neg @_{\mathbf{d}}\mathbf{G} \leftrightarrow @_{\mathbf{d}}\neg\mathbf{G}$$

the formula schema is rewritten into negation normal form in terms of the connectives $\wedge$, $\top$, $\vee$, $\bot$, $\neg$, $@$, A, and E. Step three: Using standard propositional equivalences together with the equivalences

$$A(\mathbf{G} \wedge \mathbf{H}) \leftrightarrow (A\mathbf{G} \wedge A\mathbf{H}) \qquad E(\mathbf{G} \vee \mathbf{H}) \leftrightarrow (E\mathbf{G} \vee E\mathbf{H})$$
$$A(\mathbf{G} \vee @_{\mathbf{d}}\mathbf{H}) \leftrightarrow (A\mathbf{G} \vee @_{\mathbf{d}}\mathbf{H}) \qquad E(\mathbf{G} \wedge @_{\mathbf{d}}\mathbf{H}) \leftrightarrow (E\mathbf{G} \wedge @_{\mathbf{d}}\mathbf{H})$$
$$@_{\mathbf{d}}(\mathbf{G} \wedge \mathbf{H}) \leftrightarrow (@_{\mathbf{d}}\mathbf{G} \wedge @_{\mathbf{d}}\mathbf{H}) \qquad @_{\mathbf{d}}(\mathbf{G} \vee \mathbf{H}) \leftrightarrow (@_{\mathbf{d}}\mathbf{G} \vee @_{\mathbf{d}}\mathbf{H})$$
$$A @_{\mathbf{d}}\mathbf{G} \leftrightarrow @_{\mathbf{d}}\mathbf{G} \qquad E @_{\mathbf{d}}\mathbf{G} \leftrightarrow @_{\mathbf{d}}\mathbf{G}$$
$$@_{\mathbf{d}} @_{\mathbf{e}}\mathbf{G} \leftrightarrow @_{\mathbf{e}}\mathbf{G}$$

the formula schema is rewritten such that it has the required form, except that the fourth requirement in the theorem might not be satisfied, by "pushing" occurrences of the connectives A, E, and $@$ "towards" nested occurrences of the connective $@$. The equivalences are applied in the left to right direction, that is, the left-hand sides are replaced by the right-hand sides. Step four: Using standard propositional equivalences together with the third equivalence in the second column, applied in the right to left direction, the formula schema is rewritten such that also the fourth requirement is satisfied. The fifth statement in the theorem follows from inspection of the steps in the rewrite process and Lemma 5.1 (take $\sharp$ to be A).

Definition 5.1. Consider the system $\mathbf{N}_{\mathscr{H}}$ and let a set of schematic introduction and elimination rules for the connective $\sharp$ be given where the introduction rule is non-equational and explicit. Let $\mathbf{F}$ be the formula schema

$$\bigwedge_{i=1}^{n}(\bigwedge_{j=1}^{k_i}\mathbf{H}_{ij} \rightarrow \mathbf{G}_i)$$

and let $\mathbf{I}$ be the result of rewriting $\mathbf{F}$ in accordance with Theorem 5.4. Thus, $\mathbf{I}$ is a formula schema of $\mathscr{H}$ of the form

$$\bigwedge_{i=1}^{l}(\mathbf{J}_i \vee \bigvee_{j=1}^{h_i} @_{\mathbf{b}_{ij}}\mathbf{K}_{ij}).$$

Firstly, for each $i \in \{1,\ldots,l\}$ and $j \in \{1,\ldots,h_i\}$ define $(@_{\mathbf{b}_{ij}}\mathbf{K}_{ij})^A$ as follows.

$$(@_{\mathbf{b}_{ij}}\mathbf{K}_{ij})^A = \begin{cases} \mathbf{K}_{ij} & \text{if } \mathbf{b}_{ij} = \mathbf{c} \\ @_{\mathbf{b}_{ij}}\mathbf{K}_{ij} & \text{if } \mathbf{b}_{ij} \notin \{\mathbf{c},\mathbf{d}_1,\ldots,\mathbf{d}_q\} \\ A\mathbf{K}_{ij} & \text{otherwise} \end{cases}$$

Secondly, let $\mathbf{E}$ be the formula schema

$$\bigwedge_{i=1}^{l}(\mathbf{J}_i \vee \bigvee_{j=1}^{h_i} (@_{\mathbf{b}_{ij}}\mathbf{K}_{ij})^A)$$

of $\mathscr{H}$.

Lemma 5.2. *Consider the system $\mathbf{N}_{\mathscr{H}}$ and let a set of schematic introduction and elimination rules for the connective $\sharp$ be given where the introduction rule is non-equational and explicit. Let a formula $\sharp(\overline{e},\overline{\theta})$ of $\mathscr{H}(\sharp)$ be given where $\overline{\theta}$ are formulas of $\mathscr{H}$. The formula*

$$\sharp(\overline{e},\overline{\theta}) \leftrightarrow \mathbf{E}[\overline{e},\overline{\theta}/\overline{\mathbf{e}},\overline{\boldsymbol{\theta}}]$$

is a provable equivalence in $\mathbf{N}_{\mathscr{H}(\sharp)}$ where the formula schema $\mathbf{E}$ is defined in accordance with Definition 5.1.

Proof. In what follows, $\overline{d} = d_1,\ldots,d_q$ is a list of pairwise distinct new nominals. We first prove the implication from right to left: Step one is to prove that

$$@_c(\mathbf{E}[\overline{e},\overline{\theta}/\overline{\mathbf{e}},\overline{\boldsymbol{\theta}}] \rightarrow \mathbf{I}[c,\overline{e},\overline{d},\overline{\theta}/\mathbf{c},\overline{\mathbf{e}},\overline{\mathbf{d}},\overline{\boldsymbol{\theta}}])$$

is derivable where c is an arbitrary nominal. Note that $\mathbf{E}$ has been obtained from $\mathbf{I}$ by substitution of $(@_{\mathbf{b}_{ij}}\mathbf{K}_{ij})^A$ for $@_{\mathbf{b}_{ij}}\mathbf{K}_{ij}$ for each $i \in \{1,\ldots,l\}$ and $j \in \{1,\ldots,h_i\}$, cf. Definition 5.1, whereby all occurrences of the metavariables for nominals $\mathbf{c},\mathbf{d}_1,\ldots,\mathbf{d}_q$ have disappeared. It follows that step one amounts to prove that the formula

$$@_c((@_{\mathbf{b}_{ij}}\mathbf{K}_{ij})^A[\overline{e},\overline{\theta}/\overline{\mathbf{e}},\overline{\boldsymbol{\theta}}] \rightarrow (@_{\mathbf{b}_{ij}}\mathbf{K}_{ij})[c,\overline{e},\overline{d},\overline{\theta}/\mathbf{c},\overline{\mathbf{e}},\overline{\mathbf{d}},\overline{\boldsymbol{\theta}}])$$

is derivable for each $i \in \{1,\ldots,l\}$ and $j \in \{1,\ldots,h_i\}$. There are three cases to check: If $\mathbf{b}_{ij} = \mathbf{c}$, then $(@_{\mathbf{b}_{ij}}\mathbf{K}_{ij})^A = \mathbf{K}_{ij}$ in which case the displayed formula is straightforwardly derivable. If $\mathbf{b}_{ij} \notin \{\mathbf{c},\mathbf{d}_1,\ldots,\mathbf{d}_q\}$, then $(@_{\mathbf{b}_{ij}}\mathbf{K}_{ij})^A = @_{\mathbf{b}_{ij}}\mathbf{K}_{ij}$ so the antecedent and the succedent formulas are identical. In the third case we have $\mathbf{b}_{ij} \in \{\mathbf{d}_1,\ldots,\mathbf{d}_q\}$, so $(@_{\mathbf{b}_{ij}}\mathbf{K}_{ij})^A = A\mathbf{K}_{ij}$ and therefore the displayed formula is derivable by using the rule (AE). This concludes step one. Step two consists in observing that if the formula $@_c\mathbf{I}[c,\overline{e},\overline{d},\overline{\theta}/\mathbf{c},\overline{\mathbf{e}},\overline{\mathbf{d}},\overline{\boldsymbol{\theta}}]$ is derivable from $\{@_c\mathbf{E}[\overline{e},\overline{\theta}/\overline{\mathbf{e}},\overline{\boldsymbol{\theta}}]\}$ then, by Theorem 5.4, so is $@_c\mathbf{F}[c,\overline{e},\overline{d},\overline{\theta}/\mathbf{c},\overline{\mathbf{e}},\overline{\mathbf{d}},\overline{\boldsymbol{\theta}}]$ and, by using the rule $(\sharp I)$, so is the formula $@_c\sharp(\overline{e},\overline{\theta})$.

We now prove the implication from left to right: It is straightforward to check that the formula $@_c\mathbf{E}[\overline{e},\overline{\theta}/\overline{\mathbf{e}},\overline{\boldsymbol{\theta}}]$ is derivable from $\{@_c\sharp(\overline{e},\overline{\theta})\}$ if for each $i \in \{1,\ldots,l\}$, $@_c\bot$ is derivable from the set of formulas

$$\begin{array}{l}\{@_c\sharp(\overline{e},\overline{\theta}), @_c\neg\mathbf{J}_i[\overline{e},\overline{\theta}/\overline{\mathbf{e}},\overline{\boldsymbol{\theta}}]\}\cup\\ \{@_c\neg(@_{\mathbf{b}_{i1}}\mathbf{K}_{i1})^A[\overline{e},\overline{\theta}/\overline{\mathbf{e}},\overline{\boldsymbol{\theta}}],\ldots,@_c\neg(@_{\mathbf{b}_{ih_i}}\mathbf{K}_{ih_i})^A[\overline{e},\overline{\theta}/\overline{\mathbf{e}},\overline{\boldsymbol{\theta}}]\}\end{array}$$

Now, by using the rule $(\sharp Ei)$, the formula $@_c\mathbf{F}[c,\overline{e},\overline{d},\overline{\theta}/\mathbf{c},\overline{\mathbf{e}},\overline{\mathbf{d}},\overline{\boldsymbol{\theta}}]$ is derivable from $\{@_c\sharp(\overline{e},\overline{\theta})\}$ and, so is the formula $@_c\mathbf{I}[c,\overline{e},\overline{d},\overline{\theta}/\mathbf{c},\overline{\mathbf{e}},\overline{\mathbf{d}},\overline{\boldsymbol{\theta}}]$ by Theorem 5.4. It follows that for each $i \in \{1,\ldots,l\}$, $@_c\bot$ is derivable from

$$\begin{array}{l}\{@_c\sharp(\overline{e},\overline{\theta}), @_c\neg\mathbf{J}_i[\overline{e},\overline{\theta}/\overline{\mathbf{e}},\overline{\boldsymbol{\theta}}]\}\cup\\ \{@_c\neg(@_{\mathbf{b}_{i1}}\mathbf{K}_{i1})[c,\overline{e},\overline{d},\overline{\theta}/\mathbf{c},\overline{\mathbf{e}},\overline{\mathbf{d}},\overline{\boldsymbol{\theta}}],\ldots,@_c\neg(@_{\mathbf{b}_{ih_i}}\mathbf{K}_{ih_i})[c,\overline{e},\overline{d},\overline{\theta}/\mathbf{c},\overline{\mathbf{e}},\overline{\mathbf{d}},\overline{\boldsymbol{\theta}}]\}\end{array}$$

Consider the formula

$$@_c\neg(@_{\mathbf{b}_{i1}}\mathbf{K}_{i1})[c,\overline{e},\overline{d},\overline{\theta}/\mathbf{c},\overline{\mathbf{e}},\overline{\mathbf{d}},\overline{\boldsymbol{\theta}}].$$

If it is the case that $\mathbf{b}_{i1} = \mathbf{c}$, then the displayed formula is trivially derivable from $\{@_c\neg(@_{\mathbf{b}_{i1}}\mathbf{K}_{i1})^A[\overline{e},\overline{\theta}/\overline{\mathbf{e}},\overline{\boldsymbol{\theta}}]\}$. This is also the case if $\mathbf{b}_{i1} \notin \{\mathbf{c},\mathbf{d}_1,\ldots,\mathbf{d}_q\}$ (the formulas are simply identical then). If $\mathbf{b}_{i1} \in \{\mathbf{d}_1,\ldots,\mathbf{d}_q\}$, then let $g \in \{1,\ldots,q\}$ be such that $\mathbf{b}_{i1} = \mathbf{d}_g$. It follows that $@_{d_g}\mathbf{K}_{i1}[\overline{e},\overline{\theta}/\overline{\mathbf{e}},\overline{\boldsymbol{\theta}}]$ is derivable from

$$\begin{array}{l}\{@_c\sharp(\overline{e},\overline{\theta}), @_c\neg\mathbf{J}_i[\overline{e},\overline{\theta}/\overline{\mathbf{e}},\overline{\boldsymbol{\theta}}]\}\cup\\ \{@_c\neg(@_{\mathbf{b}_{i2}}\mathbf{K}_{i2})[c,\overline{e},\overline{d},\overline{\theta}/\mathbf{c},\overline{\mathbf{e}},\overline{\mathbf{d}},\overline{\boldsymbol{\theta}}],\ldots,@_c\neg(@_{\mathbf{b}_{ih_i}}\mathbf{K}_{ih_i})[c,\overline{e},\overline{d},\overline{\theta}/\mathbf{c},\overline{\mathbf{e}},\overline{\mathbf{d}},\overline{\boldsymbol{\theta}}]\}\end{array}$$

and so is $@_cA\mathbf{K}_{i1}[\overline{e},\overline{\theta}/\overline{\mathbf{e}},\overline{\boldsymbol{\theta}}]$ by using the rule (AI). Note that the side-condition of this rule is satisfied since $\mathbf{d}_g \notin \{\mathbf{b}_{i2},\ldots,\mathbf{b}_{ih_i}\}$. But we then have $A\mathbf{K}_{i1}[\overline{e},\overline{\theta}/\overline{\mathbf{e}},\overline{\boldsymbol{\theta}}] = (@_{\mathbf{b}_{i1}}\mathbf{K}_{i1})^A[\overline{e},\overline{\theta}/\overline{\mathbf{e}},\overline{\boldsymbol{\theta}}]$. So in any of the three cases, $@_c\bot$ is derivable from

$$\begin{array}{l}\{@_c\sharp(\overline{e},\overline{\theta}), @_c\neg\mathbf{J}_i[\overline{e},\overline{\theta}/\overline{\mathbf{e}},\overline{\boldsymbol{\theta}}]\}\cup\\ \{@_c\neg(@_{\mathbf{b}_{i1}}\mathbf{K}_{i1})^A[\overline{e},\overline{\theta}/\overline{\mathbf{e}},\overline{\boldsymbol{\theta}}]\}\cup\\ \{@_c\neg(@_{\mathbf{b}_{i2}}\mathbf{K}_{i2})[c,\overline{e},\overline{d},\overline{\theta}/\mathbf{c},\overline{\mathbf{e}},\overline{\mathbf{d}},\overline{\boldsymbol{\theta}}],\ldots,@_c\neg(@_{\mathbf{b}_{ih_i}}\mathbf{K}_{ih_i})[c,\overline{e},\overline{d},\overline{\theta}/\mathbf{c},\overline{\mathbf{e}},\overline{\mathbf{d}},\overline{\boldsymbol{\theta}}]\}\end{array}$$

This step is carried out h_i times in total.

It is instructive to look at an example: Consider the language $\mathscr{H}(\prec)$ with strict implication together with the rules $(\prec I)$ and $(\prec E)$ given in the previous section. Then the formula schema **F** of Definition 5.1 is the formula schema $@_d\phi \rightarrow @_d\psi$, and the formula schema **I**, that is, the result of rewriting **F** in accordance with Theorem 5.4, is the formula schema $@_d(\neg\phi \vee \psi)$, and moreover, the formula schema **E** of Definition 5.1 is the formula schema $A(\neg\phi \vee \psi)$. Lemma 5.2 then says that for any formula $\phi \prec \psi$ of $\mathscr{H}(\prec)$, the formula $(\phi \prec \psi) \leftrightarrow A(\neg\phi \vee \psi)$ is a provable equivalence in the natural deduction system $\mathbf{N}_{\mathscr{H}(\prec)}$.

We are now ready for functional completeness.

Theorem 5.5. *(Functional completeness) Consider the system $\mathbf{N}_{\mathscr{H}}$ and let a set of schematic introduction and elimination rules for the connective $\sharp$ be given where the introduction rule is non-equational and explicit. For any formula ϕ in $\mathscr{H}(\sharp)$, there exists a formula ψ in $\mathscr{H}$ such that $\phi \leftrightarrow \psi$ is a provable equivalence in $\mathbf{N}_{\mathscr{H}(\sharp)}$.*

Proof. Induction on the number of occurrences of the connective $\sharp$ in ϕ where we use Lemma 5.2 and Lemma 5.1. We are done if $\sharp$ does not occur in ϕ. Otherwise, if $\sharp$ occurs in ϕ, then there exists an occurrence of $\sharp(\overline{e},\overline{\theta})$ in ϕ where $\overline{\theta}$ are formulas of $\mathscr{H}$. Clearly, replacing $\sharp(\overline{e},\overline{\theta})$ by $\mathbf{E}[\overline{e},\overline{\theta}/\overline{\mathbf{e}},\overline{\boldsymbol{\theta}}]$ decreases the number of occurrences of the connective $\sharp$ by one.

5.5 Discussion

There is vast literature on the proof-theory of the modal logic S5. It has turned out to be difficult to formulate natural deduction, Gentzen, and tableau systems for S5 without introducing metalinguistic machinery. The history of this problem goes back to Ohnishi and Matsumoto (1957, 1959) where a counter-example to cut-elimination is given for an otherwise very natural and straightforward Gentzen formulation of S5. See Braüner (2000) by the present author for an example of a cut-free Gentzen system for S5 where metalinguistic machinery is avoided at the expense of having to make use of a "non-local" side-condition on a derivation rule, that is, a side-condition that does not just refer to the premise of the rule, but to the whole derivation of the premise. See Mints (1992) for an example of a cut-free Gentzen formulation of S5 that makes use of metalinguistic machinery, namely indexed formulas. This system can be considered a reformulation in Gentzen style of Fitting's prefixed tableau system for S5 given in Fitting (1993).[4]

[4] This prefixed tableau system is also considered in Fitting's handbook chapter Fitting (2007) where it is shown to bear a direct relationship to a hypersequent system for S5, to be more precise, a hypersequent is a finite sequence of ordinary Gentzen sequents, and the sequents in a hypersequent play the role of the prefixes in a branch of a prefixed tableau. In Fitting (2007) this direct relationship is used to prove that completeness of the hypersequent system for S5 follows from completeness of the prefixed tableau system for S5. The direct relationship can be established since the prefix machinery for S5 is very simple as there is no accessibility relation in S5 models. It is not clear whether such a direct relationship can be established for other modal logics where the prefix machinery is more complicated.

Now, it is straightforward to check that the hybrid logic of the universal modality has the same expressive power over models as the monadic first-order logic with equality, one variable, and a countably infinite set of constants but no function symbols. That is, a model and an assignment in the usual sense for this first-order logic correspond to a model, an assignment, and a world as appropriate for the hybrid logic of the universal modality and there exist truth-preserving translations in both directions between the logics. So our functional completeness result for the hybrid logic of the universal modality can be seen as a functional completeness result for the mentioned first-order logic, just couched in modal-logical terms. It is, by the way, notable that the expressive power at the level of frames obtained by adding the difference modality to ordinary mono-modal logic is the same as the expressive power obtained by adding nominals as well as the universal modality, this is shown in Gargov and Goranko (1993). See de Rijke (1992) for a further investigation of the difference modality as an additional operator and see Goranko and Passy (1992) for an investigation of the universal modality as an additional operator.

An interesting question is whether the approach for obtaining the functional completeness result of the present chapter can be extended to other hybrid logics. As mentioned earlier, Braüner (2005c) gives a functional completeness result for a hybrid logic involving the difference modality, similar to the functional completeness result of the present chapter. It could be interesting to consider modal operators with stronger expressive power than the difference modality, for example countion modalities (a counting modality expresses that a formula is true in at least n distinct worlds, where n is some fixed number, see van der Hoek and de Rijke (1995)).

Another possible line of work concerns the natural deduction system given in Section 2.2. One technical obstacle with the introduction rule for modal operators given in Figure 2.2 of Section 2.2 is that it not only exhibits the modal operator in the conclusion, but also in the discharged assumptions. Thus, it is implicit, rather than explicit. This suggests that one should consider weakening the explicitness requirement such that the introduction rule in Figure 2.2 is covered.

Chapter 6
First-Order Hybrid Logic

In this chapter we introduce first-order hybrid logic and its proof-theory. The chapter is structured as follows. In the first section of the chapter we introduce first-order hybrid logic. In the second section we introduce a natural deduction system for first-order hybrid logic (taken from Braüner (2005b)) and in the third section we introduce an axiom system for first-order hybrid logic (also taken from Braüner (2005b)).

6.1 Introduction to First-Order Hybrid Logic

In this section we introduce the basics of first-order hybrid logic. In many cases adopt we shall the terminology of Blackburn and Marx (2002). See Fitting and Mendelsohn (1998) and Hughes and Cresswell (1996) as well as the handbook chapters (Garson 2001; Braüner and Ghilardi 2007) for the basics of first-order modal logic. The latter handbook chapter also contains a section on first-order hybrid logic.

First a couple of remarks on first-order modal logic. A basic difference between propositional modal logics and first-order modal logics is that whereas propositional symbols in propositional modal logics can have different *truth-values* in different worlds, predicates in first-order modal logics can have different *extensions* in different worlds. Thus, in first-order modal logics, predicates are relativised to worlds. This allows us to formalize natural language sentences involving predicates like for example "is a citizen of the United States". The fact that this predicate has different extensions in different worlds follows, for example, from the observation that Arnold Schwarzenegger is a citizen of the United States, but he might not have been so, for example if he had not emigrated to the United States. Predicates with different extensions in different worlds should be compared to predicates which are naturally taken to have the same extension in all worlds, one example being "is greater than five" since the extension of this predicate in any world is naturally taken to be the set of numbers greater than five. Predicates of the latter kind can be formalized in ordinary first-order logic.

T. Braüner, *Hybrid Logic and its Proof-Theory*, Applied Logic Series 37,
DOI 10.1007/978-94-007-0002-4_6,

A number of first-order modal logics involve *non-rigid designators* which are terms that can designate different individuals in different worlds. Non-rigid designators can be motivated in a number of different ways. One very instructive motivation is that non-rigid designators allow us to formalize natural language sentences involving non-rigidly designating terms like “the number of planets” and “the world champion in marathon running”. The first example term designates non-rigidly as it designates the number eight (since there are eight planets in our world), but it might have designated another number (since there might have been another number of planets if natural history had been different or if the notion of a planet had been defined differently). Similarly, the designation of the second example term is the winner of the world championship in marathon running, and the identity of the winning person is obviously also a contingent matter. The first-order hybrid logic we shall consider in this chapter involves a restricted use of non-rigid designators. In the next chapter we shall allow a more general use of non-rigid designators.

We shall now define the formal syntax of first-order hybrid logic. The first-order hybrid logic we consider is obtained by adding hybrid-logical machinery to first-order modal logic with equality, that is, first-order logic with equality extended with a modal operator. The hybrid-logical machinery involves nominals, satisfaction operators, and binders as described in Section 1.2. Besides a countably infinite set of nominals, it is assumed that a countably infinite set of ordinary first-order variables is given. The sets are assumed to be disjoint. The metavariables x, y, z, ... range over first-order variables. It is also assumed that a set of predicate symbols is given. The metavariables P, Q, R, ... range over predicate symbols. (Nominals are the only sort of propositional symbols, but 0-place predicate symbols correspond to propositional symbols in the ordinary sense.) Besides the hybrid-logical machinery described in Section 1.2, we also consider non-rigid designators. To be precise, we assume that a set of non-rigid designators is given, we let the metavariables i, j, k, ... range over non-rigid designators, and we follow Blackburn and Marx (2002) in overloading the notation for the satisfaction operator by defining a term to be either a first-order variable or an expression of the form $@_a i$ where a is a nominal and i is a non-rigid designator. Of course, the term $@_a i$ denotes the value of i at the world where a is true. Such terms are called *rigidified constants*. Formulas are defined by the grammar

$$S ::= P(t_1,\ldots,t_n) \mid t = u \mid a \mid S \wedge S \mid S \rightarrow S \mid \bot \mid \Box S \mid @_a S \mid \forall x S \mid \forall a S \mid \downarrow a S$$

where P ranges over n-place predicate symbols, t_1, ..., t_n as well as t and u range over terms, a ranges over nominals, and x ranges over ordinary first-order variables. Note that non-rigid designators only occur in connection with rigidified constants. The notions of free and bound occurrences of nominals and first-order variables are defined in the obvious way. Also, if $\bar{x}$ is a list of pairwise distinct first-order variables and $\bar{t}$ is a list of terms of the same length as $\bar{x}$, then $\psi[\bar{t}/\bar{x}]$ is the formula ψ where the terms $\bar{t}$ have been simultaneously substituted for all free occurrences of the variables $\bar{x}$. If a variable x_i in $\bar{x}$ occur free in ψ within the scope of $\forall y$ where y is any first-order variable occuring in t_i, then the variable y in ψ is renamed, and similarly, if x_i occur free in ψ within the scope of $\forall a$ or $\downarrow a$ where a is any nominal occuring in t_i, then the nominal a in ψ is renamed. An analogous definition of substitution is obtained if the lists $\bar{x}$ and $\bar{t}$ are replaced by lists of nominals. We let $\exists x\phi$ be an abbreviation for $\neg\forall x\neg\phi$ and we define the *existence predicate* by letting $E(t)$ be an abbreviation for $\exists y(y = t)$ where y is a variable distinct from any variable occuring in t.

Having defined the syntax of first-order hybrid logic, in what follows we define the formal semantics. We first define models and skeletons. Skeletons are first-order versions of the usual frames for propositional modal logic.

Definition 6.1. A *model* for first-order hybrid logic is a tuple

$$(W, R, D, \{\delta_w\}_{w\in W}, \{V_w\}_{w\in W})$$

where

1. W is a non-empty set;
2. R is a binary relation on W;
3. D is a non-empty set;
4. for each w, δ_w is a subset of D; and
5. for each w, V_w is a function that to each non-rigid designator assigns an element of D, and moreover, to each n-place predicate symbol assigns a subset of D^n.

The tuple $(W, R, D, \{\delta_w\}_{w\in W})$ is called a *skeleton* and the model is said to be *based* on this skeleton. The set δ_w is called the *domain of quantification* at the world w. A model (skeleton) has *increasing domains* if and only if $\delta_w \subseteq \delta_v$ whenever wRv, and similarly, it has *decreasing domains* if and only if $\delta_w \supseteq \delta_v$ whenever wRv.

Given a model $\mathfrak{M} = (W, R, D, \{\delta_w\}_{w\in W}, \{V_w\}_{w\in W})$, an *assignment* is a function that to each nominal assigns an element of W and to each first-order variable assigns an element of D. Given an assignment g, each term t is assigned an element $t^{\mathfrak{M},g}$ of D as follows: If t is of the form $@_a i$, then $t^{\mathfrak{M},g} = V_{g(a)}(i)$, otherwise t is a variable, in which case $t^{\mathfrak{M},g} = g(t)$. Given assignments g' and g, $g' \overset{x}{\sim} g$ means that g' agrees with g on all nominals and first-order variables, save possibly on the first-order variable x (and analogously if x is replaced by a nominal a). The relation $\mathfrak{M}, g, w \models \phi$ is defined by induction, where w is a world, g is an assignment, and ϕ is a formula of first-order hybrid logic.

$$\begin{aligned}
\mathfrak{M}, g, w \models P(t_1,\ldots,t_n) &\text{ iff } (t_1^{\mathfrak{M},g},\ldots,t_n^{\mathfrak{M},g}) \in V_w(P)\\
\mathfrak{M}, g, w \models t = u &\text{ iff } t^{\mathfrak{M},g} = u^{\mathfrak{M},g}\\
\mathfrak{M}, g, w \models a &\text{ iff } w = g(a)\\
\mathfrak{M}, g, w \models \phi \wedge \psi &\text{ iff } \mathfrak{M}, g, w \models \phi \text{ and } \mathfrak{M}, g, w \models \psi\\
\mathfrak{M}, g, w \models \phi \rightarrow \psi &\text{ iff } \mathfrak{M}, g, w \models \phi \text{ implies } \mathfrak{M}, g, w \models \psi\\
\mathfrak{M}, g, w \models \bot &\text{ iff falsum}\\
\mathfrak{M}, g, w \models \Box\phi &\text{ iff for any } v \in W \text{ such that } wRv,\ \mathfrak{M}, g, v \models \phi\\
\mathfrak{M}, g, w \models @_a\phi &\text{ iff } \mathfrak{M}, g, g(a) \models \phi\\
\mathfrak{M}, g, w \models \forall x\phi &\text{ iff for any } g' \overset{x}{\sim} g \text{ where } g'(x) \in \delta_w,\ \mathfrak{M}, g', w \models \phi\\
\mathfrak{M}, g, w \models \forall a\phi &\text{ iff for any } g' \overset{a}{\sim} g,\ \mathfrak{M}, g', w \models \phi\\
\mathfrak{M}, g, w \models \downarrow a\phi &\text{ iff } \mathfrak{M}, g', w \models \phi \text{ where } g' \overset{a}{\sim} g \text{ and } g'(a) = w
\end{aligned}$$

A formula ϕ is said to be *true* at the world w if $\mathfrak{M}, g, w \models \phi$; otherwise it is said to be *false* at w. By convention $\mathfrak{M}, g \models \phi$ means $\mathfrak{M}, g, w \models \phi$ for every world w and $\mathfrak{M} \models \phi$ means $\mathfrak{M}, g \models \phi$ for every assignment g. A formula ϕ is *valid* in a skeleton if and only if $\mathfrak{M} \models \phi$ for any model $\mathfrak{M}$ that is based on the skeleton. A formula ϕ is *valid* in a class of skeletons if and only if ϕ is valid in any skeleton in the class of skeletons in question. A formula ϕ is *valid* if and only if ϕ is valid in the class of all skeletons.

Note that we use the same notation for the binder $\forall$ and for the first-order quantifier. Also, note that the relationship $\mathfrak{M}, g, w \models E(t)$ holds if and only if $t^{\mathfrak{M},g} \in \delta_w$. Thus, the existence predicate is true of the individual designated by some term if and only if the individual in question exists. So the existence predicate behaves as desired. Let $\mathscr{O} \subseteq \{\downarrow, \forall\}$. In the present chapter we let $\mathscr{H}(\mathscr{O})$ denote the fragment of first-order hybrid logic in which the only binders are the binders in the set $\mathscr{O}$ (we use the same notation in connection with propositional hybrid logic, but no confusion should be possible).

Observe that we allow the domain of a first-order quantifier to vary from world to world. A number of other choices in the definition of a model for first-order hybrid logic should also be observed: We do not require that a predicate is false of non-existents, we do not require that a quantifier domain is non-empty and we do not require that each individual exists in some domain. In the case of first-order modal logic, most combinations of these requirements can be found in the literature. Our choices make the translation into two-sorted first-order logic very straightforward, see Section 6.1.3.

In ordinary first-order logic, the equality predicate is a designated primitive 2-place predicate symbol which is given a fixed interpretation, namely the identity relation on the domain of quantification. Note that the same pattern is followed in the above case of first-order modal and hybrid logic.

Let us take a look at a natural language sentence that can be formalized using the modal-logical machinery introduced above. Consider the sentence

Arnold Schwarzenegger is a citizen of the United States.

About Arnold Schwarzenegger, it says that he is a member of the set of persons who happen to be citizens of the United States. If the variable x stands for "Arnold Schwarzenegger" and the 1-place predicate symbol Q stands for the predicate "is a citizen of the United States", then the formula $Q(x)$ formalizes the statement. Formally, $Q(x)$ is true at a world w if and only if the designation of x belongs to the extension of the predicate symbol Q at w. The relativisation of Q to worlds formalizes that the predicate "is a citizen of the United States" has different extensions in different worlds. Consider also

Arnold Schwarzenegger is necessarily a citizen of the United States.

Of course, this statement is formalized by the formula $\Box Q(x)$ which is true at a world w if and only if for each world v accessible from w, the designation of x belongs to the extension of the predicate Q at v.

It is also instructive to take a look at a natural language sentence that can be formalized using a rigidified constant together with the $\downarrow$ binder. The example involves the term "the President of the United States" which clearly designates non-rigidly. Consider the sentence

The President of the United States is a Republican.

This sentence says something about the person who is the President of the United States, namely that the person in question is a Republican. If the non-rigid designator i stands for "the President of the United States" and the 1-place predicate symbol P stands for the predicate "is a Republican", then the formula $\downarrow aP(@_a i)$ formalizes the statement. Formally, $\downarrow aP(@_a i)$ is true at a world w if and only if the designation of i at w belongs to the extension of the predicate P at w. In Section 6.1.2 we shall give further examples of formalizations involving rigidified constants and the $\downarrow$ binder.

6.1.1 Some Remarks on Existence and Quantification

As pointed out above, the domain of a first-order quantifier is allowed to vary from world to world. This interpretation of a quantifier is called *actualist quantification* since a quantifier ranges over individuals that actually exist, that is, individuals that exist in the actual world. A different interpretation of a quantifier is obtained if it is required that the domain is constant from world to world. This is called *possibilist quantification* since the quantifier in this case ranges over individuals that possibly exist.

The difference between actualist and possibilist quantification is very clear when the modal operator is given a temporal interpretation, that is, when worlds are taken to be instants and the modal operator is interpreted using the earlier-later relation on instants. In this case actualist quantification corresponds to quantifying over things that now exist whereas possibilist quantification corresponds to quantifying over things that exist at some time. This distinction was discussed already by Prior who rejected the temporal version of possibilist quantification:

> ...even if it be true that whatever exists at any time exists at all times, there is surely no *inconsistency* in denying it, and a *logic* of time-distinctions ought to be able to proceed without assuming it. (Prior 1957, p. 30)

Since Prior, possibilist and actualist quantification has given rise to much philosophical discussion, see Fitting and Mendelsohn (1998) for an account. See also the contributions to the discussion given in Plantinga (2003) and Jager (1982).

It is straightforward to show that the famous *Barcan* formula $\forall x \Box \phi \rightarrow \Box \forall x \phi$ is valid in any decreasing domain skeleton, and moreover, it can also be shown straightforwardly that if the Barcan formula is valid in a skeleton, then the skeleton in question has decreasing domains. Thus, the class of skeletons that validates the Barcan formula is exactly the class of decreasing domain skeletons. Prior rejected the Barcan formula for the same reasons as he rejected possibilist quantification. It can also be shown straightforwardly that the class of skeletons that validates the *Converse Barcan* formula $\Box\forall x\phi \rightarrow \forall x\Box\phi$ is exactly the class of increasing domain skeletons.

First-order modal logics can be seen as combinations of two distinct logics, namely propositional modal logic and ordinary first-order logic. The two logics, propositional modal logic and ordinary first-order logic, are combined in different ways depending on the requirements on the quantifier domains. The interaction between modality and quantification is stronger with constant domains than with varying domains in the sense that the Barcan and Converse Barcan formulas (which together say that the order of quantifiers and modal operators does not matter) are both valid in the constant domain case but neither of them are valid in the varying domain case.

The semantical import of the Barcan and Converse Barcan formulas stems from the distinction between the semantics of the formulas $\forall x\Box\phi$ and $\Box\forall x\phi$. This distinction is an example of the *de re*/*de dicto* distinction. In Latin *de re* means "about the thing" and *de dicto* means "about the proposition". To explain this difference, we instantiate the formula ϕ to $P(x)$. The formula $\Box\forall xP(x)$ says that

it is necessary that each existing thing is P.

This is a *de dicto* interpretation since it says something about a proposition, namely the proposition that each existing thing is P. What it says about this proposition is that it is necessary. On the other hand, the formula $\forall x\Box P(x)$ says that

each existing thing is necessarily P.

This is a *de re* interpretation since it says something about things, namely the things that exist. What it says about these things is that each of them is necessarily P. See Fitting and Mendelsohn (1998) for a much more thorough discussion of *de re* and *de dicto*. The history of formulas like the Barcan and Converse Barcan formulas goes back to Barcan (1946).

6.1.2 Rigidified Constants

In this section, we shall give two examples of philosophical applications of rigidified constants together with the $\downarrow$ binder.

The first example has to do with the *de re*/*de dicto* distinction described in the previous section. Many natural language sentences are ambiguous as they can be given two distinct readings, a *de re* reading and a *de dicto* reading. Rigidified constants together with $\downarrow$ can be used to distinguish formally between such readings. Consider for example the sentence

The number of planets is necessarily greater than five.

which is taken from Quine (1953). On one reading, this sentence says that

> it is necessary that the number of planets is greater than five.

This is the *de dicto* reading since it says something about a proposition, namely the proposition that the number of planets is greater than five. It says about this proposition that it is necessary. However, on another reading, the sentence says that

> the number designated by the term "the number of planets" is necessarily greater than five.

This is the *de re* reading since it says something about a thing, namely a number. It says about this number that it is necessarily greater than five. Note that the *de re* reading of Quine's example sentence is naturally taken to be true (since there are eight planets and the number eight is necessarily greater than five) whereas the *de dicto* reading is naturally taken to be false (since there might have been five planets or fewer if natural history had been different or if the notion of a planet had been defined differently). The point here is that the term "the number of planets" designates non-rigidly.

In what follows, the non-rigid designator i stands for "the number of planets" and the 1-place predicate symbol P stands for the predicate "is greater than five". The formula $\Box \downarrow aP(@_a i)$ then formalizes the *de dicto* reading of Quine's sentence since this formula expresses that

> it is necessary that the thing designated by i is P.

That is, it says something about the proposition that the thing designated by i is P, namely that this proposition is necessary. Formally, $\Box \downarrow aP(@_a i)$ is true at a world w if and only if for each world v accessible from w, the designation of i at v belongs to the extension of the predicate P at v. Thus, in the *de dicto* case the predicate P and the non-rigid designator i are interpreted at the same world, namely the new world v. How about the *de re* reading of Quine's sentence? We want a formula which expresses that

> the thing designated by i is necessarily P.

That is, we want a formula which says something about the thing that i designates, namely that it is necessarily P. So, formally we want the non-rigid designator i to be interpreted at the original world w, not at the new world v where the predicate P is interpreted. It is straightforward that the formula $\downarrow a\Box P(@_a i)$ does the job since it is true at the world w exactly under the condition we want, namely under the condition that the designation of i at the world w belongs to the extension of the predicate P at each world v accessible from w. We have used the binder $\downarrow$ to indicate that the non-rigid designator i has to be interpreted at w, not v (note that this is a formally significant difference since the interpretations of i at the worlds w and v might not be the same).

Another way to formalize the *de re* and *de dicto* readings of the example sentence considered above is by using predicate abstraction and intension variables in first-order intensional logic. The *de dicto* reading of the sentence is formalized in first-order intensional logic as the formula $\Box(\lambda x P(x))(i)$ and the *de re* reading is formalized as the formula $(\lambda x \Box P(x))(i)$ where i is an intension variable (a definition of predicate abstraction and intension variables can be found in Section 7.1). See Fitting and Mendelsohn (1998) for a very thorough discussion of this and see also the handbook chapter Braüner and Ghilardi (2007).

The second example of a philosophical application of rigidified constants together with $\downarrow$ concerns a modal version of Frege's famous puzzle about the morning star and the evening star. Equality in first-order modal logic has given rise to a heated philosophical debate. This debate was initiated by a series of papers where Quine criticised quantified modal logic, see for example Quine (1953). See also the handbook chapter Lindstrøm and Segerberg (2007) for an account of Quine's criticism. Central in the debate initiated by Quine's papers is the issue of substitution of equals for equals in modal contexts. This is not the place to enter into a detailed philosophical discussion of the problem involved in substitution of equals for equals in modal contexts, so we only give a brief sketch of the problem, and we also only give a brief sketch of how a solution to the problem can be given using using appropriate hybrid-logical machinery. Now, consider the statement

> If the morning star is identical to the evening star, then it is necessary that the morning star is identical to the evening star.

which is a modal version of Frege's puzzle. This statement is naturally taken to be false (the morning star is the same celestial body as the evening star but this is a contingent fact). How can this statement be formalized in ordinary first-order modal logic with equality? An obvious candidate is the formula $x = y \to \Box(x = y)$ where the first-order variables x and y respectively stand for the terms "the morning star" and "the evening star". But this does not work since this formula is valid.

The diagnosis of the problem is that the variables x and y designate rigidly whereas the terms "the morning star" and "the evening star" designate non-rigidly. Therefore a solution to the problem is to replace the variables x and y by non-rigid designators i and j and insert hybrid-logical machinery such that the non-rigid designators are interpreted in the appropriate worlds. The resulting formula $\downarrow a(@_a i = @_a j) \to \Box \downarrow a(@_a i = @_a j)$ is not valid, as i and j designating the same object at a world w does not imply that i and j designate the same object at any world accessible from w.

To sum up, the formula $x = y \to \Box(x = y)$ is valid as it is, but it is invalid if the rigidly designating variables x and y are replaced by the non-rigid designators i and j and appropriate hybrid-logical machinery is inserted. Thereby a solution can be given to the problem of formalizing the modal version of Frege's puzzle. Another solution to the problem can be given by using predicate abstraction and intension variables in first-order intensional logic, see Fitting and Mendelsohn (1998) and Braüner and Ghilardi (2007). A radically different solution is to keep the language of ordinary first-order modal logic as it is, but instead generalize the models for first-order modal logic to encompass what are called counterpart relations. This also makes the formula invalid. The history of counterpart relations goes back to the papers Lewis (1968, 1971). After the publication of these papers, a number of generalized versions of Lewis' counterpart semantics have been introduced, one example being the semantics given in Kracht and Kutz (2002). See the discussion in Fitting (2004) where first-order intensional logic is compared to Lewis' counterpart semantics as well as to a variation of the semantics given in Kracht and Kutz (2002). Another formalization of Lewi's counterpart semantics is considered in the second part of the handbook chapter Braüner and Ghilardi (2007).

6.1.3 Translation into Two-Sorted First-Order Logic

First-order hybrid logic can be translated into two-sorted first-order logic with equality, and a fragment of two-sorted first-order logic with equality can be translated back into a fragment of first-order hybrid logic. Obviously, in the two-sorted first-order language under consideration, there is one sort for worlds and one sort for individuals. The translation from first-order hybrid logic into two-sorted first-order logic we consider in this section is an extension of the standard translation from propositional hybrid logic into one-sorted first-order logic, see Section 1.2.1. A similar translation can be found in Areces et al. (2003).

The translation we consider in this section can also be viewed as a hybridized version of a translation from a first-order modal logic into two-sorted first-order logic. There is not much literature available on translations from first-order modal logic into sorted first-order logic, two exceptions being the book van Benthem (1983) and the chapter Ohlbach et al. (2001) in *Handbook of Automated Reasoning* which beside modal logic also considers a range of other non-classical logics. In van Benthem (1983) a semantic characterisation is given of the formulas of two-sorted first-order logic which have the same expressive power as formulas of first-order modal logic. Hazen (1976) and Hodes (1984) consider a number of formulas in two-sorted first-order logic that express properties of models which are not expressible in first-order modal logic. The latter paper concentrates on a first-order version of the modal logic S5. A recent example of work in this area is Sturm and Wolter (2001) which also concerns the expressive power of a first-order version of S5. See also that paper for an overview of the area.

Now, the two-sorted first-order language under consideration in this section is defined as follows. It is assumed that a countably infinite set of first-order variables for worlds and a countably infinite set of first-order variables for individuals are given. The sets are assumed to be disjoint. The metavariables a, b, c, ... range over first-order variables for worlds and the metavariables x, y, z, ... range over first-order variables for individuals. Terms of the language are built out of variables ranging over worlds, variables ranging over individuals, and for each hybrid-logical non-rigid designator, a unary function symbol which is interpreted as a function from worlds to individuals. Thus, all terms ranging over worlds are variables and a term ranging over individuals is either a variable or of the form $i(a)$ where a is a variable ranging over worlds and i is a non-rigid designator of first-order hybrid logic. Formulas of the two-sorted first-order language are defined by the grammar

$$S ::= P^*(a,t_1,\ldots,t_n) \mid R(a,b) \mid E(a,t) \mid a=b \mid t=u \mid S \wedge S \mid S \rightarrow S \mid \bot \mid \forall x S \mid \forall a S$$

where P ranges over n-place predicate symbols of first-order hybrid logic, a and b range over variables for worlds, and $t_1, \ldots, t_n$ as well as t and u range over terms for individuals. Note that according to the grammar above, for each n-place predicate symbol P of the hybrid-logical language, there is a corresponding $(n+1)$-place predicate symbol P^* in the two-sorted first-order language. The two-sorted $(n+1)$-place predicate symbol P^* is interpreted such that it relativizes the interpretation of the corresponding hybrid-logical n-place predicate symbol P to worlds, the predicate symbol R is interpreted using the accessibility relation, and the predicate symbol E is interpreted such that it relates a world to individuals existing at that world. Note that the language contains two equality predicates and two quantifiers; an equality predicate and a quantifier for each sort. In what follows, we shall identify first-order variables for individuals with first-order variables of first-order hybrid logic and we shall identify first-order variables for worlds with nominals. Free and bound variables are defined in the obvious way and the same applies to substitution.

We first translate the first-order hybrid logic $\mathcal{H}(\downarrow,\forall)$ into two-sorted first-order logic with equality. A term t of first-order hybrid logic is translated by the translation ST defined as follows: If t is of the form $@_a i$, then $ST(t) = i(a)$, otherwise t is a variable, in which case $ST(t) = t$. Note that the translation ST of terms is not relative to a nominal. We now give the translation for formulas. Given two nominals, a and b, which do not occur in the formulas to be translated, the translations ST_a and ST_b are defined by mutual induction. We just give the translation ST_a.

$$\begin{aligned}
ST_a(P(t_1,\ldots,t_n)) &= P^*(a,ST(t_1),\ldots,ST(t_n)) \\
ST_a(t=u) &= ST(t) = ST(u) \\
ST_a(c) &= a = c \\
ST_a(\phi \wedge \psi) &= ST_a(\phi) \wedge ST_a(\psi) \\
ST_a(\phi \rightarrow \psi) &= ST_a(\phi) \rightarrow ST_a(\psi) \\
ST_a(\bot) &= \bot \\
ST_a(\Box\phi) &= \forall b(R(a,b) \rightarrow ST_b(\phi)) \\
ST_a(@_c\phi) &= ST_a(\phi)[c/a] \\
ST_a(\forall x\phi) &= \forall x(E(a,x) \rightarrow ST_a(\phi)) \\
ST_a(\forall c\phi) &= \forall c ST_a(\phi) \\
ST_a(\downarrow c\phi) &= ST_a(\phi)[a/c]
\end{aligned}$$

The definition of ST_b is obtained by exchanging a and b.

To state formally that the translation ST_a is truth-preserving, we make use of the observation that a model for first-order hybrid logic can be considered as a model for two-sorted first-order logic and vice versa.

Definition 6.2. Given a model $\mathfrak{M} = (W,R,D,\{\delta_w\}_{w\in W},\{V_w\}_{w\in W})$ for first-order hybrid logic, a model $\mathfrak{M}^* = (W,D,V^*)$ for two-sorted first-order logic is defined by letting

- $V^*(i)(w) = V_w(i)$,

- $V^*(P^*) = \{(w, d_1, \ldots, d_n) \mid (d_1, \ldots, d_n) \in V_w(P)\}$,
- $V^*(R) = R$, and
- $V^*(E) = \{(w, d) \mid d \in \delta_w\}$.

It is straightforward to see that the map $(\cdot)^*$ which maps $\mathfrak{M}$ to $\mathfrak{M}^*$ is bijective. Moreover, an assignment in the sense of first-order hybrid logic can be considered an assignment as appropriate for two-sorted first-order logic and vice versa. Note that the correspondence between first-order hybrid-logical models and two-sorted first-order models is very straightforward and simple due to the following choices made in connection with models for first-order hybrid-logic: We do not require predicates to be false of non-existents, we do not require quantifier domains to be non-empty, and we do not require that each individual exists in some domain.

Given a model $\mathfrak{M} = (W, D, V)$ for two-sorted first-order logic and an assignment g, each term t ranging over individuals is assigned an element $t^{\mathfrak{M},g}$ of D in the standard way: If t is of the form $i(a)$, then $t^{\mathfrak{M},g} = V(i)(g(a))$, otherwise t is a variable, in which case $t^{\mathfrak{M},g} = g(t)$. The relation $\mathfrak{M}, g \models \phi$ is defined by induction in the standard way, where g is an assignment and ϕ is a two-sorted first-order formula.

$$
\begin{aligned}
\mathfrak{M}, g &\models P^*(a, t_1, \ldots, t_n) \text{ iff } (g(a), t_1^{\mathfrak{M},g}, \ldots, t_n^{\mathfrak{M},g}) \in V(P^*) \\
\mathfrak{M}, g &\models R(a, b) \text{ iff } g(a)V(R)g(b) \\
\mathfrak{M}, g &\models E(a, t) \text{ iff } g(a)V(E)t^{\mathfrak{M},g} \\
\mathfrak{M}, g &\models a = b \text{ iff } g(a) = g(b) \\
\mathfrak{M}, g &\models t = u \text{ iff } t^{\mathfrak{M},g} = u^{\mathfrak{M},g} \\
\mathfrak{M}, g &\models \phi \wedge \psi \text{ iff } \mathfrak{M}, g \models \phi \text{ and } \mathfrak{M}, g \models \psi \\
\mathfrak{M}, g &\models \phi \rightarrow \psi \text{ iff } \mathfrak{M}, g \models \phi \text{ implies } \mathfrak{M}, g \models \psi \\
\mathfrak{M}, g &\models \bot \text{ iff falsum} \\
\mathfrak{M}, g &\models \forall x \phi \text{ iff for any } g' \overset{x}{\sim} g,\ \mathfrak{M}, g' \models \phi \\
\mathfrak{M}, g &\models \forall a \phi \text{ iff for any } g' \overset{a}{\sim} g,\ \mathfrak{M}, g' \models \phi
\end{aligned}
$$

The formula ϕ is said to be *true* if $\mathfrak{M}, g \models \phi$; otherwise it is said to be *false*. By convention $\mathfrak{M} \models \phi$ means $\mathfrak{M}, g \models \phi$ for every assignment g. A formula ϕ is valid if and only if $\mathfrak{M} \models \phi$ for any model $\mathfrak{M}$. We are now ready to state formally that ST_a is truth-preserving.

Proposition 6.1. *Let $\mathfrak{M}$ be a model for first-order hybrid logic. For any first-order hybrid-logical formula ϕ and any assignment g for $\mathfrak{M}$, it is the case that $\mathfrak{M}, g, g(a) \models \phi$ if and only if $\mathfrak{M}^*, g \models ST_a(\phi)$ (and the same for ST_b).*

Proof. induction on the structure of ϕ, like Proposition 1.1.

It turns out that a fragment of two-sorted first-order logic can be translated back into the first-order hybrid logic $\mathscr{H}(\forall)$. The fragment in question is obtained from the above grammar for two-sorted first-order logic by replacing the clause $\forall x S$ by the new clause $\forall x (E(a, x) \rightarrow S)$. We first translate terms. A term t of two-sorted first-order logic is translated back into first-order hybrid logic by the translation HT defined as follows: If t is of the form $i(a)$, then $HT(t) = @_a i$, otherwise t is a variable, in which case $HT(t) = t$. So HT and the translation ST given above are simply inverses to each other. A formula of the fragment above is translated by the translation given below.

$$
\begin{aligned}
HT(P^*(a,t_1,\ldots,t_n)) &= @_a P(HT(t_1),\ldots,HT(t_n))\\
HT(R(a,c)) &= @_a\Diamond c\\
HT(E(a,t)) &= @_a E(HT(t))\\
HT(a=c) &= @_a c\\
HT(t=u) &= HT(t)=HT(u)\\
HT(\phi\wedge\psi) &= HT(\phi)\wedge HT(\psi)\\
HT(\phi\rightarrow\psi) &= HT(\phi)\rightarrow HT(\psi)\\
HT(\bot) &= \bot\\
HT(\forall a\phi) &= \forall a HT(\phi)\\
HT(\forall x(E(a,x)\rightarrow\phi)) &= @_a\forall x HT(\phi)
\end{aligned}
$$

The translation *HT* is an extension of the hybrid translation from one-sorted first-order logic to propositional hybrid logic, see Section 1.2.1. The translation is truth-preserving.

Proposition 6.2. *Let $\mathfrak{M}$ be a model for first-order hybrid logic. For any formula ϕ of the above given fragment of two-sorted first-order logic and any assignment g for $\mathfrak{M}$, it is the case that $\mathfrak{M}^*, g \models \phi$ if and only if $\mathfrak{M}, g \models HT(\phi)$.*

Proof. induction on the structure of ϕ, like Proposition 1.2.

Thus, in the sense of the two propositions above, the first-order hybrid logic $\mathscr{H}(\forall)$ has the same expressive power as the fragment of two-sorted first-order logic (note that for any formula ϕ of first-order hybrid logic, the formula $ST_a(\phi)$ is in this fragment). An analogous result holds for the first-order hybrid logic $\mathscr{H}(\downarrow)$ and a bounded version of the fragment of two-sorted first-order logic, cf. Section 1.2.1. See the handbook chapter (Braüner and Ghilardi 2007, p. 577).

6.2 Natural Deduction for First-Order Hybrid Logic

In this section we shall give a natural deduction system for the first-order hybrid logic $\mathscr{H}(\mathscr{O})$. The system generalizes the natural deduction system for propositional hybrid logic given in Section 2.2. The derivation rules for the system are given in Figures 6.1, 6.2, and 6.3. All formulas in the rules are satisfaction statements. The rules given are the natural deduction rules for propositional hybrid logic given in Section 2.2 together with rules for first-order quantification, that is, $(\forall I1)$ and $(\forall E1)$, rules for first-order equality, namely $(Ref2)$ and $(Rep1)$, and moreover, rules for equality in connection with existence predicates and non-rigid designators, namely $(Nom3)$ and $(Nom4)$. It is instructive to compare the rules for first-order equality with the rules for equality in connection with nominals $(Ref1)$, $(Nom1)$, $(Nom2)$, and $(Nom3)$. The reason why the rule $(Rep1)$ is restricted to atomic formulas (which furthermore have to be different from $\bot$) is similar to the reasons why $(Nom1)$

and $(\bot 1)$ are restricted, see the remark following Proposition 2.1 in Section 2.2.2. However, the unrestricted versions of the rules are admissible, cf. Proposition 6.3.

$$\frac{@_a\phi \quad @_a\psi}{@_a(\phi \wedge \psi)}\,(\wedge I) \qquad \frac{@_a(\phi \wedge \psi)}{@_a\phi}\,(\wedge E1) \qquad \frac{@_a(\phi \wedge \psi)}{@_a\psi}\,(\wedge E2)$$

$$\frac{\begin{array}{c}[@_a\phi]\\ \vdots \\ @_a\psi\end{array}}{@_a(\phi \rightarrow \psi)}\,(\rightarrow I) \qquad \frac{@_a(\phi \rightarrow \psi) \quad @_a\phi}{@_a\psi}\,(\rightarrow E)$$

$$\frac{\begin{array}{c}[@_aE(z)]\\ \vdots \\ @_a\phi[z/x]\end{array}}{@_a\forall x\phi}\,(\forall I1)^* \qquad \frac{@_a\forall x\phi \quad @_aE(t)}{@_a\phi[t/x]}\,(\forall E1)$$

$$\frac{\begin{array}{c}[@_a\neg\phi]\\ \vdots \\ @_a\bot\end{array}}{@_a\phi}\,(\bot 1)^\star \qquad \frac{@_a\bot}{@_c\bot}\,(\bot 2)$$

$*$ z does not occur free in $@_a\forall x\phi$ or in any undischarged assumptions other than the specified occurrences of $@_aE(z)$.
$\star$ ϕ is an atomic formula different from $\bot$.

Fig. 6.1 Natural deduction rules: propositional and first-order connectives

Our natural deduction system for $\mathcal{H}(\mathcal{O})$ is obtained from the rules given in Figures 6.1, 6.2, and 6.3 by leaving out the rules for the binders that are not in the set $\mathcal{O}$. The system thus obtained will be denoted $\mathbf{N}_{\mathcal{H}(\mathcal{O})}$ (in the interest of simplicity we use the same notation as the notation used in connection with propositional hybrid logic, cf. Section 2.2). The natural deduction system $\mathbf{N}_{\mathcal{H}(\mathcal{O})}$ corresponds to the class of all skeletons, that is, the class of skeletons where no conditions are imposed on the accessibility relation or the quantifier domains.

6.2.1 Conditions on the Accessibility Relation

In what follows we shall consider natural deduction systems obtained by extending $\mathbf{N}_{\mathcal{H}(\mathcal{O})}$ with additional derivation rules corresponding to first-order conditions on the accessibility relation and the quantifier domains. These rules are generalizations of the propositional rules in Section 2.2.1 in the sense that we here not only allow conditions on the accessibility relation, but also on the quantifier domains. A two-sorted first-order formula is *geometric* if it is built out of atomic formulas of the

$$\dfrac{\begin{matrix}[@_a\Diamond c]\\ \vdots \\ @_c\phi\end{matrix}}{@_a\Box\phi}\ (\Box I)^{*} \qquad \dfrac{@_a\Box\phi \quad @_a\Diamond e}{@_e\phi}\ (\Box E)$$

$$\dfrac{@_a\phi}{@_c@_a\phi}\ (@I) \qquad \dfrac{@_c@_a\phi}{@_a\phi}\ (@E)$$

$$\dfrac{@_a\phi[c/b]}{@_a\forall b\phi}\ (\forall I2)^{\star} \qquad \dfrac{@_a\forall b\phi}{@_a\phi[e/b]}\ (\forall E2)$$

$$\dfrac{\begin{matrix}[@_a c]\\ \vdots \\ @_c\phi[c/b]\end{matrix}}{@_a\downarrow b\phi}\ (\downarrow I)^{\dagger} \qquad \dfrac{@_a\downarrow b\phi \quad @_a e}{@_e\phi[e/b]}\ (\downarrow E)$$

$*$ c does not occur free in $@_a\Box\phi$ or in any undischarged assumptions other than the specified occurrences of $@_a\Diamond c$.
$\star$ c does not occur free in $@_a\forall b\phi$ or in any undischarged assumptions.
$\dagger$ c does not occur free in $@_a\downarrow b\phi$ or in any undischarged assumptions other than the specified occurrences of $@_a c$.

Fig. 6.2 Natural deduction rules: modal and hybrid connectives

$$\dfrac{}{@_a a}\ (Ref1) \qquad \dfrac{}{@_a(t=t)}\ (Ref2)$$

$$\dfrac{@_a c \quad @_a\phi}{@_c\phi}\ (Nom1)^{*} \qquad \dfrac{@_a c \quad @_a\Diamond b}{@_c\Diamond b}\ (Nom2) \qquad \dfrac{@_a c \quad @_a E(t)}{@_c E(t)}\ (Nom3)$$

$$\dfrac{@_a c}{@_b((@_a i)=(@_c i))}\ (Nom4) \qquad \dfrac{@_a(t=u) \quad @_c\phi[t/x]}{@_c\phi[u/x]}\ (Rep1)^{*}$$

$*$ ϕ is an atomic formula different from $\bot$.

Fig. 6.3 Natural deduction rules: nominals and first-order terms

forms $R(a,c)$, $E(a,x)$, $a=c$, and $x=y$ using only the connectives $\bot$, $\wedge$, $\vee$, and $\exists$. In what follows, the metavariables S_k and S_{jk} range over atomic formulas of the mentioned forms. By a slightly modified version of the translation HT given in Section 6.1.3, atomic formulas of the mentioned forms are translated into first-order hybrid logic in a truth preserving way as follows.

$$\begin{aligned} HT(R(a,c)) &= @_a\Diamond c \\ HT(E(a,x)) &= @_aE(x) \\ HT(a=c) &= @_ac \\ HT(x=y) &= @_b(x=y) \end{aligned}$$

The nominal b in the last clause is arbitrary (we want $HT(x=y)$ to be a satisfaction statement, cf. the definition of the natural deduction rule (R_θ) below, so we simply prefix $x=y$ by an arbitrary satisfaction operator).

A *geometric theory* is a finite set of closed two-sorted first-order formulas each having the form $\forall\overline{a}\overline{x}(\phi \rightarrow \psi)$ where the formulas ϕ and ψ are geometric, $\overline{a}$ is a list $a_1,\ldots,a_l$ of variables ranging over worlds, $\overline{x}$ is a list $x_1,\ldots,x_h$ of variables ranging over individuals, and $\forall\overline{a}\overline{x}$ is an abbreviation for $\forall a_1 \ldots \forall a_l \forall x_1 \ldots \forall x_h$. It can be proved that any geometric theory is equivalent to a *basic geometric theory* which is a geometric theory in which each formula has the form

$$(*) \qquad \forall\overline{a}\overline{x}((S_1 \wedge \cdots \wedge S_n) \rightarrow \exists\overline{c}\overline{y}\bigvee_{j=1}^{m}(S_{j1} \wedge \cdots \wedge S_{jn_j}))$$

where $n,m \geq 0$ and $n_1,\ldots,n_m \geq 1$. For simplicity, we assume that the variables in each of the lists $\overline{a}$ and $\overline{x}$ are pairwise distinct, that the variables in each of $\overline{c}$ and $\overline{y}$ are pairwise distinct, and that no variable occurs in both $\overline{a}$ and $\overline{c}$ or in both $\overline{x}$ and $\overline{y}$. A formula of the form $(*)$ displayed above is a Horn clause if $\overline{c}$ and $\overline{y}$ are empty, $m=1$, and $n_m=1$.

We now give hybrid-logical natural deduction rules corresponding to a basic geometric theory. The metavariables s_k and s_{jk} range over hybrid-logical formulas of the forms $@_a\Diamond c$, $@_aE(x)$, $@_ac$, and $@_b(x=y)$. With a two-sorted first-order formula θ of the form displayed above, we associate the natural deduction derivation rule (R_θ) given in Figure 6.4 where s_k is of the form $HT(S_k)$ and s_{jk} is of the form $HT(S_{jk})$. For example, if θ is the formula

$$\forall a \forall c \forall x((R(a,c) \wedge E(a,x)) \rightarrow E(c,x))$$

then (R_θ) is the natural deduction rule

$$\dfrac{@_a\Diamond c \qquad @_aE(t) \qquad \begin{array}{c}[@_cE(t)] \\ \vdots \\ \phi\end{array}}{\phi}\ (R_\theta)$$

The formula, and hence the derivation rule, corresponds to the quantifier domains being increasing. Now, let $\mathbf{T}$ be any basic geometric theory. The natural deduction system obtained by extending $\mathbf{N}_{\mathscr{H}(\mathscr{O})}$ with the set of rules $\{(R_\theta) \mid \theta \in \mathbf{T}\}$ will be denoted $\mathbf{N}_{\mathscr{H}(\mathscr{O})} + \mathbf{T}$.

If a formula in a basic geometric theory is a Horn clause, then the rule (R_θ) given in Figure 6.4 can be replaced by the following simpler rule.

$$\dfrac{s_1[\overline{t}/\overline{x}]\ldots s_n[\overline{t}/\overline{x}] \quad \begin{matrix}[s_{11}[\overline{t}/\overline{x}]]\ldots[s_{1n_1}[\overline{t}/\overline{x}]]\\ \vdots \\ \phi\end{matrix} \quad \ldots \quad \begin{matrix}[s_{m1}[\overline{t}/\overline{x}]]\ldots[s_{mn_m}[\overline{t}/\overline{x}]]\\ \vdots \\ \phi\end{matrix}}{\phi}\,(R_\theta)^*$$

$*\,\overline{t}$ is any list of terms of the same length as $\overline{x}$, none of the nominals in $\overline{c}$ and the first-order variables in $\overline{y}$ occur in any terms in $\overline{t}$, and none of the nominals in $\overline{c}$ and the first-order variables in $\overline{y}$ occur free in ϕ or in any undischarged assumptions other than the specified occurrences of s_{jk}. (Recall that nominals are identified with first-order variables ranging over worlds and that $\overline{c}$ and $\overline{y}$ are the first-order variables existentially quantified over in the formula θ.)

Fig. 6.4 Natural deduction rules: geometric theories

$$\dfrac{s_1[\overline{t}/\overline{x}] \quad \ldots \quad s_n[\overline{t}/\overline{x}]}{s_{11}[\overline{t}/\overline{x}]}\,(R_\theta)$$

In the rule, $\overline{t}$ is any list of terms of the same length as $\overline{x}$ (note, by the way, that the rules $(Ref1)$, $(Ref2)$, $(Nom2)$, and $(Nom3)$ are all of this form). For example, if θ is the formula corresponding to increasing domains, cf. above, then the following rule will do.

$$\dfrac{@_a\Diamond c \qquad @_aE(t)}{@_cE(t)}\,(R_\theta)$$

Below is an example of a derivation in $\mathbf{N}_{\mathscr{H}} + \{\theta\}$ where θ is the formula corresponding to increasing domains and where we have used the simplified version of (R_θ).

$$\dfrac{\dfrac{\dfrac{\dfrac{\dfrac{@_a\Box\forall x\phi^3 \quad @_a\Diamond c^1}{@_c\forall x\phi}\,(\Box E) \qquad \dfrac{@_a\Diamond c^1 \quad @_aE(x)^2}{@_cE(x)}\,(R_\theta)}{@_c\phi}\,(\forall E1)}{@_a\Box\phi}\,(\Box I)^1}{@_a\forall x\Box\phi}\,(\forall I1)^2}{@_a(\Box\forall x\phi \to \forall x\Box\phi)}\,(\to I)^3$$

The nominal c is new. Note that the end-formula of the derivation is the Converse Barcan Formula prefixed by a satisfaction operator. That this formula is derivable when domains are increasing is not surprising, cf. the discussion in Section 6.1.1.

6.2.2 Some Admissible Rules

Below we state a small proposition regarding some admissible rules. Recall from Section 2.2.2 that the degree of a formula is the number of occurrences of non-nullary connectives in it.

Proposition 6.3. *The rules*

$$\dfrac{\begin{matrix}[@_a \neg\phi] \\ \vdots \\ @_a \bot\end{matrix}}{@_a \phi}\ (\bot) \qquad \dfrac{@_a c \quad @_a \phi}{@_c \phi}\ (Nom) \qquad \dfrac{@_a(t=u) \quad @_c \phi[t/x]}{@_c \phi[u/x]}\ (Rep)$$

are admissible in $\mathbf{N}_{\mathscr{H}(\mathscr{O})} + \mathbf{T}$.

Proof. The proofs that the rules $(\bot)$ and (Nom) are admissible makes use of the notion of degree, cf. above, and are straightforward extensions of the proofs in the propositional cases, see Proposition 2.1 in Section 2.2.2. The proof that (Rep) is admissible is analogous.

6.2.3 Soundness and Completeness

The aim of this section is to prove soundness and completeness. The proofs are extensions of the soundness and completeness proofs for propositional hybrid logic given in Section 2.2.3. We shall therefore skip the parts of the proofs covered in Section 2.2.3. Further references can also be found in that section. We shall need the standard substitution lemma below.

Lemma 6.1. *(Substitution lemma) Let* $\mathfrak{M}$ *be a model and let* ψ *be a formula. For any world* w *and any assignments* g *and* g' *such that* $g(a) = g'(c)$ *and* $g \overset{a}{\sim} g'$, $\mathfrak{M}, g, w \models \psi$ *if and only if* $\mathfrak{M}, g', w \models \psi[c/a]$. *Analogously, for any world* w *and any assignments* g *and* g' *such that* $g(x) = t^{\mathfrak{M},g'}$ *and* $g \overset{x}{\sim} g'$, $\mathfrak{M}, g, w \models \psi$ *if and only if* $\mathfrak{M}, g', w \models \psi[t/x]$.

Proof. Straightforward induction on the structure of ψ, like the propositional case, Lemma 2.1.

A model $\mathfrak{M}$ for first-order hybrid logic is called a $\mathbf{T}$-*model* if and only if $\mathfrak{M}^* \models \theta$ for every formula θ in $\mathbf{T}$ (recall that $\mathfrak{M}^*$ is the two-sorted first-order model corresponding to $\mathfrak{M}$). Being a $\mathbf{T}$-model is really a property of the skeleton on which the model $\mathfrak{M}$ is based, the reason being that the formulas in $\mathbf{T}$ do not contain predicate symbols beside R, $=$, and E.

Theorem 6.1. *(Soundness) Let* ϕ *be a satisfaction statement and let* Γ *be a set of satisfaction statements. The first statement below implies the second statement.*

1. ϕ *is derivable from* Γ *in* $\mathbf{N}_{\mathscr{H}(\mathscr{O})} + \mathbf{T}$.
2. *For any* $\mathbf{T}$*-model* $\mathfrak{M}$ *and any assignment* g*, if, for any formula* $\psi \in \Gamma$, $\mathfrak{M}, g \models \psi$*, then* $\mathfrak{M}, g \models \phi$.

Proof. Induction on the structure of the derivation of ϕ, like the propositional case, Theorem 2.1.

In what follows, we shall prove completeness.

Definition 6.3. A set of satisfaction statements Γ in $\mathcal{H}(\mathcal{O})$ is $\mathbf{N}_{\mathcal{H}(\mathcal{O})}+\mathbf{T}$-*inconsistent* if and only if $@_a\bot$ is derivable from Γ in $\mathbf{N}_{\mathcal{H}(\mathcal{O})}+\mathbf{T}$ for some nominal a and Γ is $\mathbf{N}_{\mathcal{H}(\mathcal{O})}+\mathbf{T}$-*consistent* if and only if Γ is not $\mathbf{N}_{\mathcal{H}(\mathcal{O})}+\mathbf{T}$-inconsistent. Moreover, Γ is *maximal* $\mathbf{N}_{\mathcal{H}(\mathcal{O})}+\mathbf{T}$-*consistent* if and only if Γ is $\mathbf{N}_{\mathcal{H}(\mathcal{O})}+\mathbf{T}$-consistent and any set of satisfaction statements in $\mathcal{H}(\mathcal{O})$ that properly extends Γ is $\mathbf{N}_{\mathcal{H}(\mathcal{O})}+\mathbf{T}$-inconsistent.

We shall frequently omit the reference to $\mathcal{H}(\mathcal{O})$ and $\mathbf{N}_{\mathcal{H}(\mathcal{O})}+\mathbf{T}$ where no confusion can occur. The definition above leads us to the lemma below.

Lemma 6.2. *If a set of satisfaction statements Γ is consistent, then for every satisfaction statement $@_a\phi$, either $\Gamma\cup\{@_a\phi\}$ is consistent or $\Gamma\cup\{@_a\neg\phi\}$ is consistent.*

Proof. Straightforward, like the propositional case, Lemma 2.2.

Now a Lindenbaum lemma.

Lemma 6.3. *(Lindenbaum lemma) Let $\overline{\mathcal{H}(\mathcal{O})}$ be the hybrid logic obtained by extending the set of nominals in $\mathcal{H}(\mathcal{O})$ with a countably infinite set of new nominals and a countably infinite set of new first-order variables. Let $\phi_1, \phi_2, \phi_3, \ldots$ be an enumeration of all satisfaction statements in $\overline{\mathcal{H}(\mathcal{O})}$. For every $\mathbf{N}_{\mathcal{H}(\mathcal{O})}+\mathbf{T}$-consistent set of satisfaction statements Γ, a maximal $\mathbf{N}_{\overline{\mathcal{H}(\mathcal{O})}}+\mathbf{T}$-consistent set of satisfaction statements $\Gamma^*\supseteq\Gamma$ is defined as follows. Firstly, Γ^0 is defined to be Γ. Secondly, Γ^{n+1} is defined by induction. If $\Gamma^n\cup\{\phi_{n+1}\}$ is $\mathbf{N}_{\overline{\mathcal{H}(\mathcal{O})}}+\mathbf{T}$-inconsistent, then Γ^{n+1} is defined to be Γ^n. Otherwise Γ^{n+1} is defined to be*

1. *$\Gamma^n\cup\{\phi_{n+1}, @_a\psi[z/x], @_aE(z)\}$ if ϕ_{n+1} is of the form $@_a\exists x\psi$;*
2. *$\Gamma^n\cup\{\phi_{n+1}, @_b\psi, @_a\Diamond b\}$ if ϕ_{n+1} is of the form $@_a\Diamond\psi$;*
3. *$\Gamma^n\cup\{\phi_{n+1}, @_b\psi[b/c], @_ab\}$ if ϕ_{n+1} is of the form $@_a\downarrow c\psi$;*
4. *$\Gamma^n\cup\{\phi_{n+1}, @_a\psi[b/c]\}$ if ϕ_{n+1} is of the form $@_a\exists c\psi$;*
5. *$\Gamma^n\cup\{\phi_{n+1}, @_e\bigvee_{j=1}^m(s_{j1}\wedge\cdots\wedge s_{jn_j})[\overline{d},\overline{b}/\overline{a},\overline{c}][\overline{t},\overline{z}/\overline{x},\overline{y}]\}$ if there exists a formula in $\mathbf{T}$ of the form $\forall\overline{a}\overline{x}((S_1\wedge\cdots\wedge S_n)\rightarrow\exists\overline{c}\overline{y}\bigvee_{j=1}^m(S_{j1}\wedge\cdots\wedge S_{jn_j}))$ such that $m\geq 1$ and $\phi_{n+1}=@_e(s_1\wedge\cdots\wedge s_n)[\overline{d}/\overline{a}][\overline{t}/\overline{x}]$ for some terms $\overline{t}$ and some nominals $\overline{d}$ and e; and*
6. *$\Gamma^n\cup\{\phi_{n+1}\}$ if none of the clauses above apply.*

In clause 1, z is a new first-order variable that does not occur in Γ^n or ϕ_{n+1}, in clause 2, 3, and 4, b is a new nominal that does not occur in Γ^n or ϕ_{n+1}, in clause 5, $\overline{b}$ is a list of new nominals such that none of the nominals in $\overline{b}$ occur in Γ^n or ϕ_{n+1}, and similarly, $\overline{z}$ is a list of new first-order variables such that none of the variables in $\overline{z}$ occur in Γ^n or ϕ_{n+1}. Finally, Γ^* is defined to be $\cup_{n\geq 0}\Gamma^n$.

Proof. Firstly, Γ^0 is $\mathbf{N}_{\mathcal{H}(\mathcal{O})}+\mathbf{T}$-consistent by definition and hence also $\mathbf{N}_{\overline{\mathcal{H}(\mathcal{O})}}+\mathbf{T}$-consistent. Secondly, to check that the consistency of Γ^n implies the consistency of Γ^{n+1}, we need to check the first five clauses in the definition of Γ^{n+1}. We only cover clause 1 and clause 5.

- If ϕ_{n+1} is of the form $@_a\exists x\psi$, then assume conversely that $@_f\bot$ is derivable from $\Gamma^n \cup \{\phi_{n+1}, @_a\psi[z/x], @_aE(z)\}$. Then $@_a\neg\psi[z/x]$ is derivable from the set $\Gamma^n \cup \{\phi_{n+1}, @_aE(z)\}$ and therefore $@_a\forall x\neg\psi$ is derivable from $\Gamma^n \cup \{\phi_{n+1}\}$ by the rule $(\forall I1)$. But then $@_a\bot$ is derivable from $\Gamma^n \cup \{\phi_{n+1}\}$ as $\phi_{n+1} = @_a\exists x\psi$.
- If there exists a formula $\forall\overline{a}\overline{x}((S_1 \wedge \cdots \wedge S_n) \rightarrow \exists\overline{c}\overline{y}\bigvee_{j=1}^m (S_{j1} \wedge \cdots \wedge S_{jn_j}))$ in **T** such that $m \geq 1$ and $\phi_{n+1} = @_e(s_1 \wedge \cdots \wedge s_n)[\overline{d}/\overline{a}][\overline{t}/\overline{x}]$ for some terms $\overline{t}$ and some nominals $\overline{d}$ and e, then assume conversely that $@_f\bot$ is derivable from the set $\Gamma^n \cup \{\phi_{n+1}, @_e \bigvee_{j=1}^m (s_{j1} \wedge \cdots \wedge s_{jn_j})[\overline{d},\overline{b}/\overline{a},\overline{c}][\overline{t},\overline{z}/\overline{x},\overline{y}]\}$. Then it is the case that the formula $@_e \bigwedge_{j=1}^m \neg(s_{j1} \wedge \cdots \wedge s_{jn_j})[\overline{d},\overline{b}/\overline{a},\overline{c}][\overline{t},\overline{z}/\overline{x},\overline{y}]$ is derivable from the set $\Gamma^n \cup \{\phi_{n+1}\}$, and hence, it is the case that the formula $@_e\bot$ is derivable from $\Gamma^n \cup \{\phi_{n+1}, s_{j1}[\overline{d},\overline{b}/\overline{a},\overline{c}][\overline{t},\overline{z}/\overline{x},\overline{y}], \ldots, s_{jn_j}[\overline{d},\overline{b}/\overline{a},\overline{c}][\overline{t},\overline{z}/\overline{x},\overline{y}]\}$ for any j where $1 \leq j \leq m$. But $s_1[\overline{d}/\overline{a}][\overline{t}/\overline{x}]$, ..., $s_n[\overline{d}/\overline{a}][\overline{t}/\overline{x}]$ are derivable from $\Gamma^n \cup \{\phi_{n+1}\}$. Therefore $@_e\bot$ is derivable from $\Gamma^n \cup \{\phi_{n+1}\}$ by the rule (R_θ).

We conclude that each Γ^n is consistent which trivially implies the consistency of Γ^*. It is straightforward to check that furthermore Γ^* is maximal consistent. See also the propositional case, Lemma 2.3.

Below we shall define a canonical model. First a small lemma.

Lemma 6.4. *Let Δ be a maximal consistent set of satisfaction statements. Let $\sim_\Delta$ be the binary relation on the set of nominals defined by the convention that $a \sim_\Delta a'$ if and only if $@_a a' \in \Delta$ and let $\sim_\Delta$ be the binary relation on the set of terms defined by the convention that $t \sim_\Delta t'$ if and only if for some nominal b, $@_b(t = t') \in \Delta$ (note that the notation $\sim_\Delta$ is overloaded). Then the defined relations are equivalence relations with the following properties.*

1. *If $a \sim_\Delta a'$ and $@_aE(t) \in \Delta$, then $@_{a'}E(t) \in \Delta$.*
2. *If $a \sim_\Delta a'$, $c \sim_\Delta c'$, and $@_a\Diamond c \in \Delta$, then $@_{a'}\Diamond c' \in \Delta$.*
3. *If $a \sim_\Delta a'$, $t_1 \sim_\Delta t_1'$, ..., $t_n \sim_\Delta t_n'$, and $@_aP(t_1, \ldots, t_n) \in \Delta$, then $@_{a'}P(t_1', \ldots, t_n') \in \Delta$.*
4. *If $a \sim_\Delta a'$, then $(@_a i) \sim_\Delta (@_{a'} i)$.*

Proof. A straightforward extension of the proof in the propositional case, Lemma 2.4, where Lemma 6.2 is used.

Given a nominal a, we let $[a]$ denote the equivalence class of a with respect to $\sim_\Delta$ (and analogously if the nominal a is replaced by a first-order term t). We now define a canonical model.

Definition 6.4. (Canonical model) Let Δ be a maximal $\mathbf{N}_{\overline{\mathcal{H}(\mathcal{O})}} + \mathbf{T}$-consistent set of satisfaction statements. A model $\mathfrak{M}^\Delta = (W^\Delta, R^\Delta, D^\Delta, \{\delta_w^\Delta\}_{w \in W^\Delta}, \{V_w^\Delta\}_{w \in W^\Delta})$ and an assignment g^Δ for $\mathfrak{M}^\Delta$ are defined as follows.

- $W^\Delta = \{[a] \mid a \text{ is a nominal of } \overline{\mathcal{H}(\mathcal{O})}\}$.
- $R^\Delta = \{([a], [c]) \mid @_a\Diamond c \in \Delta\}$.
- $D^\Delta = \{[t] \mid t \text{ is a term of } \overline{\mathcal{H}(\mathcal{O})}\}$.
- $\delta_{[a]}^\Delta = \{[t] \mid @_aE(t) \in \Delta\}$.

- $V^{\Delta}_{[a]}(i) = [@_a i]$.
- $V^{\Delta}_{[a]}(P) = \{([t_1], \ldots, [t_n]) \mid @_a P(t_1, \ldots, t_n) \in \Delta\}$.
- $g^{\Delta}(a) = [a]$.
- $g^{\Delta}(x) = [x]$.

Note that properties of the relations mentioned in Lemma 6.4 imply that the model $\mathfrak{M}^{\Delta}$ is well-defined. Given the Lindenbaum lemma and the definition of a canonical model, we just need one small lemma before we are ready to prove a truth lemma.

Lemma 6.5. *Let ϕ be a satisfaction statement of the hybrid logic $\mathscr{H}(\mathscr{O})$, and let x and y be first-order variables such that y does not occur in ϕ. Let ϕ' be ϕ where each occurrence of x that is not free has been replaced by y. Then ϕ' is derivable from $\{\phi\}$ and ϕ is derivable from $\{\phi'\}$ in $\mathbf{N}_{\mathscr{H}(\mathscr{O})} + \emptyset$. An analogous result is obtained if the first-order variables x and y are replaced by nominals as appropriate.*

Proof. Induction on the degree of ϕ, like the propositional case, Lemma 2.5.

Now the truth lemma.

Lemma 6.6. *(Truth lemma) Let Γ be a $\mathbf{N}_{\mathscr{H}(\mathscr{O})} + \mathbf{T}$-consistent set of satisfaction statements. Then for any satisfaction statement $@_a\phi$, it is the case that $@_a\phi \in \Gamma^*$ if and only if $\mathfrak{M}^{\Gamma^*}, g^{\Gamma^*}, [a] \models \phi$.*

Proof. First note that $t^{\mathfrak{M}^{\Delta}, g^{\Delta}} = [t]$ for any term t. Induction on the degree of ϕ. We only consider the case where ϕ is of the form $\forall x\theta$.

Assume that $@_a\forall x\theta \in \Gamma^*$. We then have to prove that $\mathfrak{M}^{\Gamma^*}, g, [a] \models \theta$ for any $g \overset{x}{\sim} g^{\Gamma^*}$ such that $g(x) \in \delta^{\Gamma^*}_{[a]}$, that is, such that $g(x) = [t]$ for some term t where $@_a E(t) \in \Gamma^*$. If t is a first-order variable, then let θ' be θ where each occurrence of t that is not free has been replaced by some first-order variable that does not occur in $@_a\theta$, and similarly, if t is a rigidified constant, that is, if it is of the form $@_c i$, then let θ' be θ where each occurrence of the nominal c that is not free has been replaced by some nominal that does not occur in $@_a\theta$. Then $@_a\forall x\theta' \in \Gamma^*$ as $@_a(\forall x\theta \rightarrow \forall x\theta')$ is derivable by Lemma 6.5. So $@_a\theta'[t/x] \in \Gamma^*$ by the rule $(\forall E1)$. By induction we get $\mathfrak{M}^{\Gamma^*}, g^{\Gamma^*}, [a] \models \theta'[t/x]$ and therefore $\mathfrak{M}^{\Gamma^*}, g, [a] \models \theta'$ by Lemma 6.1. But $@_a(\theta' \rightarrow \theta)$ is derivable by Lemma 6.5 and therefore valid by Theorem 6.1, so $\mathfrak{M}^{\Gamma^*}, g, [a] \models \theta$.

On the other hand, assume that $\mathfrak{M}^{\Gamma^*}, g^{\Gamma^*}, [a] \models \forall x\theta$. If $@_a\neg\forall x\theta \in \Gamma^*$, then also $@_a\exists x\neg\theta \in \Gamma^*$ as $@_a(\neg\forall x\theta \rightarrow \exists x\neg\theta)$ is derivable. Therefore by definition of Γ^*, there exists a first-order variable z such that $@_a\neg\theta[z/x] \in \Gamma^*$ and $@_a E(z) \in \Gamma^*$. Now, let $g \overset{x}{\sim} g^{\Gamma^*}$ such that $g(x) = [z]$. So $g(x) \in \delta^{\Gamma^*}_{[a]}$ as $@_a E(z) \in \Gamma^*$. Therefore by assumption $\mathfrak{M}^{\Gamma^*}, g, [a] \models \theta$ and hence $\mathfrak{M}^{\Gamma^*}, g^{\Gamma^*}, [a] \models \theta[z/x]$ by Lemma 6.1. Therefore $@_a\theta[z/x] \in \Gamma^*$ by induction. We conclude that $@_a\neg\forall x\theta \notin \Gamma^*$ and hence $@_a\forall x\theta \in \Gamma^*$ by Lemma 6.2.

See also the propositional case, Lemma 2.6.

Now we need only one lemma before we can prove completeness.

Lemma 6.7. *Let Γ be a $\mathbf{N}_{\mathcal{H}(\mathcal{O})}+\mathbf{T}$-consistent set of satisfaction statements. Then the model $\mathfrak{M}^{\Gamma^*}$ is a $\mathbf{T}$-model.*

Proof. If $\theta \in \mathbf{T}$, θ has the form $\forall \overline{a}\overline{x}((S_1 \wedge \cdots \wedge S_n) \rightarrow \exists \overline{c}\overline{y} \bigvee_{j=1}^{m} (S_{j1} \wedge \cdots \wedge S_{jn_j}))$ where $\overline{a} = a_1,\ldots,a_l$ and $\overline{x} = x_1,\ldots,x_h$. Assume g is an assignment for a hybrid-logical model $\mathfrak{M}^{\Gamma^*}$ such that $(\mathfrak{M}^{\Gamma^*})^*, g \models S_1, \ldots, (\mathfrak{M}^{\Gamma^*})^*, g \models S_n$. Let $g(a_1) = [d_1], \ldots, g(a_l) = [d_l]$ and $g(x_1) = [t_1], \ldots, g(x_h) = [t_h]$. Then it is the case that $s_1[\overline{d}/\overline{a}][\overline{t}/\overline{x}],\ldots,s_n[\overline{d}/\overline{a}][\overline{t}/\overline{x}] \in \Gamma^*$ by the definition of a canonical model. If $m \geq 1$, then by definition of Γ^* there exists a list of nominals $\overline{b}$ and a list of first-order variables $\overline{z}$ such that $@_e \bigvee_{j=1}^{m} (s_{j1} \wedge \cdots \wedge s_{jn_j})[\overline{d},\overline{b}/\overline{a},\overline{c}][\overline{t},\overline{z}/\overline{x},\overline{y}] \in \Gamma^*$ since it is the case that $@_e(s_1 \wedge \cdots \wedge s_n)[\overline{d}/\overline{a}][\overline{t}/\overline{x}] \in \Gamma^*$ where e is an arbitrary nominal. Therefore it is the case that $@_e(s_{j1} \wedge \cdots \wedge s_{jn_j})[\overline{d},\overline{b}/\overline{a},\overline{c}][\overline{t},\overline{z}/\overline{x},\overline{y}] \in \Gamma^*$ and hence $s_{j1}[\overline{d},\overline{b}/\overline{a},\overline{c}][\overline{t},\overline{z}/\overline{x},\overline{y}],\ldots,s_{jn_j}[\overline{d},\overline{b}/\overline{a},\overline{c}][\overline{t},\overline{z}/\overline{x},\overline{y}] \in \Gamma^*$ for some j where it is the case that $1 \leq j \leq m$. But then it follows from the definition of a canonical model that $(\mathfrak{M}^{\Gamma^*})^*, g \models \exists \overline{c}\overline{y} \bigvee_{j=1}^{m} (S_{j1} \wedge \cdots \wedge S_{jn_j})$. On the other hand, if $m = 0$, then $@_e \bot \in \Gamma^*$ by the rule (R_θ) which contradicts the consistency of Γ^*. See also the propositional case, Lemma 2.7.

Theorem 6.2. *(Completeness) Let ϕ be a satisfaction statement and let Γ be a set of satisfaction statements. The second statement below implies the first statement.*

1. *ϕ is derivable from Γ in $\mathbf{N}_{\mathcal{H}(\mathcal{O})}+\mathbf{T}$.*
2. *For any $\mathbf{T}$-model $\mathfrak{M}$ and any assignment g, if, for any formula $\psi \in \Gamma$, $\mathfrak{M}, g \models \psi$, then $\mathfrak{M}, g \models \phi$.*

Proof. Analogous to the propositional case, Theorem 2.2.

6.2.4 Normalization

In this section we give reduction rules for the natural deduction system $\mathbf{N}_{\mathcal{H}(\mathcal{O})}+\mathbf{T}$ and we give a normalization theorem. First some conventions. If a premise of a rule has one of the forms $@_a\Diamond c$, $@_aE(t)$, $@_ac$, or $@_b(t = u)$, then it is called a *relational premise*, and similarly, if the conclusion of a rule has one of these forms, then it is called a *relational conclusion*. Moreover, if an assumption discharged by a rule has one of the mentioned forms, then it is called a *relationally discharged assumption*. The premise of the form $@_a\phi$ in the rule $(\rightarrow E)$ is called *minor*. A premise of an elimination rule that is neither minor nor relational is called *major*.

A *maximum formula* in a derivation is a formula occurrence that is both the conclusion of an introduction rule and the major premise of an elimination rule. Maximum formulas can be removed by applying *proper reductions*. The rules for proper reductions are the propositional proper reduction rules given in Section 2.2.4 together with the rule below. In what follows, we let $\pi[\overline{t}/\overline{x}]$ be the derivation π where each formula occurrence ψ has been replaced by $\psi[\overline{t}/\overline{x}]$.

$(\forall I1)$ followed by $(\forall E1)$

$$\dfrac{\dfrac{\begin{array}{c}[@_a E(z)]\\ \vdots\,\pi_1\\ @_a\phi[z/x]\end{array}}{@_a\forall x\phi} \qquad \begin{array}{c}\vdots\,\pi_2\\ @_aE(t)\end{array}}{@_a\phi[t/x]} \quad\rightsquigarrow\quad \begin{array}{c}\vdots\,\pi_2\\ @_aE(t)\\ \vdots\,\pi_1[t/z]\\ @_a\phi[t/x]\end{array}$$

We also need reduction rules in connection with the (R_θ) derivation rules. A *permutable formula* in a derivation is a formula occurrence that is both the conclusion of a (R_θ) rule and the major premise of an elimination rule. Permutable formulas in a derivation can be removed by applying *permutative reductions*. The rule for permutative reductions in the case where the elimination rule has two premises is given below. By convention, S is an abbreviation for $[\overline{t}/\overline{x}]$ and T is an abbreviation for $[\overline{b}/\overline{c}][\overline{z}/\overline{y}]$ where the nominals in the list $\overline{b}$ are pairwise distinct and new, and similarly, the first-order variables in the list $\overline{z}$ are pairwise distinct and new.

$$\dfrac{\dfrac{\begin{array}{c}\vdots\,\tau_1\\ s_1S\end{array}\ \ldots\ \begin{array}{c}\vdots\,\tau_n\\ s_nS\end{array} \qquad \begin{array}{c}[s_{11}S]\ldots[s_{1n_1}S]\\ \vdots\,\pi_1\\ \phi\end{array} \quad\ldots\quad \begin{array}{c}[s_{m1}S]\ldots[s_{mn_m}S]\\ \vdots\,\pi_m\\ \phi\end{array}}{\phi} \qquad \begin{array}{c}\vdots\,\pi\\ \theta\end{array}}{\psi}$$

$$\rightsquigarrow$$

$$\dfrac{\begin{array}{c}\vdots\,\tau_1\\ s_1S\end{array}\ \ldots\ \begin{array}{c}\vdots\,\tau_n\\ s_nS\end{array} \qquad \dfrac{\begin{array}{c}[s_{11}TS]\ldots[s_{1n_1}TS]\\ \vdots\,\pi_1T\\ \phi\end{array} \quad \begin{array}{c}\vdots\,\pi\\ \theta\end{array}}{\psi} \quad\ldots\quad \dfrac{\begin{array}{c}[s_{m1}TS]\ldots[s_{mn_m}TS]\\ \vdots\,\pi_mT\\ \phi\end{array} \quad \begin{array}{c}\vdots\,\pi\\ \theta\end{array}}{\psi}}{\psi}$$

The case where the elimination rule has only one premise is obtained by deleting all instances of the derivation π from the reduction rule.

A derivation is *normal* if it contains no maximum or permutable formula.

Theorem 6.3. *(Normalization) Any derivation in* $\mathbf{N}_{\mathcal{H}(\mathcal{O})}+\mathbf{T}$ *can be rewritten to a normal derivation by repeated applications of proper and permutative reductions.*

Proof. The proof of this theorem is analogous to the proof of the propositional normalization theorem, Theorem 2.3 (where the preceding definitions and lemmas, namely Definition 2.3, Lemma 2.8, Definition 2.4, Lemma 2.9, Definition 2.5, and Lemma 2.10, are extended as appropriate).

The first step of the theorem is to prove that any derivation can be rewritten to a derivation in which all maximum or permutable formulas are of the form $@_a\Diamond c$ or $@_aE(t)$. The second step of the theorem is to prove that any derivation which is the result of the first step can be rewritten to a derivation in which all maximum or permutable formulas are of the form $@_a\neg c$ or $@_a\neg(t=u)$. The third step of the theorem is to prove that any derivation which is the result of the second step can be rewritten to a normal derivation.

6.2.5 The Form of Normal Derivations

The definition below is the same as in the case of propositional hybrid logic, that is, Definition 2.6, but note that it is now applied to first-order hybrid logic.

Definition 6.5. A *branch* in a derivation π is a non-empty list $\phi_1,\ldots,\phi_n$ of formula occurrences in π with the following properties.

1. For each $i<n$, ϕ_i stands immediately above ϕ_{i+1}.
2. ϕ_1 is an assumption, or a relational conclusion, or the conclusion of a (R_θ) rule with zero non-relational premises.
3. ϕ_n is either the end-formula, or a minor or relational premise.
4. For each $i<n$, ϕ_i is not a minor or relational premise.

Lemma 6.8. *Any formula occurrence in a derivation π belongs to a branch in π.*

Proof. Induction on the structure of π, like the propositional case, Lemma 2.11.

Lemma 6.9. *Let $\beta=\phi_1,\ldots,\phi_n$ be a branch in a normal derivation in $\mathbf{N}_{\mathcal{H}(\mathcal{O})}+\mathbf{T}$. Then there exists a formula occurrence ϕ_i in β, called the* minimum formula *in β, such that*

1. *for each $j<i$, ϕ_j is the major premise of an elimination rule, or the non-relational premise of an instance of $(Nom1)$ or $(Rep1)$, or the premise of an instance of the rule $(\bot2)$, or a non-relational premise of an instance of a (R_θ) rule;*
2. *if $i\neq n$, then ϕ_i is a premise of an introduction rule or the premise of an instance of the rule $(\bot1)$; and*
3. *for each j, where $i<j<n$, ϕ_j is a premise of an introduction rule, or the non-relational premise of an instance of $(Nom1)$ or $(Rep1)$, or a non-relational premise of an instance of a (R_θ) rule.*

Proof. A straightforward extension of the proof in the propositional case, Lemma 2.12.

We shall now prove a theorem where we make use of the definition below.

Definition 6.6. The notion of a *subformula* is defined by the conventions that

- ϕ is a subformula of ϕ;
- if ψ is an atomic formula and $\psi[t/x]$ is a subformula of ϕ, then so is $\psi[u/x]$;
- if $\psi \wedge \theta$ or $\psi \rightarrow \theta$ is a subformula of ϕ, then so are ψ and θ;
- if $@_a\psi$ or $\Box\psi$ is a subformula of ϕ, then so is ψ;
- if $\downarrow a\psi$ or $\forall a\psi$ is a subformula of ϕ, then so is $\psi[c/a]$ for any nominal c; and
- if $\forall x\psi$ is a subformula of ϕ, then so is $\psi[t/x]$ for any term t.

A formula $@_a\phi$ is a *quasi-subformula* of a formula $@_c\psi$ if and only if ϕ is a subformula of ψ.

Regarding the definition of subformulas, it might be asked why $\psi[u/x]$ is a subformula of $\psi[t/x]$ but $@_a\psi$ is only a quasi-subformula of $@_b\psi$ (where ψ is atomic). The reason is pragmatic: A clause making $@_a\psi$ a subformula of $@_b\psi$ could have been included for the sake of systematicity, but the resulting quasi-subformula property, that is, the theorem below, would then have been less informative. Now the theorem.

Theorem 6.4. *(Quasi-subformula property) Let π be a normal derivation of ϕ from Γ in $\mathbf{N}_{\mathscr{H}(\mathscr{O})} + \mathbf{T}$. Moreover, let θ be a formula occurrence in π such that*

1. *θ is not an assumption discharged by an instance of the rule $(\perp 1)$ where the discharged assumption is the major premise of an instance of $(\rightarrow E)$;*
2. *θ is not an occurrence of $@_a\perp$ in a branch whose first formula is an assumption discharged by an instance of the rule $(\perp 1)$ where the discharged assumption is the major premise of an instance of $(\rightarrow E)$; and*
3. *θ is not an occurrence of $@_a\perp$ in a branch whose first formula is the conclusion of a (R_θ) rule with zero non-relational premises.*

Then θ is a quasi-subformula of ϕ, or of some formula in Γ, or of some relational premise, or of some relational conclusion, or of some relationally discharged assumption.

Proof. A straightforward extension of the proof in the propositional case, Theorem 2.4.

Note that if the formula occurrence θ is not covered by one of the three exceptions, then it is a quasi-subformula of ϕ, or of some formula in Γ, or of a formula of one of the forms $@_a\Diamond c$, $@_aE(t)$, $@_ac$, and $@_b(x = y)$.

6.3 Axiom Systems for First-Order Hybrid Logic

In this section we shall give a sound and complete Hilbert-style axiom system for the first-order hybrid logic $\mathscr{H}(\mathscr{O})$. The axiom system is comprised of all instances of theorems of propositional logic together with the axioms and rules in Figure 6.5. Our axiom system for the first-order hybrid logic $\mathscr{H}(\mathscr{O})$, which will be denoted $\mathbf{A}_{\mathscr{H}(\mathscr{O})}$, is obtained by leaving out axioms and rules for the binders that are not in

$(Distr_{\rightarrow})$	$@_a(\phi \rightarrow \psi) \leftrightarrow (@_a\phi \rightarrow @_a\psi)$
$(\bot)$	$@_a\bot \rightarrow \bot$
$(Scope)$	$@_a@_b\phi \leftrightarrow @_b\phi$
$(Ref1)$	$@_a a$
$(Ref2)$	$t = t$
$(Transfer)$	$@_a(t = u) \rightarrow @_c(t = u)$
$(Intro)$	$(a \wedge \phi) \rightarrow @_a\phi$
$(Nom4)$	$@_a c \rightarrow (@_a i) = (@_c i)$
(Rep)	$(t = u \wedge \phi[t/x]) \rightarrow \phi[u/x]$
$(\forall E1)$	$(\forall x\phi \wedge E(t)) \rightarrow \phi[t/x]$
$(\Box E)$	$(\Box\phi \wedge \Diamond e) \rightarrow @_e\phi$
$(\downarrow E)$	$(\downarrow b\phi \wedge e) \rightarrow @_e\phi[e/b]$
$(\forall E2)$	$\forall b\phi \rightarrow \phi[e/b]$

$$\frac{\phi \rightarrow \psi \qquad \phi}{\psi}\,(MP) \qquad \frac{\phi}{@_a\phi}\,(N_@) \qquad \frac{@_a\phi}{\phi}\,(Name)^*$$

$$\frac{(\psi \wedge E(z)) \rightarrow \phi[z/x]}{\psi \rightarrow \forall x\phi}\,(\forall I1)^\star \qquad \frac{(\psi \wedge \Diamond c) \rightarrow @_c\phi}{\psi \rightarrow \Box\phi}\,(\Box I)^\dagger$$

$$\frac{\psi \rightarrow \phi[c/b]}{\psi \rightarrow \forall b\phi}\,(\forall I2)^\ddagger \qquad \frac{(\psi \wedge c) \rightarrow @_c\phi[c/b]}{\psi \rightarrow \downarrow b\phi}\,(\downarrow I)^\circ$$

$*$ a does not occur free in ϕ.
$\star$ z does not occur free in $\forall x\phi$ or ψ.
$\dagger$ c does not occur free in ϕ or ψ.
$\ddagger$ c does not occur free in $\forall b\phi$ or ψ.
$\circ$ c does not occur free in $\downarrow b\phi$ or ψ.

Fig. 6.5 Hilbert-style axioms and rules for first-order hybrid logic

$\mathscr{O}$. The first-order axiom system $\mathbf{A}_{\mathscr{H}}$ is an extension of the propositional axiom system with the same name given in Figure 2.11 of Section 2.5.

The rule $(\forall I1)$ in Figure 6.5 corresponds to the natural deduction introduction rule with the same name given in Figure 6.1 of Section 6.2, and similarly, the axiom $(\forall E1)$ corresponds to the natural deduction rule with the same name. This remark also applies to the other cases where Hilbert-style axioms and rules have the same names as natural deduction rules. As remarked in Section 2.5, the rule $(\forall I1)$ and the axiom $(\forall E1)$ are the standard Hilbert-style axiomatic machinery for quantifiers in free logic (where they have different names).

Note that ϕ in the axiom (Rep) can be any formula, also a formula that involves modal operators. Thus, we allow substitution of equals for equals in modal contexts, so for example the formula $t = u \rightarrow \Box(t = u)$ is derivable. This is justified by the fact that terms designate rigidly, that is, a term designates the same object in all worlds. Thus, the terms t and u designating the same object at a world w imply that t and u designate the same object in any world, in particular any world accessible from w. This gives rise to a philosophical discussion; see the example with Frege's puzzle in Section 6.1.2.

The axiom system $\mathbf{A}_{\mathscr{H}}$ is sound and complete with respect to the first-order semantics given in Section 6.1. Soundness is a straightward induction proof analogous to Theorem 2.7. Completeness is analogous to the proof of Theorem 2.8 but here we make use of the complete natural deduction system $\mathbf{N}_{\mathscr{H}}$ for first-order hybrid logic given in Section 6.2.

It is straightforward to modify soundness and completeness to encompass rules corresponding to a basic geometric theory $\mathbf{T}$. First, a Hilbert-style rule (R_θ) is associated with each formula θ in the basic geometric theory $\mathbf{T}$, analogous to the propositional case described in Section 2.5. Second, a skeleton, cf. Definition 6.1, is called a $\mathbf{T}$-*skeleton* if and only if for every model $\mathfrak{M}$ for first-order hybrid logic which is based on the skeleton in question and every formula $\theta \in \mathbf{T}$, it is the case that $\mathfrak{M}^* \models \theta$. The notion of validity is then relativised to the class of first-order $\mathbf{T}$-skeletons and the axiom system $\mathbf{A}_{\mathscr{H}}$ is extended with the set of rules $\{(R_\theta) \mid \theta \in \mathbf{T}\}$.

Chapter 7
Intensional First-Order Hybrid Logic

In this chapter we introduce intensional first-order hybrid logic and its proof-theory. This chapter is structured as follows. In the first section of the chapter we introduce intensional first-order hybrid logic, including two different kinds of models—standard models and generalized models. In the second section we introduce a natural deduction system which is complete with respect to generalized models and we then show how to extend it with a further rule such that a system which is complete with respect to standard models is obtained. In the third section we discuss models for intensional first-order hybrid logic where intension functions are allowed to be partial. All the results of this chapter are taken from Braüner (2008).

7.1 Introduction to Intensional First-Order Hybrid Logic

In the first-order hybrid logic we gave in the previous chapter we only allowed non-rigid designators in a restricted form, namely in connection with rigidified constants, that is, terms of the form $@_a i$ where i is a non-rigid designator and a is a nominal (and where $@_a i$ denotes the value of the non-rigid designator i at the world at which the nominal a is true). In the present chapter we will be more general; we will allow non-rigid designators to be bound by quantifiers and we will allow non-rigid designators to occur as arguments to predicates. Thus, non-rigid designators are now considered to be variables on equal terms with the ordinary first-order variables of the previous chapter, but whereas the ordinary first-order variables designate individual objects, the non-rigidly designating variables (directly or indirectly) designate functions from worlds to objects. Such functions are called *intensions* (some authors call them *individual concepts*).

Rigidified constants constitute the "interface" between the two different types of semantic values—objects and intensions—in the sense that a rigidified constant allows the function designated by an intension variable to be applied to an argument, that is, a world, whereby an object is obtained. To be more precise, from a mathematical point of view, an intension is just a relation of a particular kind between

T. Braüner, *Hybrid Logic and its Proof-Theory*, Applied Logic Series 37,
DOI 10.1007/978-94-007-0002-4_7, © Springer Science+Business Media B.V. 2011

the set of worlds and the set of objects, namely what we usually call the graph of a function, and given such a relation together with a world, rigidified constants are the only built-in machinery in the logic that allows us to perform the mathematical operation we usually call applying a function to an argument, thereby obtaining an object.

Above we described the logic under consideration in the present chapter as an extension of first-order hybrid logic with certain first-order intensional machinery, namely intensional quantifiers and intensional arguments to predicates. Another way to view the logic under consideration is as a hybridization of first-order intensional logic, which is ordinary modal logic extended with first-order intensional machinery. A number of versions of first-order intensional logic can be found in the literature, for example Melvin Fitting's First-Order Intensional Logic given in Fitting (2004) and also considered in Braüner and Ghilardi (2007). See Hughes and Cresswell (1996), the handbook chapter Garson (2001), and the papers (Scott 1970; Parks 1976) for other versions. See also Fitting (2002) for a treatment of higher-order intensional logic. The history of first-order intensional logic goes back to the work of Montague (1974) and Gallin (1975). See the handbook chapter Lindstrøm and Segerberg (2007) for a historical account of intensional logic.

We will now extend the formal syntax and semantics of first-order hybrid logic with first-order intensional machinery. See Section 6.1 where the formal syntax and semantics of the plain first-order hybrid logic are defined. Where appropriate, we do not repeat the conventions and definitions that are identical.

First the syntax. It is assumed that a countably infinite set of intension variables is given. The metavariables $i, j, k, \ldots$ range over intension variables. It is assumed that the set of intension variables is disjoint from the set of nominals as well as the set of ordinary first-order variables, that is, variables for objects. An object term is either a first-order variable for objects or an expression of the form $@_a i$ where a is a nominal and i is an intension variable. A term is either an object term or an intension variable. Note that until now, the intension variables play the same roles as the non-rigid designators of plain first-order hybrid logic, except that we require that there are infinitely many intension variables, the reason being that we shall include intension quantifiers in the formulas. Predicate symbols are typed, that is, it is not only specified which arity a predicate symbol has, it is also specified which type each argument place has (whether it is for an object term or an intension term). Following the paper Fitting (2004), the types of an n-place predicate symbol are specified by a list $T_1, \ldots, T_n$ where $T_i \in \{O, I\}$ for each T_i (the letter O stands for object and the letter I stands for intension). Formulas are defined by the grammar

$$S ::= P(t_1,\ldots,t_n) \mid t = u \mid i = j \mid a \mid S \wedge S \mid S \rightarrow S \mid \bot \mid \Box S \mid @_a S \mid \downarrow a S \mid \forall a S \mid \forall x S \mid \forall i S$$

where P ranges over n-place predicate symbols and $t_1, \ldots, t_n$ range over terms of the respective types $T_1, \ldots, T_n$ specified for P, t and u range over object terms, i and j range over intension variables, a ranges over nominals, and x ranges over object variables. The definition of substitution is extended in accordance with the extension of the language with intension quantifiers. Note that we have an equality predicate for intensions, that is, the predicate $i = j$, this kind of equality is called *synonymy*.

Having defined the syntax, we now define the semantics. We first define what we call standard models. Later we shall define a more general kind of model.

Definition 7.1. A *standard model* for intensional first-order hybrid logic is a tuple

$$(W, R, D_O, D_I, \{V_w\}_{w \in W})$$

where

1. W is a non-empty set;
2. R is a binary relation on W;
3. D_O is a non-empty set;
4. D_I is a non-empty set of functions from W to D_O; and
5. for each w, V_w is a function that to each n-place predicate symbol P assigns a subset of $D_{T_1} \times \cdots \times D_{T_n}$ where $T_1, \ldots, T_n$ are the types specified for P.

The set D_O is called the *domain of object quantification* and the set D_I is called the *domain of intension quantification*.

The standard models defined above are the same as Fitting's models for First-Order Intensional Logic, cf. Fitting (2004). Note that the domains of object quantification and intension quantification are both taken to be constant. Compare to the first-order hybrid logic given in Section 6.1 where the domain of object quantification is relativized to worlds (this difference is, however, not directly of significance to the main issue of the present chapter, and in any case, a varying domain semantics can be embedded into a constant domain semantics by introducing a designated primitive 1-place predicate whose extension at any world is the set of things taken to exist at that world, see Braüner and Ghilardi (2007, p. 560).

Given a standard model $\mathfrak{M} = (W, R, D_O, D_I, \{V_w\}_{w \in W})$, an *assignment* is a function that to each nominal assigns an element of W, to each object variable assigns an element of D_O, and to each intension variable assigns an element of D_I. Given an assignment g, each object term t is assigned an element $t^{\mathfrak{M},g}$ of D_O as follows: If t is of the form $@_a i$, then $t^{\mathfrak{M},g} = g(i)(g(a))$, otherwise t is a variable, in which case $t^{\mathfrak{M},g} = g(t)$. Similarly, each intension term t, which has to be a variable, is assigned an element $t^{\mathfrak{M},g}$ of D_I by $t^{\mathfrak{M},g} = g(t)$. Given assignments g' and g, $g' \overset{x}{\sim} g$ means that g' agrees with g on all nominals and variables, save possibly on the object variable x (and analogously if x is replaced by an intension variable i or a nominal a). The relation $\mathfrak{M}, g, w \models \phi$ is defined in the same way as in the case of plain first-order hybrid logic, cf. Section 6.1, except that the clause for object quantification is replaced by

$$\mathfrak{M}, g, w \models \forall x \phi \text{ iff for any } g' \overset{x}{\sim} g,\ \mathfrak{M}, g', w \models \phi$$

since the domain of object quantification is here taken to be constant, and moreover, the clauses

$$\mathfrak{M}, g, w \models i = j \text{ iff } i^{\mathfrak{M},g} = j^{\mathfrak{M},g}$$

$$\mathfrak{M}, g, w \models \forall i \phi \text{ iff for any } g' \overset{i}{\sim} g,\ \mathfrak{M}, g', w \models \phi$$

for intension equality and intension quantification are added. A formula ϕ is *valid* in a class of standard models if and only if $\mathfrak{M} \models \phi$ for any standard model $\mathfrak{M}$ in the class in question. A formula ϕ is *valid* with respect to standard models if and only if ϕ is valid in the class of all standard models. The notion of *satisfiability* is defined accordingly.

Let $\mathcal{O} \subseteq \{\downarrow, \forall\}$. We let $\mathcal{H}(\mathcal{O})$ denote the fragment of intensional first-order hybrid logic in which the only binders are the binders in the set $\mathcal{O}$ (we use the same notation in connection with propositional and plain first-order hybrid logic).

With the aim of clarifying the difference between objectual and intensional predication, we shall return to an example sentence from Section 6.1, namely the sentence

The President of the United States is a Republican.

which is about the person who is the President of the United States. Thus, it is about an individual object. The statement was formalized by the formula $\downarrow aP(@_a i)$ where the non-rigid designator i stands for "the President of the United States" and the predicate symbol P stands for the predicate "is a Republican". In the terminology of the present chapter, the predicate symbol P is objectual. On the other hand, consider the sentence

The President of the United States is an important concept in politics.

This sentence is not about the person who happens to be the President of the United States, rather it is about the concept of the President of the United States. What the sentence says about this concept, is that it is politically important. If the intensional 1-place predicate symbol R stands for the predicate "is a politically important concept", then the formula $R(i)$ formalizes the statement in question. Formally, $R(i)$ is true at a world w if and only if the extension of R at w contains the intension, that is, the function, designated by i. Clearly, the statement is true, but if "the President of the United States" is replaced by for example "the world champion in marathon running", then it becomes false.

Fitting's First-Order Intensional Logic, cf. Fitting (2004), includes *predicate abstraction* which for any formula ϕ allows a new formula $(\lambda x\phi)(i)$ to be formed. Free occurrences of the object variable x in ϕ are bound in $(\lambda x\phi)(i)$. The semantics is as follows.

$$\mathfrak{M}, g, w \models (\lambda x\phi)(i) \text{ iff } \mathfrak{M}, g', w \models \phi \text{ where } g' \overset{x}{\sim} g \text{ and } g'(x) = g(i)(w)$$

Predicate abstractions play a role similar to that of rigidified constants: They constitute the interface between the two different types of semantic values—objects and intensions. It is instructive to compare the formalization of the first example sentence above (the sentence saying that the President of the United States is a Republican) with a formalization using predicate abstraction. Using a rigidified constant together with the $\downarrow$ binder, the example sentence was formalized as $\downarrow aP(@_a i)$, but using predicate abstraction, it can be formalized as $(\lambda xP(x))(i)$. That these two different formalizations are possible is no coincidence: If intensional first-order hybrid logic is extended with predicate abstraction, then any formula $(\lambda x\phi)(i)$ is equivalent to $\downarrow a\phi[@_a i/x]$ where the nominal a is new. Hence, predicate abstractions are eliminable in the context of appropriate hybrid-logical machinery, namely rigidified constants and the $\downarrow$ binder. See Section 6.1.2 that presents two philosophical applications of this hybrid-logical machinery where alternatively predicate abstraction together with intension variables could have been used.[1]

Note that the intension quantifiers in the semantics above range over elements of the set D_I, which is an arbitrary non-empty subset of the set of all functions from W to D_O. An alternative semantics can be obtained by letting D_I be the set of all functions from W to D_O. Contrary to the original semantics, this alternative semantics validates the formula $\Box\exists xP(x) \to \exists i\Box \downarrow aP(@_a i)$. Roughly, this formula says that if an object is associated with each accessible world, then there exists an intension which maps each accessible world to the object associated with it. A criticism often raised against this property of being able to make an intension out of any association of objects with worlds is that the choices of objects in such an intension need not in any sense be coherent, contrary to what is intuitively expected. In general, modal and hybrid logics along the lines of the alternative logic are unaxiomatisable (but it should be mentioned that no proof is available of unaxiomatisability of the alternative logic described here). See Hughes and Cresswell (1996) and Garson (2001) for proofs of unaxiomatisability of modal versions of such logics.

7.1.1 Generalized Models

The kind of models considered previously in the present chapter are the standard models from Definition 7.1. In this section we shall consider a more general kind of models (but the syntax in unchanged).

Definition 7.2. A *generalized model* for intensional first-order hybrid logic is a tuple

$$(W, R, D_O, D_I, \mathscr{E}, \{V_w\}_{w\in W})$$

where

1. W is a non-empty set;
2. R is a binary relation on W;
3. D_O is a non-empty set;
4. D_I is a non-empty set;

---

[1] Note that predicate abstraction plays a role in a modal version of Herbrand's theorem given in Fitting (1996). The role of predicate abstraction in this work is to enable appropriate Skolemization of formulas involving objectual quantifiers within the scope of modal operators, for example, the formula $\Box\exists xP(x)$ is Skolemized as $\Box(\lambda xP(x))(i)$. In the case of ordinary first-order logic, Herbrand's theorem gives rise to a semi-decision procedure by a reduction to the search for a tautology in a countably infinite set of propositional formulas. A similar result can be proved in the modal case for a first-order modal logic involving non-rigidly designating constant and function symbols as well as predicate abstraction, see Fitting (1996, 1999). Clearly, the formula $\Box\exists xP(x)$ could instead have been Skolemized as $\Box \downarrow aP(@_a i)$ using a rigidified constant and the $\downarrow$ binder. Skolemization of modal formulas using hybrid-logical machinery is a hithertho unexplored line of work.

5. $\mathscr{E}$ is a function that to each element of D_I assigns a function from W to D_O; and
6. for each w, V_w is a function that to each n-place predicate symbol P assigns a subset of $D_{T_1} \times \cdots \times D_{T_n}$ where $T_1, \cdots, T_n$ are the types specified for P.

In this more general kind of model, the set D_I is an arbitrary non-empty set, not necessarily a set of functions from W to D_O, but there is a function $\mathscr{E}$ which to each element of D_I assigns a function from W to D_O. Thus, elements of D_I "encode" functions from W to D_O. It is important to note that this encoding need not be unique, that is, it is not required that $\mathscr{E}$ is injective.

The interpretation of terms we earlier gave for standard models is unchanged, except that we let $(@_a i)^{\mathfrak{M},g} = \mathscr{E}(g(i))(g(a))$ instead of $(@_a i)^{\mathfrak{M},g} = g(i)(g(a))$ (note that intension variables still designate functions from W to D_O, but now indirectly via the $\mathscr{E}$ function). The interpretation of formulas, that is, the relation $\mathfrak{M}, g, w \models \phi$, is unchanged. The definitions of validity and satisfiability are relativized to generalized models instead of standard models.

There is a close correspondence between standard models and generalized models where the function $\mathscr{E}$ is injective, as is witnessed by the following proposition.

Proposition 7.1. *A formula of intensional first-order hybrid logic is satisfiable with respect to standard models if and only if it is satisfiable with respect to generalized models where the function $\mathscr{E}$ is injective.*

Proof. We only sketch the proof. It is straightforward to turn a standard model into a generalized model: Let the generalized version of D_I be the same as the standard version of D_I and let $\mathscr{E}$ be the identity function. Conversely, a generalized model where $\mathscr{E}$ is injective is straightforward to turn into a standard model: Let the standard version of D_I be the image of the generalized version of D_I under $\mathscr{E}$ and adjust V_w accordingly. Thus, a generalized model where $\mathscr{E}$ is injective corresponds to a standard model and vice versa.

Thus, from a mathematical point of view it does not matter whether one considers standard models or generalized models with injective $\mathscr{E}$ function (although the difference may matter from a philosophical point of view). It is in the sense of the proposition that generalized models are more general than standard models.

In what follows we shall show that under certain conditions one can dispense with the requirement in Proposition 7.1 that $\mathscr{E}$ is injective.

Definition 7.3. Let $\mathfrak{M} = (W, R, D_O, D_I, \mathscr{E}, \{V_w\}_{w \in W})$ be a generalized model for intensional first-order hybrid logic where $|D_O| > 1$. Let $e_1, e_2 \in D_O$ such that $e_1 \neq e_2$. A new generalized model $\mathfrak{M}' = (W', R', D_O, D_I, \mathscr{E}', \{V'_w\}_{w \in W'})$ is defined by letting

- $W' = \{(1,w) \mid w \in W\} \cup \{(2,f) \mid f \in D_I\}$,
- $R' = \{((1,w),(1,v)) \mid (w,v) \in R\}$,
- $\mathscr{E}'(g)((1,w)) = \mathscr{E}(g)(w)$,
- $\mathscr{E}'(g)((2,f)) = \begin{cases} e_1 \text{ if } g = f \\ e_2 \text{ otherwise,} \end{cases}$
- $V'_{(1,w)}(P) = V_w(P)$, and

- $V'_{(2,f)}(P) = \emptyset$.

It is straightforward to check that the function $\mathscr{E}'$ is injective. Moreover, given an assignment g for $\mathfrak{M}$, an assignment g' for $\mathfrak{M}'$ is defined by letting $g'(a) = (1, g(a))$ for any nominal a, by letting $g'(x) = g(x)$ for any object variable x, and by letting $g'(i) = g(i)$ for any intension variable i.

The trick in the definition above is to add new inaccessible worlds such that distinct elements of D_I encode distinct functions from worlds to D_O.[2]

Lemma 7.1. *Let* $\mathfrak{M} = (W, R, D_O, D_I, \mathscr{E}, \{V_w\}_{w \in W})$ *be a generalized model for intensional first-order hybrid logic where* $|D_O| > 1$. *For any formula* ϕ *of the intensional first-order hybrid logic* $\mathscr{H}(\downarrow)$, *any assignment* g, *and any world* w, *it is the case that* $\mathfrak{M}, g, w \models \phi$ *if and only if* $\mathfrak{M}', g', (1, w) \models \phi$.

Proof. Induction on the structure of ϕ. We make use of the observation that $t^{\mathfrak{M}', g'} = t^{\mathfrak{M}, g}$ for any object term t.

Proposition 7.2. *If a formula of the intensional first-order hybrid logic* $\mathscr{H}(\downarrow)$ *is satisfiable with respect to generalized models where* $|D_O| > 1$, *then it is satisfiable with respect to generalized models where the function* $\mathscr{E}$ *is injective.*

Proof. Follows from Lemma 7.1.

The requirement in the proposition above that $|D_O| > 1$ is indispensable, that is, the proposition does not hold without this requirement. The formula

$$\forall x \forall y (x = y) \wedge \exists i \exists j (i \neq j)$$

is a counter example since it is satisfiable with respect to generalized models, but not with respect to generalized models with injective $\mathscr{E}$ function (and hence not standard models, cf. Proposition 7.1). This formula says that $|D_O| = 1$ and $|D_I| > 1$ which together imply that $\mathscr{E}$ is not injective. The counter example depends on the presence of intension equality, but if one does not want to involve intension equality, then $i \neq j$ should be replaced by $P(i) \wedge \neg P(j)$. The requirement in the proposition that formulas do not contain the very expressive $\forall$ binder is also indispensable. The formula

$$\exists i \exists j (\forall a (@_a i = @_a j) \wedge i \neq j)$$

is a counter example. It simply says that the function $\mathscr{E}$ is not injective. Again, $i \neq j$ can be replaced by $P(i) \wedge \neg P(j)$. The point in the counter example is that injectivity of $\mathscr{E}$ (and hence the negation of injectivity) can be expressed in the object language if the binder $\forall$ is allowed. In the light of the results of the next section this is no surprise, since the binder $\forall$ gives us full first-order expressive power on models, cf. Proposition 7.3 and Proposition 7.4, and injectivity of $\mathscr{E}$ is a first-order property of models, cf. Proposition 7.5.

[2] A similar trick was suggested to the author by Melvin Fitting in a discussion on First-Order Intensional Logic (personal communication).

7.1.2 Translation into Three-Sorted First-Order Logic

In Section 6.1.3 it was shown that the plain first-order hybrid logic can be translated into two-sorted first-order logic with equality. In a similar way intensional first-order hybrid logic can be translated into three-sorted first-order logic with equality. There is one sort for worlds, one sort for objects, and one sort for intensions.

The three-sorted first-order language under consideration here is defined as follows. It is assumed that countably infinite sets of first-order variables for respectively worlds, objects, and intensions are given. The three sets are assumed to be pairwise disjoint. As in the plain first-order hybrid logic of Section 6.1.3, the metavariables a, b, c, ... range over variables for worlds, and x, y, z, ... range over variables for objects. The metavariables i, j, k, ... range over variables for intensions. There is only one function symbol, namely the 2-place function symbol ℓ which is of type objects and whose argument places are of types intensions and worlds respectively. Thus, a term for worlds is a variable, a term for intensions is a variable, and a term for objects is either a variable or of the form $\ell(i,a)$ where i is a variable for intensions and a is a variable for worlds. Formulas of the three-sorted first-order language are defined by the grammar

$$S ::= P^*(a,t_1,\dots,t_n) \mid R(a,b) \mid a=b \mid t=u \mid i=j \mid S\wedge S \mid S\rightarrow S \mid \perp \mid \forall xS \mid \forall aS \mid \forall iS$$

where P ranges over n-place predicate symbols of intensional first-order hybrid logic and t_1, ..., t_n range over terms of the respective types T_1, ..., T_n specified for P, a and b range over variables for worlds, t and u range over terms for objects, i and j range over variables for intensions, and x ranges over variables for objects. As in Section 6.1.3, we identify first-order variables for objects with object variables of first-order hybrid logic and we identify first-order variables for worlds with nominals. Similarly, we identify first-order variables for intensions with intension variables of intensional first-order hybrid logic.

We now give the translation which translates the intensional first-order hybrid logic $\mathcal{H}(\downarrow,\forall)$ into three-sorted first-order logic with equality. A term t of intensional first-order hybrid logic is translated by the translation ST defined as follows: If t is of the form $@_a i$, then $ST(t) = \ell(i,a)$, otherwise t is a variable, in which case $ST(t) = t$. The translation ST_a of formulas is defined in the same way as in the case of plain first-order hybrid logic, cf. Section 6.1.3, except that the clause for object quantification is replaced by the clause

$$ST_a(\forall x\phi) = \forall x ST_a(\phi)$$

and the clauses

$$\begin{aligned} ST_a(i=j) &= ST(i) = ST(j) \\ ST_a(\forall i\phi) &= \forall i ST_a(\phi) \end{aligned}$$

for intension equality and intension quantification are added. The translation ST_b is modified analogously.

To state formally that the translation given above is truth-preserving, we make use of the observation that a generalized model for intensional first-order hybrid logic can be considered as a model for three-sorted first-order logic and vice versa.

Definition 7.4. Let $\mathfrak{M} = (W, R, D_O, D_I, \mathscr{E}, \{V_w\}_{w \in W})$ be a generalized model for intensional first-order hybrid logic. We define a three-sorted first-order model $\mathfrak{M}^* = (W, D_O, D_I, V^*)$ by letting

- $V^*(\ell)(f, w) = \mathscr{E}(f)(w)$,
- $V^*(P^*) = \{(w, d_1, \ldots, d_n) \mid (d_1, \ldots, d_n) \in V_w(P)\}$, and
- $V^*(R) = R$.

It is straightforward to see that the map $(\cdot)^*$ which maps $\mathfrak{M}$ to $\mathfrak{M}^*$ is bijective. Moreover, an assignment in the sense of intensional first-order hybrid logic can be considered an assignment as appropriate for three-sorted first-order logic and vice versa.

Given a model $\mathfrak{M} = (W, D_O, D_I, V)$ for three-sorted first-order logic and an assignment g, each object term t is assigned an element $t^{\mathfrak{M},g}$ of D_O in the standard way: If t is of the form $\ell(i,a)$, then $t^{\mathfrak{M},g} = V(\ell)(g(i), g(a))$, otherwise t is a variable, in which case $t^{\mathfrak{M},g} = g(t)$. Similarly, each intension term t, which has to be a variable, is assigned an element $t^{\mathfrak{M},g}$ of D_I by $t^{\mathfrak{M},g} = g(t)$. Given a model $\mathfrak{M}$ for three-sorted first-order logic, the relation $\mathfrak{M}, g \models \phi$ is defined by induction in the standard way, where g is an assignment for three-sorted first-order logic and ϕ is a three-sorted first-order formula, that is, the relation is defined as described in Section 6.1.3, except that the clause for the predicate E is removed and the clauses

$$\mathfrak{M}, g, w \models i = j \text{ iff } i^{\mathfrak{M},g} = j^{\mathfrak{M},g}$$
$$\mathfrak{M}, g, w \models \forall i \phi \text{ iff for any } g' \overset{i}{\sim} g,\ \mathfrak{M}, g', w \models \phi$$

for intension equality and intension quantification are added. We are now ready to state formally that the translation is truth-preserving.

Proposition 7.3. *Let $\mathfrak{M}$ be a generalized model for intensional first-order hybrid logic. For any formula ϕ of intensional first-order hybrid logic and any assignment g for $\mathfrak{M}$, it is the case that $\mathfrak{M}, g, g(a) \models \phi$ if and only if $\mathfrak{M}^*, g \models ST_a(\phi)$ (and the same for ST_b).*

Proof. Induction on the structure of ϕ.

It follows that validity with respect to generalized models for intensional first-order hybrid logic can be simulated by validity in three-sorted first-order logic.

Theorem 7.1. *Any formula ϕ of intensional first-order hybrid logic is valid with respect to generalized models if and only if the first-order formula $ST_a(\phi)$ is valid.*

Proof. By Proposition 7.3.

Three-sorted first-order logic can be translated back into $\mathscr{H}(\forall)$. We first translate object and intension terms of three-sorted first-order logic. Such a term t is translated back into intensional first-order hybrid logic by the translation HT defined as follows: If t is of the form $\ell(i,a)$, then $HT(t) = @_a i$, otherwise t is a variable, in which case $HT(t) = t$. So HT and the translation ST given above are simply inverses to each other. A formula is translated by the translation given below.

$$\begin{aligned}
HT(P^*(a,t_1,\ldots,t_n)) &= @_a P(HT(t_1),\ldots,HT(t_n)) \\
HT(R(a,c)) &= @_a \Diamond c \\
HT(a = c) &= @_a c \\
HT(t = u) &= HT(t) = HT(u) \\
HT(\phi \wedge \psi) &= HT(\phi) \wedge HT(\psi) \\
HT(\phi \rightarrow \psi) &= HT(\phi) \rightarrow HT(\psi) \\
HT(\bot) &= \bot \\
HT(\forall x \phi) &= \forall x HT(\phi) \\
HT(\forall i \phi) &= \forall i HT(\phi) \\
HT(\forall a \phi) &= \forall a HT(\phi)
\end{aligned}$$

Compare to the translation *HT* given in Section 6.1.3. The translation above is truth-preserving.

Proposition 7.4. *Let $\mathfrak{M}$ be a generalized model for intensional first-order hybrid logic. For any formula ϕ of three-sorted first-order logic and any assignment g for $\mathfrak{M}$, it is the case that $\mathfrak{M}^*, g \models \phi$ if and only if $\mathfrak{M}, g \models HT(\phi)$.*

Proof. Induction on the structure of ϕ.

Thus, in the sense of Proposition 7.3 and Proposition 7.4, the intensional first-order hybrid logic $\mathscr{H}(\forall)$ has the same expressive power as three-sorted first-order logic. An analogous result holds for the intensional first-order hybrid logic $\mathscr{H}(\downarrow)$ and a bounded version of three-sorted first-order logic, cf. Section 1.2.1.

The map $(\cdot)^*$ from Definition 7.4 gives a bijective correspondence between generalized models for intensional first-order hybrid logic and models for three-sorted first-order logic. What about standard models for intensional first-order hybrid logic? Instead of standard models, we shall consider generalized models where the function $\mathscr{E}$ is injective, cf. Proposition 7.1. It turns out that there is a bijective correspondence between generalized models with injective $\mathscr{E}$ function and models for three-sorted first-order logic that satisfy a certain first-order condition, as is witnessed by the small proposition below.

Proposition 7.5. *A generalized model $\mathfrak{M}$ for intensional first-order hybrid logic has injective $\mathscr{E}$ function if and only if $\mathfrak{M}^* \models \forall i \forall j (\forall a (\ell(i,a) = \ell(j,a)) \rightarrow i = j)$.*

Proof. Trivial.

Thus, validity with respect to generalized models with injective $\mathscr{E}$ function (and hence standard models) can be simulated by validity in three-sorted first-order logic.

Theorem 7.2. *Any formula ϕ of intensional first-order hybrid logic is valid with respect to generalized models with injective $\mathscr{E}$ function (and hence standard models) if and only if the first-order formula $\forall i \forall j (\forall a (\ell(i,a) = \ell(j,a)) \rightarrow i = j) \rightarrow ST_a(\phi)$ is valid.*

Proof. By Proposition 7.3 and Proposition 7.5 (and also Proposition 7.1).

7.2 Natural Deduction for Intensional First-Order Hybrid Logic

In the present section we shall give a natural deduction system for the intensional first-order hybrid logic $\mathcal{H}(\mathcal{O})$. The rules for the system are obtained from the rules for plain first-order hybrid logic given in Figures 6.1, 6.2, and 6.3 of Section 6.2 by removing the rules $(\forall I1)$ and $(\forall E1)$ (given in Figure 6.1) and the rule $(Nom3)$ (given in Figure 6.3) and by adding the new rules for object quantification as well as the rules for intension quantifiers and intension equality given in Figure 7.1. The system for $\mathcal{H}(\mathcal{O})$ is obtained by leaving out the rules for the binders that are not in the set $\mathcal{O}$. The system thus obtained will be denoted $\mathbf{N}_{\mathcal{H}(\mathcal{O})}$ (we use the same notation in connection with propositional and plain first-order hybrid logic).

$$\frac{@_a\phi[z/x]}{@_a\forall x\phi}\,(\forall I1)^{*} \qquad \frac{@_a\forall x\phi}{@_a\phi[t/x]}\,(\forall E1)$$

$$\frac{@_a\phi[j/i]}{@_a\forall i\phi}\,(\forall I3)^{\star} \qquad \frac{@_a\forall i\phi}{@_a\phi[k/i]}\,(\forall E3)$$

$$\frac{}{@_a(i=i)}\,(Ref3) \qquad \frac{@_a(j=k) \quad @_c\phi[j/i]}{@_c\phi[k/i]}\,(Rep2)^{\dagger}$$

$*$ z does not occur free in $@_a\forall x\phi$ or in any undischarged assumptions.
$\star$ j does not occur free in $@_a\forall i\phi$ or in any undischarged assumptions.
$\dagger$ ϕ is an atomic formula different from $\bot$.

Fig. 7.1 New natural deduction rules for intensional first-order hybrid logic

It can be verified that all the results for plain first-order hybrid logic given in Sections 6.2.2, 6.2.4, and 6.2.5 can be adapted to the natural deduction system $\mathbf{N}_{\mathcal{H}(\mathcal{O})}$ for intensional first-order hybrid logic. Thus, a normalization theorem can be proved for the system such that normal derivations satisfy the quasi-subformula property. Moreover, the natural deduction system $\mathbf{N}_{\mathcal{H}(\mathcal{O})}$ for intensional first-order hybrid logic can be extended with additional derivation rules corresponding to first-order conditions on the accessibility relation expressed by geometric theories, see Section 6.2.1 for the plain first-order case (the rules will not involve conditions on the quantifier domains since both of the quantifier domains—the domain of object quantification and the domain of intension quantification—are taken to be constant).

7.2.1 Soundness and Completeness: Generalized Models

The natural deduction system $\mathbf{N}_{\mathscr{H}(\mathscr{O})}$ for intensional first-order hybrid logic is sound and complete with respect to generalized models, which is straightforward to prove by adapting the soundness and completeness proof for plain first-order hybrid logic given in Section 6.2.3. We only give a selection of the adapted versions of the results given in Section 6.2.3. First the Lindenbaum lemma.

Lemma 7.2. *(Lindenbaum lemma) Let $\overline{\mathscr{H}(\mathscr{O})}$ be the hybrid logic obtained by extending the set of nominals in $\mathscr{H}(\mathscr{O})$ with a countably infinite set of new nominals, a countably infinite set of new object variables, and a countably infinite set of new intension variables. Let ϕ_1, ϕ_2, ϕ_3, ... be an enumeration of all satisfaction statements in $\overline{\mathscr{H}(\mathscr{O})}$. For every $\mathbf{N}_{\mathscr{H}(\mathscr{O})}$-consistent set of satisfaction statements Γ, a maximal $\mathbf{N}_{\overline{\mathscr{H}(\mathscr{O})}}$-consistent set of satisfaction statements $\Gamma^* \supseteq \Gamma$ is defined as follows. Firstly, Γ^0 is defined to be Γ. Secondly, Γ^{n+1} is defined by induction. If $\Gamma^n \cup \{\phi_{n+1}\}$ is $\mathbf{N}_{\overline{\mathscr{H}(\mathscr{O})}}$-inconsistent, then Γ^{n+1} is defined to be Γ^n. Otherwise Γ^{n+1} is defined to be*

1. *$\Gamma^n \cup \{\phi_{n+1}, @_b\psi, @_a\Diamond b\}$ if ϕ_{n+1} is of the form $@_a\Diamond\psi$;*
2. *$\Gamma^n \cup \{\phi_{n+1}, @_b\psi[b/c], @_ab\}$ if ϕ_{n+1} is of the form $@_a \downarrow c\psi$;*
3. *$\Gamma^n \cup \{\phi_{n+1}, @_a\psi[b/c]\}$ if ϕ_{n+1} is of the form $@_a\exists c\psi$;*
4. *$\Gamma^n \cup \{\phi_{n+1}, @_a\psi[z/x]\}$ if ϕ_{n+1} is of the form $@_a\exists x\psi$;*
5. *$\Gamma^n \cup \{\phi_{n+1}, @_a\psi[j/i]\}$ if ϕ_{n+1} is of the form $@_a\exists i\psi$; and*
6. *$\Gamma^n \cup \{\phi_{n+1}\}$ if none of the clauses above apply.*

In clauses 1–3, b is a new nominal that does not occur in Γ^n or ϕ_{n+1}, in clause 4, z is a new object variable that does not occur in Γ^n or ϕ_{n+1}, and in clause 5, j is a new intension variable that does not occur in Γ^n or ϕ_{n+1}. Finally, Γ^* is defined to be $\cup_{n\geq 0}\Gamma^n$.

Proof. Firstly, Γ^0 is $\mathbf{N}_{\mathscr{H}(\mathscr{O})}$-consistent by definition and hence also $\mathbf{N}_{\overline{\mathscr{H}(\mathscr{O})}}$-consistent. Secondly, to check that the consistency of Γ^n implies the consistency of Γ^{n+1}, we need to check the first five clauses in the definition of Γ^{n+1}. We only cover clause 5.

If ϕ_{n+1} is of the form $@_a\exists i\psi$, then assume conversely that $@_f\bot$ is derivable from $\Gamma^n \cup \{\phi_{n+1}, @_a\psi[j/i]\}$. Then $@_a\neg\psi[j/i]$ is derivable from $\Gamma^n \cup \{\phi_{n+1}\}$ and therefore $@_a\forall i\neg\psi$ is derivable from $\Gamma^n \cup \{\phi_{n+1}\}$ by the rule $(\forall I2)$. But then $@_a\bot$ is derivable from $\Gamma^n \cup \{\phi_{n+1}\}$ as $\phi_{n+1} = @_a\exists i\psi$.

We conclude that each Γ^n is consistent which trivially implies the consistency of Γ^*. It is straightforward to check that furthermore Γ^* is maximal consistent.

Below we shall define a canonical generalized model. First a small lemma.

Lemma 7.3. *Let Δ be a maximal consistent set of satisfaction statements. Let $\sim_\Delta$ be the binary relation on the set of nominals defined by the convention that $a \sim_\Delta a'$ if and only if $@_aa' \in \Delta$ and let $\sim_\Delta$ be the binary relation on the set of terms of either type defined by the convention that $t \sim_\Delta t'$ if and only if for some nominal b,*

$@_b(t = t') \in \Delta$ *(note that the notation* $\sim_\Delta$ *is overloaded). Then the defined relations are equivalence relations with the following properties.*

1. *If* $a \sim_\Delta a'$, $c \sim_\Delta c'$, *and* $@_a\Diamond c \in \Delta$, *then* $@_{a'}\Diamond c' \in \Delta$.
2. *If* $a \sim_\Delta a'$, $t_1 \sim_\Delta t'_1, \ldots, t_n \sim_\Delta t'_n$, *and* $@_aP(t_1, \ldots, t_n) \in \Delta$, *then* $@_{a'}P(t'_1, \ldots, t'_n) \in \Delta$.
3. *If* $a \sim_\Delta a'$ *and* $i \sim_\Delta i'$, *then* $(@_a i) \sim_\Delta (@_{a'} i')$.

Proof. Use the appropriate derivation rules together with a version of Lemma 6.2 for intensional first-order hybrid logic.

Given a nominal a, we let $[a]$ denote the equivalence class of a with respect to $\sim_\Delta$, and analogously if the nominal a is replaced by a term t of either type. We now define a canonical generalized model.

Definition 7.5. (Canonical generalized model) Let Δ be a set of satisfaction statements which is maximal $\mathbf{N}_{\overline{\mathscr{H}(\mathscr{O})}}$-consistent. We define a generalized model $\mathfrak{M}^\Delta = (W^\Delta, R^\Delta, D^\Delta_O, D^\Delta_I, \mathscr{E}^\Delta, \{V^\Delta_w\}_{w \in W^\Delta})$ and an assignment g^Δ for $\mathfrak{M}^\Delta$ as follows.

- $W^\Delta = \{[a] \mid a \text{ is a nominal of } \overline{\mathscr{H}(\mathscr{O})}\}$.
- $R^\Delta = \{([a],[c]) \mid @_a\Diamond c \in \Delta\}$. $D^\Delta_O = \{[t] \mid t \text{ is an object term of } \overline{\mathscr{H}(\mathscr{O})}\}$.
- $D^\Delta_I = \{[i] \mid i \text{ is an intension variable of } \overline{\mathscr{H}(\mathscr{O})}\}$.
- $\mathscr{E}^\Delta([i])([a]) = [@_a i]$.
- $V^\Delta_{[a]}(P) = \{([t_1], \ldots, [t_n]) \mid @_aP(t_1, \ldots, t_n) \in \Delta\}$.
- $g^\Delta(a) = [a]$.
- $g^\Delta(x) = [x]$.
- $g^\Delta(i) = [i]$.

Now the truth lemma.

Lemma 7.4. *(Truth lemma) Let* Γ *be a* $\mathbf{N}_{\mathscr{H}(\mathscr{O})}$*-consistent set of satisfaction statements. Then for any satisfaction statement* $@_a\phi$, *it is the case that* $@_a\phi \in \Gamma^*$ *if and only if* $\mathfrak{M}^{\Gamma^*}, g^{\Gamma^*}, [a] \models \phi$.

Proof. Induction on the degree of ϕ. We only consider the case where ϕ is of the form $\forall i\theta$.

Assume that $@_a\forall i\theta \in \Gamma^*$. We then have to prove that $\mathfrak{M}^{\Gamma^*}, g, [a] \models \theta$ for any $g \stackrel{i}{\sim} g^{\Gamma^*}$. Let $g(i) = [j]$ for some intension variable j and let θ' be θ where each occurrence of j that is not free has been replaced by some intension variable that does not occur in $@_a\theta$. Then $@_a\forall i\theta' \in \Gamma^*$ as $@_a(\forall i\theta \rightarrow \forall i\theta')$ is derivable, cf. an adapted version of Lemma 6.5. So $@_a\theta'[j/i] \in \Gamma^*$ by the rule $(\forall E2)$. By induction we get $\mathfrak{M}^{\Gamma^*}, g^{\Gamma^*}, [a] \models \theta'[j/i]$ and therefore $\mathfrak{M}^{\Gamma^*}, g, [a] \models \theta'$, cf. an adapted version of Lemma 6.1. But $@_a(\theta' \rightarrow \theta)$ is derivable cf. an adapted version of Lemma 6.5 and therefore valid cf. soundness, so $\mathfrak{M}^{\Gamma^*}, g, [a] \models \theta$. On the other hand, assume that $\mathfrak{M}^{\Gamma^*}, g^{\Gamma^*}, [a] \models \forall i\theta$. If $@_a\neg\forall i\theta \in \Gamma^*$, then also $@_a\exists i\neg\theta \in \Gamma^*$ as $@_a(\neg\forall i\theta \rightarrow \exists i\neg\theta)$ is derivable. Therefore by definition of Γ^*, there exists an intension variable j such that $@_a\neg\theta[j/i] \in \Gamma^*$. Now, let $g \stackrel{i}{\sim} g^{\Gamma^*}$ such that $g(i) = [j]$. Then by assumption $\mathfrak{M}^{\Gamma^*}, g, [a] \models \theta$ and hence $\mathfrak{M}^{\Gamma^*}, g^{\Gamma^*}, [a] \models \theta[j/i]$, cf. an adapted version of Lemma 6.1. Therefore $@_a\theta[j/i] \in \Gamma^*$ by induction. We conclude that $@_a\neg\forall i\theta \notin \Gamma^*$ and hence $@_a\forall i\theta \in \Gamma^*$, cf. a version of Lemma 6.2 for intensional first-order hybrid logic.

Now completeness.

Theorem 7.3. *(Completeness) Let ϕ be a satisfaction statement and let Γ be a set of satisfaction statements. The second statement below implies the first statement.*

1. *ϕ is derivable from Γ in $\mathbf{N}_{\mathscr{H}(\mathscr{O})}$.*
2. *For any generalized model $\mathfrak{M}$ and any assignment g, if, for any formula $\psi \in \Gamma$, $\mathfrak{M}, g \models \psi$, then $\mathfrak{M}, g \models \phi$.*

Proof. Clearly, we are done if Γ is inconsistent, so assume that Γ is consistent. Now, assume that ψ is not derivable from Γ and let $\psi = @_a\phi$. Then $\Gamma \cup \{@_a\neg\phi\}$ is consistent. Let $\Delta = (\Gamma \cup \{@_a\neg\phi\})^*$ according to Lemma 7.2, and consider the generalized model $\mathfrak{M}^\Delta$ and the assignment g^Δ. By Lemma 7.4, $\mathfrak{M}^\Delta, g^\Delta \models \theta$ for any formula $\theta \in \Gamma$, and also $\mathfrak{M}^\Delta, g^\Delta \models @_a\neg\phi$. But this contradicts with the second statement in the theorem.

7.2.2 Soundness and Completeness: Standard Models

In the preceeding section we proved that the natural deduction system $\mathbf{N}_{\mathscr{H}(\mathscr{O})}$ is sound and complete with respect to generalized models. Now, by negating the first displayed formula in the last paragraph of Section 7.1.1, we obtain a formula of $\mathscr{H}$, namely the formula

$$\forall x \forall y(x = y) \rightarrow \forall i \forall j(i = j),$$

which is valid with respect to standard models, but not valid with respect to generalized models, and therefore not derivable in $\mathbf{N}_{\mathscr{H}}$ since this system is sound. Clearly, the formula is not derivable in $\mathbf{N}_{\mathscr{H}(\mathscr{O})}$ whatever set $\mathscr{O}$ of binders is chosen, hence, no system $\mathbf{N}_{\mathscr{H}(\mathscr{O})}$ is complete with respect to standard models. Another example of a formula which is valid with respect to standard models, but not valid with respect to generalized models, is obtained by negating the second displayed formula in the last paragraph of Section 7.1.1. This is the formula

$$\forall i \forall j(\forall a(@_a i = @_a j) \rightarrow i = j).$$

Note that this formula involves the $\forall$ binder.

A natural deduction system for $\mathscr{H}(\mathscr{O})$ which is sound and complete with respect to standard models can be obtained by adding one of the rules of Figure 7.2 to $\mathbf{N}_{\mathscr{H}(\mathscr{O})}$. The choice of rule depends on whether or not the binder $\forall$ is included in $\mathscr{O}$. In the case where $\forall$ is not included in $\mathscr{O}$, we add $(Ext1)$ to $\mathbf{N}_{\mathscr{H}(\mathscr{O})}$, and in the case where $\forall$ is included in $\mathscr{O}$, we add $(Ext2)$ to $\mathbf{N}_{\mathscr{H}(\mathscr{O})}$. Note that none of these rules are sound with respect to arbitrary generalized models: The rule $(Ext1)$ is only sound with respect to generalized models where $|D_O| = 1$ implies $|D_I| = 1$, and similarly, the rule (*Ext*2) is only sound with respect to generalized models where the function $\mathscr{E}$ is injective. Note also that both of the rules are sound with respect to standard models. Completeness of $\mathbf{N}_{\mathscr{H}(\mathscr{O})}$ extended with (*Ext*1) with respect to standard models is proved via completeness with respect to generalized models where $|D_O| = 1$ implies $|D_I| = 1$, and similarly, completeness of $\mathbf{N}_{\mathscr{H}(\mathscr{O})}$ extended with (*Ext*2) with respect to standard models is proved via completeness with respect to generalized models where $\mathscr{E}$ is injective. The proofs are similar to the proof that $\mathbf{N}_{\mathscr{H}(\mathscr{O})}$ is complete with respect to generalized models, but there are a couple of essential differences which we point out below.

$$\frac{@_b(x = y)}{@_b(i = j)}\ (Ext1)^* \qquad \frac{@_b(@_a i = @_a j)}{@_b(i = j)}\ (Ext2)^\star$$

$*$ x and y do not occur free in any undischarged assumptions.
$\star$ a is different from b and does not occur free in any undischarged assumptions.

Fig. 7.2 Natural deduction rules for standard models

The first essential difference is in the Lindenbaum lemma, Lemma 7.2. Recall that in this lemma, a consistent set of satisfaction statements Γ^0 is extended to a maximal consistent set of satisfaction statements $\Gamma^* = \cup_{n \geq 0} \Gamma^n$ where Γ^{n+1} is defined by induction in terms of a number of clauses. We here add a new clause in the definition of Γ^{n+1}. In the case where $\forall$ is not included in $\mathscr{O}$ (thus, $\mathbf{N}_{\mathscr{H}(\mathscr{O})}$ has been extended with (*Ext*1)) we add the clause

$$\Gamma^n \cup \{\phi_{n+1}, @_b(x \neq y)\} \text{ if } \phi_{n+1} \text{ is of the form } @_b(i \neq j)$$

where x and y are new object variables. This will force $|D_O| = 1$ to imply that $|D_I| = 1$ in the canonical generalized model, Definition 7.5, when the maximal consistent set used in the canonical generalized model is defined in accordance with the Lindenbaum lemma. In the case where $\forall$ is included in $\mathscr{O}$ (thus, $\mathbf{N}_{\mathscr{H}(\mathscr{O})}$ has been extended with (*Ext*2)) we add the clause

$$\Gamma^n \cup \{\phi_{n+1}, @_b(@_a i \neq @_a j)\} \text{ if } \phi_{n+1} \text{ is of the form } @_b(i \neq j)$$

where a is a new nominal. This will force the function $\mathscr{E}^\Delta$ in the canonical generalized model to be injective. The second essential difference is in the completeness theorem, Theorem 7.3. In the case where $\forall$ is not included in $\mathscr{O}$, we build into Theorem 7.3 applications of Proposition 7.1 and Proposition 7.2, where both propositions are generalized to sets of formulas (rather than single formulas). In the case where $\forall$ is included in $\mathscr{O}$, we build into Theorem 7.3 an application of Proposition 7.1, again generalized to sets of formulas. The remaining parts of the completeness proof are essentially the same.

Note: By modifying the completeness proof for $\mathbf{N}_{\mathscr{H}(\forall)}$ extended with $(Ext2)$, it follows that $\mathbf{N}_{\mathscr{H}}$ extended with $(Ext2)$ is complete with respect to standard models. Thus, we have two different natural deduction systems for $\mathscr{H}$ which are sound and complete with respect to standard models: The first system is $\mathbf{N}_{\mathscr{H}}$ extended with $(Ext1)$ and the second system is $\mathbf{N}_{\mathscr{H}}$ extended with $(Ext2)$. At first sight this seems peculiar, but the intuitive reason is that even though the rule $(Ext2)$ semantically is stronger than $(Ext1)$, this difference in semantic expressive power cannot be expressed within the relatively weak object language $\mathscr{H}$, that is, within this language the second system cannot derive more formulas than the first system. A similar remark applies if $\mathscr{H}$ is replaced by $\mathscr{H}(\downarrow)$.

7.3 Partial Intensions

In the models we have considered previously in the present chapter, intensions are total functions from worlds to objects. It is arguable that for some purposes it is more appropriate to take intensions to be partial functions from worlds to objects, see Fitting (2006b). In this section we discuss models for intensional first-order hybrid logic where intensions are allowed to be partial functions.

It is surprising that if the total functions in the definitions of standard and generalized models considered previously are replaced by partial functions, and the interpretations of terms and formulas are adjusted accordingly, then standard and generalized models validate the same formulas not involving the $\forall$ binder. To be more precise, the definition of standard models, Definition 7.1, is adjusted such that D_I is a non-empty set of partial functions from W to D_O and the interpretation of terms is adjusted such that an object term of the form $@_a i$ is assigned the element $t^{\mathfrak{M},g} = g(i)(g(a))$ of D_O just in case $g(a)$ is in the domain of the partial function $g(i)$, otherwise the object term is said to be undefined. The interpretation of formulas is adjusted such that an atomic formula $P(t_1,\ldots,t_n)$ or $t = u$ is false if one of the terms involved is an undefined object term. The definition of generalized models, Definition 7.2, and the associated interpretations of terms and formulas are adjusted analogously.

Given the adjustments described above, it is straightforward to check that Proposition 7.1 still holds. Moreover, Definition 7.3 can be adjusted such that the requirement that $|D_O| > 1$ is removed, which makes it possible to remove the same requirement in Lemma 7.1 and Proposition 7.2. It follows that a formula not involving the binder $\forall$ is satisfiable with respect to partial standard models if and only if it is satisfiable with respect to partial generalized models. Thus, the same formulas are validated as long as the $\forall$ binder is not allowed. However, there is a difference if the binder $\forall$ is allowed, which is witnessed by the formula

$$\forall i \forall j(\forall a(@_a i = @_a j) \rightarrow i = j)$$

we considered earlier in the paper. This formula is valid with respect to partial standard models, but as observed earlier in the paper, it is not valid with respect to generalized models, and hence not with respect to partial generalized models either.

An axiom system for first-order intensional logic, complete with respect to partial standard models, can be found in Fitting (2006b).

Chapter 8
Intuitionistic Hybrid Logic

In this chapter we introduce intuitionistic hybrid logic and its proof-theory. Intuitionistic hybrid logic is hybrid modal logic over an intuitionistic logic basis instead of a classical logical basis. The chapter is structured as follows. In the first section of the chapter we introduce intuitionistic hybrid logic (this is taken from Braüner and de Paiva (2006)). In the second section we introduce a natural deduction system for intuitionistic hybrid logic (taken from Braüner and de Paiva (2006)) and in the third and fourth sections we introduce axiom systems for intuitionistic and paraconsistent hybrid logic (taken from Braüner and de Paiva (2006)). In the last section we discuss certain other work, namely a Curry-Howard interpretation of intuitionistic hybrid logic.

8.1 Introduction to Intuitionistic Hybrid Logic

The formulas of our intuitionistic hybrid logic are the same as those of the classical hybrid logic $\mathcal{H}$ defined in Section 1.2, except that the connectives $\vee$ and $\Diamond$ are taken to be primitive, the reason being that they are not intuitionistically definable in terms of the other connectives (contrary to the classical case). The connectives $\neg$, $\top$, and $\leftrightarrow$ are defined by the conventions that $\neg\phi$ is an abbreviation for $\phi \rightarrow \bot$, $\top$ is an abbreviation for $\neg\bot$, and $\phi \leftrightarrow \psi$ is an abbreviation for $(\phi \rightarrow \psi) \wedge (\psi \rightarrow \phi)$.

Now, as indicated above, intuitionistic hybrid logic is obtained by replacing the classical logic basis of hybrid modal logic by an intuitionistic logic basis. Thus, two logics are combined, namely intuitionistic logic (which by the standard Kripke semantics is interpreted in terms of a set of "states of knowledge" equipped with a partial order, called the "epistemic" partial order, with respect to which the interpretations of propositional symbols are monotone, that is, the interpretations of propositional symbols are preserved by the partial order) and hybrid modal logic (where modal operators, nominals, and satisfaction operators are interpreted in terms of a set of possible worlds equipped with an accessibility relation, cf. the classical semantics described earlier).

T. Braüner, *Hybrid Logic and its Proof-Theory*, Applied Logic Series 37, DOI 10.1007/978-94-007-0002-4_8,

The main intuition behind our combined semantics is that we want to give an intuitionistic reading of hybrid modal logic where a distinction is made between the way of reasoning and what the reasoning is about, that is, we want to reason intuitionistically about time, space, states in a computer, or whatever the subject-matter is. The principle that logical reasoning should not depend on what the reasoning is about is expressed many places in the logical literature; one of them is the following quotation by J.A. Robinson.

> The correctness of a piece of reasoning, ... does not depend on what the reasoning is about (we can see that the conclusion *all epiphorins are turpy* follows from the premises *all epiphorins are febrids* and *all febrids are turpy*, without understanding all the words) so much as on how the reasoning is done; on the pattern of relationships between the various constituent ideas rather than on the actual ideas themselves. (Robinson 1979, p. 1)

Following this principle, we keep the intuitionistic states of knowledge separate from the modal possible worlds (representing times, locations, states in a computer, or something else). Consequently, we keep the epistemic partial order separate from the interpretation of the hybrid-logical machinery as well as the accessibility relation involved in interpreting modal operators. This is contrary to a number of intuitionistic modal logics where the epistemic partial order and the modal accessibility relation are relations on the same set, thus, in these logics the way of reasoning is not kept distinct from what the reasoning is about, see Section 8.1.2.

Our distinction between the way of reasoning and what the reasoning is about explains why the natural deduction system given in the next section has the property that if it is extended with the natural deduction rule corresponding to the excluded middle (technically, we just modify the rule for $\perp$ as appropriate), then we get back all the derivable formulas of the classical natural deduction system $\mathbf{N}_{\mathcal{H}}$ from Section 2.2 and the modal operator $\Diamond$ becomes definable in terms of $\Box$ ($\Diamond$ becomes equivalent to $\neg\Box\neg$) and also $\vee$ becomes definable as usual. So only the excluded middle has to be added to get the classical modality from the intuitionistic modalities, which we think is significant from a philosophical point of view. In this sense the modalities do not have a constructive component of their own. This gives a very transparent relationship between proof-systems for intuitionistic and classical hybrid logic.

In what follows, we give our formal semantics for intuitionistic hybrid logic. This semantics is an extension of a semantics for intuitionistic modal logic which was introduced in a tense-logical version in Ewald (1986). We first define models.

Definition 8.1. A *model* for intuitionistic hybrid logic is a tuple

$$(W, \leq, \{D_w\}_{w\in W}, \{\sim_w\}_{w\in W}, \{R_w\}_{w\in W}, \{V_w\}_{w\in W})$$

where

1. W is a non-empty set partially ordered by $\leq$;
2. for each w, D_w is a non-empty set such that $w \leq v$ implies $D_w \subseteq D_v$;
3. for each w, $\sim_w$ is an equivalence relation on D_w such that $w \leq v$ implies $\sim_w \subseteq \sim_v$;

4. for each w, R_w is a binary relation on D_w such that $w \leq v$ implies $R_w \subseteq R_v$; and
5. for each w, V_w is a function that to each ordinary propositional symbol p assigns a subset of D_w such that $w \leq v$ implies $V_w(p) \subseteq V_v(p)$.

It is assumed that if $d \sim_w d'$, $e \sim_w e'$, and dR_we, then $d'R_we'$, and similarly, if $d \sim_w d'$ and $d \in V_w(p)$, then $d' \in V_w(p)$. The tuple $(W, \leq, \{D_w\}_{w \in W}, \{\sim_w\}_{w \in W}, \{R_w\}_{w \in W})$ is called a *frame* for intuitionistic hybrid logic and the model is said to be *based* on this frame.

As explained above, the elements of the set W are states of knowledge and for any such state w, the set D_w is the set of possible worlds known in the state of knowledge w, the relation $\sim_w$ is the set of known identities between possible worlds, the relation R_w is the set of known relationships between possible worlds, and the set $V_w(p)$ is the set of possible worlds at which p is known to be true. Note that the definition requires that the epistemic partial order $\leq$ preserves all these kinds of knowledge, that is, if an advance to a greater state of knowledge is made, then what is known is preserved.

We now give a simple example of a model for intuitionistic hybrid logic. We first specify that $W = \{w, v, u\}$ and that $\leq$ is the reflexive and transitive closure of the relation $\{(w,v),(v,u)\}$. Thus, w, v, and u are successively greater states of knowledge. It remains to specify what is known at each of the three states of knowledge. To keep things as simple as possible, we ignore the equivalence relation $\sim_w$ and we ignore all other propositional symbols than p. The remaining parts of the example model are specified below where there is one column for each of state of knowledge.

$D_w = \{d, e\}$	$D_v = \{d, e, f\}$	$D_u = \{d, e, f\}$
$R_w = \{(d, e)\}$	$R_v = \{(d, e), (d, f)\}$	$R_u = \{(d, e), (d, f)\}$
$V_w(p) = \{e\}$	$V_v(p) = \{e\}$	$V_u(p) = \{e, f\}$

Note that each column is a notational variant of a model for classical propositional hybrid logic, cf. Definition 1.1 of Section 1.2. Using this observation, we can depict the example model for intuitionistic hybrid logic as

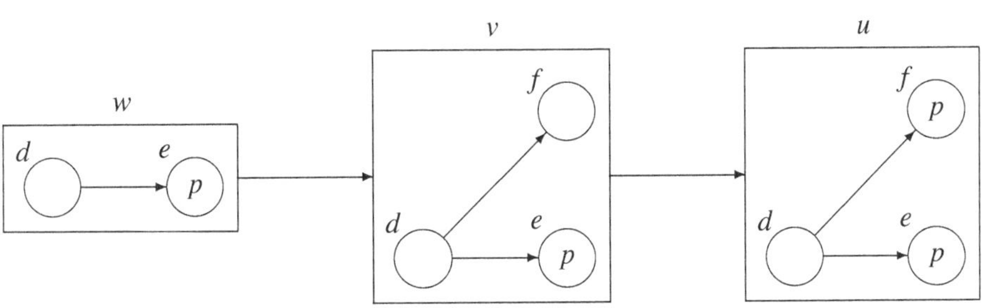

where each state of knowledge is represented by a box containing a model for classical propositional hybrid logic, depicted in the same way as the example model following Definition 1.1 (beware that it is the possible worlds d, e, and f above that are worlds in the sense of Definition 1.1, not the states of knowledge w, v, and u). Two states of knowledge being related by the epistemic partial order is indicated by an arrow, but arrows generated by reflexivity and transitivity are omitted. Inportant remark: If a propositional symbol is absent in a circle representing a possible world, then it means that it is not known whether the propositional symbol is true at the possible world in question (it does not mean that the propositional symbol is false as it does in the classical case). Read from left to right, the depiction above says as follows: In state w it is known that the possible world d has one successor, e, and that p is true at e, in state v it is moreover known that d has a second successor, f, and in state u it is furthermore known that p is true at f.

Given a model $\mathfrak{M} = (W, \leq, \{D_w\}_{w\in W}, \{\sim_w\}_{w\in W}, \{R_w\}_{w\in W}, \{V_w\}_{w\in W})$ and an element w of W, a *w-assignment* is a function g that to each nominal assigns an element of D_w. Given a w-assignment g', $g' \overset{a}{\sim} g$ means that g' agrees with g on all nominals save possibly a. Note that if g is an *w-assignment* and $w \leq v$, then g is also a *v-assignment* (this is used in the clauses below for implication and the $\Box$ operator). The relation $\mathfrak{M}, g, w, d \models \phi$ is defined by induction, where w is an element of W, g is a w-assignment, d is an element of D_w, and ϕ is a formula.

$$
\begin{array}{rl}
\mathfrak{M}, g, w, d \models p & \text{iff } d \in V_w(p) \\
\mathfrak{M}, g, w, d \models a & \text{iff } d \sim_w g(a) \\
\mathfrak{M}, g, w, d \models \phi \wedge \psi & \text{iff } \mathfrak{M}, g, w, d \models \phi \text{ and } \mathfrak{M}, g, w, d \models \psi \\
\mathfrak{M}, g, w, d \models \phi \vee \psi & \text{iff } \mathfrak{M}, g, w, d \models \phi \text{ or } \mathfrak{M}, g, w, d \models \psi \\
\mathfrak{M}, g, w, d \models \phi \rightarrow \psi & \text{iff for all } v \geq w, \\
& \quad \mathfrak{M}, g, v, d \models \phi \text{ implies } \mathfrak{M}, g, v, d \models \psi \\
\mathfrak{M}, g, w, d \models \bot & \text{iff falsum} \\
\mathfrak{M}, g, w, d \models \Box\phi & \text{iff for all } v \geq w, \text{ for all } e \in D_v, \\
& \quad dR_v e \text{ implies } \mathfrak{M}, g, v, e \models \phi \\
\mathfrak{M}, g, w, d \models \Diamond\phi & \text{iff for some } e \in D_w,\ dR_w e \text{ and } \mathfrak{M}, g, w, e \models \phi \\
\mathfrak{M}, g, w, d \models @_a\phi & \text{iff } \mathfrak{M}, g, w, g(a) \models \phi
\end{array}
$$

By convention $\mathfrak{M}, g, w \models \phi$ means $\mathfrak{M}, g, w, d \models \phi$ for every element d of D_w and $\mathfrak{M} \models \phi$ means $\mathfrak{M}, g, w \models \phi$ for every element w of W and every w-assignment g. A formula ϕ is *valid* in a frame if and only if $\mathfrak{M} \models \phi$ for any model $\mathfrak{M}$ that is based on the frame. A formula ϕ is *valid* in a class of frames if and only if ϕ is valid in every frame in the class of frames in question. A formula ϕ is *valid* if and only if ϕ is valid in the class of all frames.

We let $\mathcal{H}^{\mathcal{I}}$ denote intuitionistic hybrid logic. Note the difference in the interpretations of the two modal operators: The interpretation of the $\Box$ operator involves quantification over accessible states of knowledge whereas the interpretation of $\Diamond$ does not. This is the case since the modal operators correspond to quantifiers in intuitionistic first-order logic where the interpretation of the $\forall$ quantifier involves the accessibility relation whereas the interpretation of $\exists$ does not, see Section 8.1.3.

An example of a formula valid in the classical hybrid-logical semantics, but not valid in the intuitionistic semantics, is $@_a b \vee @_a \neg b$. The fact that this formula is not intuitionistically valid is not a surprise since it corresponds to the formula $a = b \vee \neg a = b$ in the first-order correspondence language we introduce below, and in intuitionistic first-order logic we do not have a general excluded middle. (Incidentally, this formula *would* be valid if for any w, the relation $\sim_w$ is taken to be the identity on the set D_w. Thus, the relation $\sim_w$ is needed.)

Another example is the formula $@_a\phi \leftrightarrow \neg @_a \neg \phi$. This formula should not be valid intuitionistically, but it should be valid classically and it is actually taken as an axiom for classical hybrid logic by some authors.

The semantics satisfy the following important propositions.

Proposition 8.1. *(Monotonicity) If* $\mathfrak{M}, g, w, d \models \phi$ *and* $w \leq v$, *then* $\mathfrak{M}, g, v, d \models \phi$.

Proof. Induction on the structure of ϕ.

Proposition 8.2. *If* $\mathfrak{M}, g, w, d \models \phi$ *and* $d \sim_w d'$, *then* $\mathfrak{M}, g, w, d' \models \phi$.

Proof. Induction on the structure of ϕ.

8.1.1 Relation to Many-Valued Semantics

The intuitionistic semantics for hybrid logic given above is related to the many-valued semantics for hybrid logic given in Hansen et al. (2008). In that paper the two-valued basis of classical hybrid logic is generalized to a many-valued basis involving a truth-value space having the structure of a finite Heyting algebra. A notable feature of the many-valued semantics is that it allows formulas as well as the accessibility relation to take on many truth-values, that is, the many-valued interpretations of the modal operators $\Box$ and $\Diamond$ generalizes the classical two-valued interpretation, making use of the many-valued accessibility relation.

To be more specific, let $\mathscr{T}$ denote a fixed finite Heyting algebra. As part of this, $\mathscr{T}$ has join and meet operations (denoted $\sqcup$ and $\sqcap$), and also, it has smallest and largest elements (denoted $\bot$ and $\top$). Moreover, for any elements y and z of $\mathscr{T}$, there is a greatest element x of $\mathscr{T}$ satisfying $y \sqcap x \leq z$. The element x is the relative pseudo-complement of y with respect to z (denoted $y \Rightarrow z$). A many-valued model for hybrid logic is then a tuple (W, R, V), where W is a set (the worlds), R is a function from $W \times W$ to $\mathscr{T}$ (the many-valued accessibility relation), and V is a function from $W \times \{p, q, r, \ldots\}$ to $\mathscr{T}$. As in the classical two-valued case, an assignment is a function g from $\{a, b, c, \ldots\}$ to W. The function V is inductively extended to all formulas as follows.

$$
\begin{aligned}
V(w,a) &= \begin{cases} \top \text{ if } g(a) = w \\ \bot \text{ else} \end{cases} \\
V(w, \phi \wedge \psi) &= V(w,\phi) \sqcap V(w,\psi) \\
V(w, \phi \vee \psi) &= V(w,\phi) \sqcup V(w,\psi) \\
V(w, \phi \rightarrow \psi) &= V(w,\phi) \Rightarrow V(w,\psi) \\
V(w, \bot) &= \bot \\
V(w, \Box\phi) &= \sqcap\{R(w,v) \Rightarrow V(v,\phi) \mid v \in W\} \\
V(w, \Diamond\phi) &= \sqcup\{R(w,v) \sqcap V(v,\phi) \mid v \in W\} \\
V(w, @_a\phi) &= V(g(a), \phi)
\end{aligned}
$$

A formula ϕ is valid if and only if $V(w,\phi) = \top$ for any model (W,R,V) and any assignment g. Note that if the fixed finite Heyting algebra $\mathscr{T}$ is the two-valued Heyting algebra containing only the elements $\bot$ and $\top$, then validity not surprisingly coincides with the classical notion of validity considered in the first section of the present chapter. It is straightforward to check that if a formula is valid with respect to the many-valued semantics, whatever fixed finite Heyting algebra $\mathscr{T}$ is considered, then it is also valid with respect to the classical two-valued semantics.

Note that the many-valued semantics assigns to a nominal the truth-value $\top$ in exactly one world, and $\bot$ in all other worlds. This is in agreement with the classical two-valued semantics for hybrid logic in which a nominal refers to a unique world.

Now, if the notion of an intuitionistic model in Definition 8.1 above is restricted such that W is finite, D_w is constant, that is, the same for any w, and $\sim_w$ is the identity relation, then an intuitionistic model, which accordingly can be written as $(W,\leq,D,\{R_w\}_{w\in W},\{V_w\}_{w\in W})$, corresponds to a many-valued model as follows. We first note that the $\leq$-closed subsets of W ordered by inclusion constitute a finite Heyting algebra, which we take to be the space $\mathscr{T}$ of truth-values. A many-valued model (D,R^*,V^*) is then defined by letting

- $R^*(d,e) = \{w \in W \mid dR_we\}$ and
- $V^*(d,p) = \{w \in W \mid d \in V_w(p)\}$.

It can be proved that for any formula ϕ, $V^*(d,\phi) = \{w \in W \mid \mathfrak{M},g,w,d \models \phi\}$. Thus, with the mentioned restriction on the notion of an intuitionistic model, the intuitionistic semantics can be simulated by the many-valued semantics.

There is also a correspondence in the opposite direction, enabling the many-valued semantics to be simulated by the intuitionistic semantics. Given a finite Heyting algebra $\mathscr{T}$ and a many-valued model (D,R,V), a restricted intuitionistic model $\mathfrak{M} = (W,\subseteq,D,\{R^*_w\}_{w\in W},\{V^*_w\}_{w\in W})$ can be defined by letting

- $W = \{w \mid w$ is a proper prime filter in $\mathscr{T}\}$,
- dR^*_we if and only if $R(d,e) \in w$, and
- $d \in V^*_w(p)$ if and only if $V(d,p) \in w$.

It can be proved that for any formula ϕ, $\mathfrak{M},g,w,d \models \phi$ if and only if $V(d,\phi) \in w$. Thus, with the above restriction on intuitionistic models, the intuitionistic semantics is equivalent to the many-valued semantics for hybrid logic.

It is an open question how the many-valued and the intuitionistic semantics are related if the restriction on intuitionistic models is removed.[1] The above described equivalence between the two semantics for hybrid logic is taken from Hansen et al. (2008). It is an extension of a similar equivalence between an intuitionistic and a many-valued semantics for ordinary modal logic which originally was given in Fitting (1992b). In the latter paper, the epistemic worlds of the semantics are thought of as experts and the epistemic partial order is thought of as a relation of dominance

[1] The fact that in the intuitionistic semantics based on Definition 8.1, nominals are interpreted using a family $\{\sim_w\}_{w\in W}$ of equivalence relations, not identity, seems to imply that in an equivalent many-valued semantics, nominals should be allowed to take on arbitrary truth-values, not just top and bottom.

between experts: One expert dominates another one if whatever the first expert says is true is also said to be true by the second expert.

The above implies that if a formula is valid with respect to the intuitionistic semantics for hybrid logic, then it is also valid with respect to the many-valued semantics, whatever finite Heyting algebra is chosen as the fixed truth-value space (as the many-valued semantics can be simulated by the intuitionistic semantics in the above sense and the set $\{\top\}$ is a proper prime filter). Since intuitionistic validity implies many-valued validity, and many-valued validity implies classical validity, many-valued hybrid logics are logics between classical hybrid logic and intuitionistic hybrid logic. This is similar to the fuzzy hybrid logics presented in Galmiche and Salhi (to appear) which constitute a linearly ordered set of logics between classical hybrid logic and intuitionistic hybrid logic.

8.1.2 Relation to Birelational Semantics

In the intuitionistic semantics for hybrid logic considered hithertho in the present chapter, the epistemic partial order is separate from the interpretation of the hybrid-logical machinery as well as the accessibility relation involved in interpreting modal operators. This semantics is an extension of a semantics for intuitionistic modal logic which has been considered in a number of places, in particular Ewald (1986), Simpson (1994), and Gabbay et al. (2003). This semantics, which we shall refer to as the Kripke semantics for intuitionistic modal logic, is different from a number of semantics for intuitionistic modal logics where the epistemic partial order and the modal accessibility relation are relations on the same set. Such semantics are called *birelational* semantics.

It is a remarkable fact that there is a birelational semantics that validates exactly the same modal-logical formulas as the Kripke semantics. This applies to a number of ordinary intuitionistic modal logics, including intuitionistic K and intuitionistic $\mathsf{S5}$, see the accounts given in Simpson (1994) and Gabbay et al. (2003). Note that the Kripke semantics for hybrid logic considered in the present chapter is a hybrid-logical extension of K.

It is also a remarkable fact that the finite model property (if a formula is falsifiable, then it is falsifiable by a finite model) is not satisfied relative to the Kripke semantics, but it is satisfied relative to the birelational semantics. Thus, via the finite model property, the birelational semantics can serve as a vehicle for proving decidability. This technique of proving that an intuitionistic modal logic is decidable has been used several places, see in particular Simpson (1994) where it is used in connection with a number of ordinary intuitionistic modal logics, including intuitionistic K and intuitionistic $\mathsf{S5}$. See also Gabbay et al. (2003) where the proof of the finite model property is somewhat different.

A simple counterexample to the finite model property relative to the Kripke semantics is the formula $\Box\neg\neg\phi \rightarrow \neg\neg\Box\phi$. This formula is only falsifiable by infinite Kripke models, whatever intuitionistic K, intuitionistic $\mathsf{S5}$, or a number of other intuitionistic modal logics are considered, cf. Simpson (1994). The intuitive reason why the finite model property holds for birelational models, but fails for Kripke models, is that there are more birelational models than Kripke models, in the sense that every Kripke model can in a truth-preserving way be "encoded" as a birelational model, but not vice versa, cf. Simpson (1994). The origin of the counterexample above is Ono and Suzuki (1988).

Now, back to hybrid logic. In Chadha et al. (2006), a birelational semantics for intuitionistic S5 has been extended with satisfaction operators, and it has been proved that the birelational semantics satisfies the finite model property, from which it follows that intuitionistic S5 with satisfaction operators is decidable. See Section 8.5 for the background of this work. It is not clear how to add nominals to the birelational semantics, the problem being that if nominals are given their obvious interpretation, namely singleton sets, then the interpretation of nominals cannot be preserved by the partial order, thus, monotonicity is violated.

8.1.3 Translation into Intuitionistic First-Order Logic

Intuitionistic hybrid logic can be translated into intuitionistic first-order logic with equality. The first-order language under consideration is the same as the language considered in Section 1.2.1 in connection with classical first-order logic, except that the connectives $\vee$ and $\exists$ are taken to be primitive since they are not intuitionistically definable in terms of the other connectives. The connectives $\neg$, $\top$, and $\leftrightarrow$ are defined as in intuitionistic hybrid logic.

As in the classical case, the translations ST_a and ST_b are defined by mutual induction.

$$\begin{aligned}
ST_a(p) &= p^*(a) \\
ST_a(c) &= a = c \\
ST_a(\phi \wedge \psi) &= ST_a(\phi) \wedge ST_a(\psi) \\
ST_a(\phi \vee \psi) &= ST_a(\phi) \vee ST_a(\psi) \\
ST_a(\phi \rightarrow \psi) &= ST_a(\phi) \rightarrow ST_a(\psi) \\
ST_a(\bot) &= \bot \\
ST_a(\Box\phi) &= \forall b(R(a,b) \rightarrow ST_b(\phi)) \\
ST_a(\Diamond\phi) &= \exists b(R(a,b) \wedge ST_b(\phi)) \\
ST_a(@_c\phi) &= ST_a(\phi)[c/a]
\end{aligned}$$

The definition of ST_b is obtained by exchanging a and b. Note that the translation above is identical to the extended standard translation from classical hybrid logic into classical first-order logic with equality that we have considered in Section 1.2.1, except that clauses for the new connectives $\vee$ and $\Diamond$ are added. However, it is important to note that the formulas involved in the translations are interpreted differently, that is, in the classical case they are interpreted classically whereas in the intuitionistic case they are interpreted intuitionistically.

A model for intuitionistic hybrid logic can be considered as a model for intuitionistic first-order logic with equality and vice versa.

Definition 8.2. Let a model $\mathfrak{M} = (W, \leq, \{D_w\}_{w\in W}, \{\sim_w\}_{w\in W}, \{R_w\}_{w\in W}, \{V_w\}_{w\in W})$ for intuitionistic hybrid logic be given. A model for intuitionistic first-order logic with equality $\mathfrak{M}^* = (W, \leq, \{D_w\}_{w\in W}, \{V_w^*\}_{w\in W})$ is defined by letting

- $V_w^*(p^*) = V_w(p)$,
- $V_w^*(R) = R_w$, and
- $V_w^*(=) = \sim_w$.

The map $(\cdot)^*$ is bijective. See Simpson (1994) for the simpler correspondence between modal-logical models and intuitionistic first-order models without equality. The definition of a model for intuitionistic first-order logic with equality can be found in Troelstra and van Dalen (1988), but it should be straightforward to read it off from the definition above.[2] Moreover, if nominals of hybrid logic are identified with first-order variables, then a w-assignment in the sense of intuitionistic hybrid logic can be considered as a w-assignment in the sense of intuitionistic first-order logic and vice versa.

Given a model $\mathfrak{M} = (W, \leq, \{D_w\}_{w\in W}, \{V_w\}_{w\in W})$ for intuitionistic first-order logic, the relation $\mathfrak{M}, g, w \models \phi$ is defined by induction, where w is an element of W, g is a w-assignment, and ϕ is a first-order formula.

$$
\begin{aligned}
\mathfrak{M}, g, w \models p^*(a) &\text{ iff } g(a) \in V_w(p^*)\\
\mathfrak{M}, g, w \models R(a,b) &\text{ iff } g(a)V_w(R)g(b)\\
\mathfrak{M}, g, w \models a = b &\text{ iff } g(a)V_w(=)g(b)\\
\mathfrak{M}, g, w \models \phi \wedge \psi &\text{ iff } \mathfrak{M}, g, w \models \phi \text{ and } \mathfrak{M}, g, w \models \psi\\
\mathfrak{M}, g, w \models \phi \vee \psi &\text{ iff } \mathfrak{M}, g, w \models \phi \text{ or } \mathfrak{M}, g, w \models \psi\\
\mathfrak{M}, g, w \models \phi \to \psi &\text{ iff for all } v \geq w,\\
&\qquad \mathfrak{M}, g, v \models \phi \text{ implies } \mathfrak{M}, g, v \models \psi\\
\mathfrak{M}, g, w \models \bot &\text{ iff falsum}\\
\mathfrak{M}, g, w \models \forall a\phi &\text{ iff for all } v \geq w, \text{ for all } g' \overset{a}{\sim} g,\\
&\qquad g'(a) \in D_v \text{ implies } \mathfrak{M}, g, v \models \phi[g']\\
\mathfrak{M}, g, w \models \exists a\phi &\text{ iff for some } g' \overset{a}{\sim} g,\ g'(a) \in D_w \text{ and } \mathfrak{M}, g, w \models \phi[g']
\end{aligned}
$$

By convention $\mathfrak{M} \models \phi$ means $\mathfrak{M}, g, w \models \phi$ for every element w of W and every w-assignment g. We are now ready to state formally that truth is preserved in the sense of the proposition below.

Proposition 8.3. *Let $\mathfrak{M}$ be a model for intuitionistic hybrid logic and let ϕ be a hybrid-logical formula in which the nominals a and b do not occur. For any element*

[2] In the model for intuitionistic hybrid logic, and in the model for intuitionistic first-order logic as well, there is a set D_w for each state of knowledge w (subject to the monotonicity requirement that $D_w \subseteq D_v$ whenever $w \leq v$). One might instead consider having a constant set like in the constant domain semantics for first-order modal logic, cf. Section 6.1.1. In Görnemann (1971) such a constant semantics for intuitionistic first-order logic was axiomatized by adding to an axiom system for the usual semantics the axiom $\forall a(\phi \vee \psi) \to (\phi \vee \forall a\psi)$, where the variable a does not occur in ϕ. It would be interesting to investigate whether such a constant version of intuitionistic hybrid logic could be given (whatever proof-theoretic machinery is chosen, it is not clear whether the completeness proof of Section 8.2.3 can be modified as appropriate).

w of W and any w-assignment g, $\mathfrak{M}, g, w, g(a) \models \phi$ *if and only if* $\mathfrak{M}^*, g, w \models ST_a(\phi)$ *(and the same for* ST_b*).*

Proof. Induction on the structure of ϕ.

Note that in the proposition above the formulas are interpreted intuitionistically, not classically. Thus, our semantics for intuitionistic hybrid logic can be considered as obtained by changing the interpretation of the first-order metalanguage from classical to intuitionistic. The introduction of ordinary intuitionistic modal logics via the standard translation has been considered a number of places, see in particular Gabbay et al. (2003), which concentrates on intuitionistic versions of K and S5, denoted respectively **FS** and **MIPC**. This idea can be traced back to Bull (1966) which considers **MIPC**, that is, intuitionistic S5.

Also following this idea, de Paiva (2006) introduces an intuitionistic version of the description logic $\mathcal{ALC}$ via the standard translation to intuitionistic first-order logic. Bozzato et al. (2009) also follow this recipe, but in this paper intuitionistic first-order logic is extended with what is called the Kuroda Principle $\forall a \neg\neg\phi \rightarrow \neg\neg\forall a\phi$. In the modal semantics this principle corresponds to any epistemic state having at least one finite element in its future, where a finite element is an element with no elements in its future. It is in Bozzato et al. (2009) conjectured that the Kuroda Principle imply that the resulting modal logic satisfies the finite model property (but this is not proved in the paper).

8.2 Natural Deduction for Intuitionistic Hybrid Logic

In this section we shall give a natural deduction system for the intuitionistic hybrid logic $\mathcal{H}^{\mathcal{I}}$. See Girard et al. (1989) for an introduction to natural deduction with a slant towards intuitionistic logic. The derivation rules for the system are given in Figures 8.1 and 8.2. All formulas in the rules are satisfaction statements. The system will be denoted $\mathbf{N}_{\mathcal{H}\mathcal{I}}$. The system $\mathbf{N}_{\mathcal{H}\mathcal{I}}$ is an intuitionistic version of the natural deduction system for classical hybrid logic $\mathbf{N}_{\mathcal{H}}$ given in Figures 2.2 and 2.3 of Section 2.2. The intuitionistic system can be seen as obtained from the classical system by deleting the rules $(\bot 1)$ and $(\bot 2)$ and instead adding the rule $(\bot E)$ as well as rules for the connectives $\vee$ and $\Diamond$.

8.2.1 Conditions on the Accessibility Relation

In what follows we shall consider natural deduction systems obtained by extending $\mathbf{N}_{\mathcal{H}\mathcal{I}}$ with additional derivation rules corresponding to first-order conditions on the accessibility relation. A first-order formula is *geometric* if it is built out of atomic formulas of the form $R(a,c)$ and $a = c$ using only the connectives $\bot$, $\wedge$, $\vee$, and $\exists$. In what follows, the metavariables S_k and S_{jk} range over atomic formulas of the

$$\frac{@_a\phi \qquad @_a\psi}{@_a(\phi \wedge \psi)}\,(\wedge I) \qquad \frac{@_a(\phi \wedge \psi)}{@_a\phi}\,(\wedge E1) \qquad \frac{@_a(\phi \wedge \psi)}{@_a\psi}\,(\wedge E2)$$

$$\frac{@_a\phi}{@_a(\phi \vee \psi)}\,(\vee I1) \qquad \frac{@_a\psi}{@_a(\phi \vee \psi)}\,(\vee I2) \qquad \frac{@_a(\phi \vee \psi) \qquad \begin{array}{c}[@_a\phi]\\ \vdots \\ \theta\end{array} \qquad \begin{array}{c}[@_a\psi]\\ \vdots \\ \theta\end{array}}{\theta}\,(\vee E)$$

$$\frac{\begin{array}{c}[@_a\phi]\\ \vdots \\ @_a\psi\end{array}}{@_a(\phi \rightarrow \psi)}\,(\rightarrow I) \qquad \frac{@_a(\phi \rightarrow \psi) \qquad @_a\phi}{@_a\psi}\,(\rightarrow E)$$

$$\frac{@_a\bot}{\theta}\,(\bot E)$$

$$\frac{@_a\phi}{@_c@_a\phi}\,(@I) \qquad \frac{@_c@_a\phi}{@_a\phi}\,(@E)$$

$$\frac{@_e\phi \qquad @_a\Diamond e}{@_a\Diamond\phi}\,(\Diamond I) \qquad \frac{@_a\Diamond\phi \qquad \begin{array}{c}[@_c\phi]\ [@_a\Diamond c]\\ \vdots \\ \theta\end{array}}{\theta}\,(\Diamond E)^*$$

$$\frac{\begin{array}{c}[@_a\Diamond c]\\ \vdots \\ @_c\phi\end{array}}{@_a\Box\phi}\,(\Box I)^\star \qquad \frac{@_a\Box\phi \qquad @_a\Diamond e}{@_e\phi}\,(\Box E)$$

$*$ c does not occur in $@_a\Diamond\phi$, in θ, or in any undischarged assumptions other than the specified occurrences of $@_c\phi$ and $@_a\Diamond c$.
$\star$ c does not occur in $@_a\Box\phi$ or in any undischarged assumptions other than the specified occurrences of $@_a\Diamond c$.

Fig. 8.1 Natural deduction rules: Connectives

$$\frac{}{@_a a}\,(Ref) \qquad \frac{@_a c \qquad @_a\phi}{@_c\phi}\,(Nom1)^* \qquad \frac{@_a c \qquad @_a\Diamond b}{@_c\Diamond b}\,(Nom2)$$

$*$ ϕ is a propositional symbol (ordinary or a nominal).

Fig. 8.2 Natural deduction rules: Nominals

mentioned forms. Atomic formulas of the mentioned forms can be translated into hybrid logic in a truth-preserving way as follows.

$$HT(R(a,c)) = @_a\Diamond c$$
$$HT(a = c) = @_a c$$

A *geometric theory* is a finite set of closed first-order formulas, each having the form $\forall\overline{a}(\phi \rightarrow \psi)$ where the formulas ϕ and ψ are geometric, $\overline{a}$ is a list $a_1,\dots,a_l$ of first-order variables, and $\forall\overline{a}$ is an abbreviation for $\forall a_1 \dots \forall a_l$. It can be proved, cf. Simpson (1994), that any geometric theory is intuitionistically equivalent to a *basic geometric theory* which is a geometric theory in which each formula has the form

$$(*) \qquad \forall\overline{a}((S_1 \wedge \dots \wedge S_n) \rightarrow \exists\overline{c} \bigvee_{j=1}^{m} (S_{j1} \wedge \dots \wedge S_{jn_j}))$$

where $n,m \geq 0$ and $n_1,\dots,n_m \geq 1$. For simplicity, we assume that the variables in the list $\overline{a}$ are pairwise distinct, that the variables in $\overline{c}$ are pairwise distinct, and that no variable occurs in both $\overline{a}$ and $\overline{c}$.

We now give hybrid natural deduction rules corresponding to a basic geometric theory. The metavariables s_k and s_{jk} range over hybrid-logical formulas of the forms $@_a\Diamond c$ and $@_a c$. With a first-order formula θ of the form $(*)$ displayed above, we associate the natural deduction derivation rule (R_θ) given in Figure 8.3 where s_k is of the form $HT(S_k)$ and s_{jk} is of the form $HT(S_{jk})$. Now, let $\mathbf{T}$ be any basic geometric theory. The natural deduction system obtained by extending $\mathbf{N}_{\mathscr{H}\mathscr{I}}$ with the set of rules $\{(R_\theta) \mid \theta \in \mathbf{T}\}$ will be denoted $\mathbf{N}_{\mathscr{H}\mathscr{I}} + \mathbf{T}$.

$$\dfrac{s_1 \quad \dots \quad s_n \qquad \begin{matrix}[s_{11}]\dots[s_{1n_1}]\\ \vdots \\ \phi\end{matrix} \qquad \dots \qquad \begin{matrix}[s_{m1}]\dots[s_{mn_m}]\\ \vdots \\ \phi\end{matrix}}{\phi}\,(R_\theta)^*$$

$*$ None of the nominals in $\overline{c}$ occur in ϕ or in any of the undischarged assumptions other than the specified occurrences of s_{jk}. (Recall that nominals are identified with first-order variables and that $\overline{c}$ are the first-order variables existentially quantified over in the formula θ.)

Fig. 8.3 Natural deduction rules: Geometric theories

It is straightforward to check that if a formula in a basic geometric theory is a Horn clause, then the rule (R_θ) given in Figure 8.3 can be replaced by a simpler rule, exactly as in the classical case described in Section 2.2.1.

8.2.2 An Admissible Rule

Below we state a small proposition regarding an admissible rule. Recall from Section 2.2.2 that the degree of a formula is the number of occurrences of non-nullary connectives in it.

Proposition 8.4. *The rule*

$$\frac{@_a c \qquad @_a \phi}{@_c \phi} \; (Nom)$$

is admissible.

Proof. The proof makes use of the notion of degree, cf. above, and is a straightforward extension of the proof in the propositional cases, see Proposition 2.1 in Section 2.2.2.

8.2.3 Soundness and Completeness

Having given the Kripke semantics and the natural deduction system, we are now ready to prove soundness and completeness.

A model $\mathfrak{M}$ for intuitionistic hybrid logic is called a **T**-*model* if and only if $\mathfrak{M}^* \models \theta$ for every formula θ in **T** (recall that $\mathfrak{M}^*$ is the model for intuitionistic first-order logic corresponding to $\mathfrak{M}$). Compare to the definition of a classical **T**-model after Lemma 2.1 in Section 2.2.3 and note that if the natural deduction system is classical, then the condition imposed on the accessibility relation is the classical interpretation of the geometric theory **T**, and if the natural deduction system is intuitionistic, then the condition imposed is the intuitionistic interpretation of **T**, that is, **T** is interpreted as a statement in intuitionistic first-order logic. Thus, the condition on the accessibility relation in the intuitionistic proof-system can be seen as obtained by changing the interpretation of **T** from classical to intuitionistic (see also the remarks concluding Section 8.1.3).

Theorem 8.1. *(Soundness) Let ψ be a satisfaction statement and let Γ be a set of satisfaction statements. The first statement below implies the second statement.*

1. *ψ is derivable from Γ in $\mathbf{N}_{\mathcal{HI}} + \mathbf{T}$.*
2. *For any **T**-model $\mathfrak{M}$, any element w of W, and any w-assignment g, if, for any formula $\theta \in \Gamma$, $\mathfrak{M}, g, w \models \theta$, then $\mathfrak{M}, g, w \models \psi$.*

Proof. Induction on the structure of the derivation of ψ where we make use of Proposition 8.1.

In what follows, we shall give a Henkin-type proof of completeness. In the interest of simplicity, we shall often omit reference to the basic geometric theory **T** and to the natural deduction system $\mathbf{N}_{\mathcal{HI}} + \mathbf{T}$.

Definition 8.3. A set of satisfaction statements Γ is *inconsistent* if and only if $@_a\bot$ is derivable from Γ for some nominal a and Γ is *consistent* if and only if Γ is not inconsistent.

Let $\mathbf{C}$ be any countably infinite set of nominals and let $\mathtt{L}(\mathbf{C})$ denote the set of hybrid-logical formulas built using the nominals in $\mathbf{C}$. Moreover, let $\mathbf{C}_0$ denote the set of nominals in the language defined earlier, thus, $\mathtt{L}(\mathbf{C}_0)$ denotes the language we have considered hitherto.

Definition 8.4. Let $\mathbf{C}$ and $\mathbf{E}$ be disjoint countably infinite sets of nominals. A set of satisfaction statements Γ in the language $\mathtt{L}(\mathbf{C} \cup \mathbf{E})$ is $\mathbf{E}$-saturated if and only if

1. Γ is consistent;
2. if ϕ is derivable from Γ, then $\phi \in \Gamma$;
3. if $@_a(\phi \vee \psi) \in \Gamma$, then $@_a\phi \in \Gamma$ or $@_a\psi \in \Gamma$;
4. if $@_a\Diamond\phi \in \Gamma$, then for some nominal e in $\mathbf{E}$, $@_e\phi \in \Gamma$ and $@_a\Diamond e \in \Gamma$; and
5. if for some formula $\forall\overline{a}((S_1 \wedge \ldots \wedge S_n) \rightarrow \exists\overline{c} \bigvee_{j=1}^m (S_{j1} \wedge \ldots \wedge S_{jn_j}))$ in $\mathbf{T}$ where $m \geq 1$, it is the case that $@_e(s_1 \wedge \ldots \wedge s_n)[\overline{d}/\overline{a}] \in \Gamma$, then for some list $\overline{b}$ of nominals in $\mathbf{E}$, $@_e \bigvee_{j=1}^m (s_{j1} \wedge \ldots \wedge s_{jn_j})[\overline{d},\overline{b}/\overline{a},\overline{c}] \in \Gamma$.

Recall from Section 8.2.1 that the metavariables S_k range over atomic first-order formulas of the forms $R(a,c)$ and $a = c$, and that s_k is the hybrid-logical translation of any such formula S_k, that is, $s_k = HT(S_k)$. An analogous remark applies to S_{jk}. Clause 5 in the definition above ensures that the canonical model of Definition 8.5 is a $\mathbf{T}$-model, cf. Lemma 8.3, where $\mathbf{T}$ is the fixed basic geometric theory. We are now ready for a saturation lemma.

Lemma 8.1. *(Saturation lemma) Let* $\mathbf{C}$ *and* $\mathbf{E}$ *be disjoint countably infinite sets of nominals and let* $\phi_1, \phi_2, \phi_3, \ldots$ *be an enumeration of all satisfaction statements in* $\mathtt{L}(\mathbf{C} \cup \mathbf{E})$. *Let* Γ *be a set of satisfaction statements in* $\mathtt{L}(\mathbf{C})$ *and let* ψ *be a satisfaction statement in* $\mathtt{L}(\mathbf{C})$ *such that* ψ *is not derivable from* Γ. *An* $\mathbf{E}$*-saturated set of satisfaction statements* $\Gamma^* \supseteq \Gamma$ *from which* ψ *is not derivable is defined as follows. Firstly,* Γ^0 *is defined to be* Γ. *Secondly,* Γ^{n+1} *is defined by induction. If* ψ *is derivable from* $\Gamma^n \cup \{\phi_{n+1}\}$, *then* Γ^{n+1} *is defined to be* Γ^n. *Otherwise* Γ^{n+1} *is defined to be*

1. $\Gamma^n \cup \{\phi_{n+1}, @_a\theta\}$ *if* ϕ_{n+1} *is of the form* $@_a(\theta \vee \chi)$ *and* ψ *is not derivable from* $\Gamma^n \cup \{\phi_{n+1}, @_a\theta\}$;
2. $\Gamma^n \cup \{\phi_{n+1}, @_a\chi\}$ *if* ϕ_{n+1} *is of the form* $@_a(\theta \vee \chi)$ *and the first clause does not apply;*
3. $\Gamma^n \cup \{\phi_{n+1}, @_e\theta, @_a\Diamond e\}$ *if* ϕ_{n+1} *is of the form* $@_a\Diamond\theta$;
4. $\Gamma^n \cup \{\phi_{n+1}, @_e \bigvee_{j=1}^m (s_{j1} \wedge \ldots \wedge s_{jn_j})[\overline{d},\overline{b}/\overline{a},\overline{c}]\}$ *if there exists a formula in* $\mathbf{T}$ *of the form* $\forall\overline{a}((S_1 \wedge \ldots \wedge S_n) \rightarrow \exists\overline{c} \bigvee_{j=1}^m (S_{j1} \wedge \ldots \wedge S_{jn_j}))$ *such that* $m \geq 1$ *and* $\phi_{n+1} = @_e(s_1 \wedge \ldots \wedge s_n)[\overline{d}/\overline{a}]$ *for some nominals* $\overline{d}$ *and* e; *and*
5. $\Gamma^n \cup \{\phi_{n+1}\}$ *if none of the first four clauses apply.*

In clause 3, e *is a nominal in* $\mathbf{E}$ *that does not occur in* Γ^n *or* ϕ_{n+1}, *and similarly, in clause 4,* $\overline{b}$ *is a list of nominals in* $\mathbf{E}$ *such that none of the nominals in* b *occur in* Γ^n *or* ϕ_{n+1}. *Finally,* Γ^* *is defined to be* $\cup_{n\geq 0}\Gamma^n$.

Proof. Firstly, ψ is not derivable from Γ^0 by definition. Secondly, to check that the non-derivability of ψ from Γ^n implies the non-derivability of ψ from Γ^{n+1}, we need to check each of the clauses in the definition of Γ^{n+1}. The first and fifth clauses are trivial. For the second clause, the derivability of ψ from $\Gamma^n \cup \{\phi_{n+1}, @_a\chi\}$ implies the derivability of ψ from $\Gamma^n \cup \{\phi_{n+1}, @_a\theta\}$ since the first clause does not apply, therefore ψ is derivable from $\Gamma^n \cup \{\phi_{n+1}\}$ by the rule $(\vee E)$. For the third clause, the derivability of ψ from $\Gamma^n \cup \{\phi_{n+1}, @_e\theta, @_a\Diamond e\}$ implies the derivability of ψ from $\Gamma^n \cup \{\phi_{n+1}\}$ by the rule $(\Diamond E)$. The fourth clause is analogous to the third clause. We conclude that ψ is not derivable from Γ^*. It is straightforward to check that Γ^* is **E**-saturated.

The canonical model given below is similar to a canonical model for first-order intuitionistic logic given in Troelstra and van Dalen (1988).

Definition 8.5. (Canonical model) Let $\mathbf{C}_1, \mathbf{C}_2, \mathbf{C}_3, \ldots$ be pairwise disjoint countably infinite sets of nominals disjoint from $\mathbf{C}_0$ and let $\mathbf{C}_n^* = \cup_{1 \leq i \leq n} \mathbf{C}_i$ where $n \geq 1$. Let Γ be a consistent set of satisfaction statements in the language $\mathrm{L}(\mathbf{C}_0)$. A model

$$\mathfrak{M}^\Gamma = (W^\Gamma, \subseteq, \{D_w^\Gamma\}_{w \in W^\Gamma}, \{\sim_w^\Gamma\}_{w \in W^\Gamma}, \{R_w^\Gamma\}_{w \in W^\Gamma}, \{V_w^\Gamma\}_{w \in W^\Gamma})$$

and for each $w \in W^\Gamma$, a w-assignment g_w^Γ, is defined as follows.

- $W^\Gamma = \{\Delta \supseteq \Gamma \mid \text{for some } n,\ \Delta \subseteq \mathrm{L}(\mathbf{C}_0 \cup \mathbf{C}_n^*) \text{ and } \Delta \text{ is } \mathbf{C}_n^*\text{-saturated}\}$.
- $D_\Delta^\Gamma = \mathbf{C}_0 \cup \mathbf{C}_n^*$ where Δ is $\mathbf{C}_n^*$-saturated.
- $\sim_\Delta^\Gamma = \{(a,c) \mid @_a c \in \Delta\}$.
- $R_\Delta^\Gamma = \{(a,c) \mid @_a \Diamond c \in \Delta\}$.
- $V_\Delta^\Gamma(p) = \{a \mid @_a p \in \Delta\}$.
- $g_\Delta^\Gamma(a) = a$ where $a \in D_\Delta^\Gamma$.

Note that it follows from Lemma 8.1 that W^Γ is non-empty. It is straightforward to check the other requirements $\mathfrak{M}^\Gamma$ has to satisfy to be a model for intuitionistic hybrid logic. Given the saturation lemma and the definition of a canonical model, we are ready to prove a truth lemma.

Lemma 8.2. *(Truth lemma) For any $\Delta \in W^\Gamma$ and any satisfaction statement $@_a\phi$ in $\mathrm{L}(D_\Delta^\Gamma)$, $@_a\phi \in \Delta$ if and only if $\mathfrak{M}^\Gamma, g_\Delta^\Gamma, \Delta, a \models \phi$.*

Proof. Induction on the degree of ϕ. We only consider the case where ϕ is of the form $\Box\theta$; the other cases are simpler.

Assume that $@_a\Box\theta \in \Delta$. Let $\Lambda \supseteq \Delta$ and $aR_\Lambda^\Gamma e$, that is, $@_a\Diamond e \in \Lambda$. Then $@_e\theta \in \Lambda$ by the rule $(\Box E)$ which by the induction hypothesis implies that $\mathfrak{M}^\Gamma, g_\Lambda^\Gamma, \Lambda, e \models \theta$. It follows that $\mathfrak{M}^\Gamma, g_\Delta^\Gamma, \Delta, a \models \Box\theta$.

Assume that $@_a\Box\theta \notin \Delta$. Assume that Δ is $\mathbf{C}_n^*$-saturated and let $e \in \mathbf{C}_{n+1}^*$. Then the formula $@_e\theta$ is not derivable from $\Delta \cup \{@_a\Diamond e\}$ for otherwise we could derive $@_a\Box\theta$ from Δ by the rule $(\Box I)$. According to Lemma 8.1, there exists a $\mathbf{C}_{n+2}^*$-saturated extension Λ of $\Delta \cup \{@_a\Diamond e\}$ such that $@_e\theta$ is not derivable from Λ. It follows by the induction hypothesis that $\mathfrak{M}^\Gamma, g_\Lambda^\Gamma, \Lambda, e \models \theta$ is not the case. This contradicts $\mathfrak{M}^\Gamma, g_\Delta^\Gamma, \Delta, a \models \Box\theta$ since $@_a\Diamond e \in \Lambda$ implies that $aR_\Lambda^\Gamma e$.

We only need one more lemma before we can prove completeness.

Lemma 8.3. *Let Γ be a consistent set of satisfaction statements. Then the canonical model $\mathfrak{M}^{\Gamma}$ is a* **T***-model.*

Proof. If $\theta \in \mathbf{T}$, then θ has the form $\forall\overline{a}((S_1 \wedge \cdots \wedge S_n) \rightarrow \exists\overline{c} \bigvee_{j=1}^{m} (S_{j1} \wedge \cdots \wedge S_{jn_j}))$ where $\overline{a} = a_1, \ldots, a_l$. Assume that Δ is an element of W^{Γ} and g is a Δ-assignment for an intuitionistic hybrid-logical model $\mathfrak{M}^{\Gamma}$ such that $(\mathfrak{M}^{\Gamma})^*, g, \Delta \models S_1$, ..., $(\mathfrak{M}^{\Gamma})^*, g, \Delta \models S_n$. So $g(a_1) = d_1$, ..., $g(a_l) = d_l$ for some nominals d_1, ..., d_l in D^{Γ}_{Δ}. Then $s_1[\overline{d}/\overline{a}], \ldots, s_n[\overline{d}/\overline{a}] \in \Delta$ by the definition of a canonical model. If $m \geq 1$, then it follows from Δ being saturated that there exists a list of nominals $\overline{b}$ such that $@_e \bigvee_{j=1}^{m} (s_{j1} \wedge \cdots \wedge s_{jn_j})[\overline{d}, \overline{b}/\overline{a}, \overline{c}] \in \Delta$ where e is an arbitrary nominal. Therefore $@_e(s_{j1} \wedge \cdots \wedge s_{jn_j})[\overline{d}, \overline{b}/\overline{a}, \overline{c}] \in \Delta$ and hence $s_{j1}[\overline{d}, \overline{b}/\overline{a}, \overline{c}], \ldots, s_{jn_j}[\overline{d}, \overline{b}/\overline{a}, \overline{c}] \in \Delta$ for some j where $1 \leq j \leq m$. But then it follows from the definition of a canonical model that $(\mathfrak{M}^{\Gamma})^*, \Delta \models \exists\overline{c} \bigvee_{j=1}^{m} (S_{j1} \wedge \cdots \wedge S_{jn_j})$. On the other hand, if $m = 0$, then $@_e\bot \in \Delta$ by the rule (R_θ) which contradicts the consistency of Δ.

Now the completeness theorem.

Theorem 8.2. *(Completeness) Let ψ be a satisfaction statement and let Γ be a set of satisfaction statements. The second statement below implies the first statement.*

1. *ψ is derivable from Γ in $\mathbf{N}_{\mathscr{H}\mathscr{I}} + \mathbf{T}$.*
2. *For any* **T***-model $\mathfrak{M}$, any element w of W, and any w-assignment g, if, for any formula $\theta \in \Gamma$, $\mathfrak{M}, g, w \models \theta$, then $\mathfrak{M}, g, w \models \psi$.*

Proof. Assume that ψ is not derivable from Γ. Consider the canonical model $\mathfrak{M}^{\Gamma}$ and let Λ be a $\mathbf{C}_1^*$ saturated extension of Γ from which ψ is not derivable, cf. Lemma 8.1. It follows from Lemma 8.2 that $\mathfrak{M}^{\Gamma}, g^{\Gamma}_{\Lambda}, \Lambda \models \psi$ is not the case but it also follows from Lemma 8.2 that for any $\theta \in \Gamma$, $\mathfrak{M}^{\Gamma}, g^{\Gamma}_{\Lambda}, \Lambda \models \theta$ is the case. But this contradicts the second statement in the theorem since $\mathfrak{M}^{\Gamma}$ is a **T**-model by Lemma 8.3

8.2.4 Normalization

In what follows we give reduction rules for the natural deduction system $\mathbf{N}_{\mathscr{H}\mathscr{I}} + \mathbf{T}$ and we prove a normalization theorem. First some conventions. If a premise of a rule has the form $@_a c$ or $@_a\Diamond c$, then it is called a *relational premise*, and similarly, if the conclusion of a rule has the form $@_a c$ or $@_a\Diamond c$, then it is called a *relational conclusion*. Moreover, if an assumption discharged by a rule has the form $@_a\Diamond c$, then it is called a *relationally discharged assumption*. The premise of the form $@_a\phi$ in the rule $(\rightarrow E)$ is called *minor* and the premises of the form θ in the rules $(\vee E)$, $(\Diamond E)$, and (R_θ) are called *parametric premises*. A premise of an elimination rule that is neither minor, relational, or parametric is called *major*.

A *maximum formula* in a derivation is a formula occurrence that is both the conclusion of an introduction rule and the major premise of an elimination rule. Maximum formulas can be removed by applying *proper reductions*. The rules for proper reductions are as follows. We have omitted the reduction rules involving the connectives $\wedge$, $\rightarrow$, $@$, and $\Box$ which can be found in Section 2.2.4.

$(\vee I1)$ followed by $(\vee E)$ (analogously in the case of $(\vee I2)$)

$$\dfrac{\dfrac{\begin{matrix}\vdots\,\pi_1\\ @_a\phi\end{matrix}}{@_a(\phi\vee\psi)} \quad \begin{matrix}[@_a\phi]\\ \vdots\,\pi_2\\ \theta\end{matrix} \quad \begin{matrix}[@_a\psi]\\ \vdots\,\pi_3\\ \theta\end{matrix}}{\theta} \quad \rightsquigarrow \quad \begin{matrix}\vdots\,\pi_1\\ @_a\phi\\ \vdots\,\pi_2\\ \theta\end{matrix}$$

$(\Diamond I)$ followed by $(\Diamond E)$

$$\dfrac{\dfrac{\begin{matrix}\vdots\,\pi_1\\ @_e\phi\end{matrix} \quad \begin{matrix}\vdots\,\pi_2\\ @_a\Diamond e\end{matrix}}{@_a\Diamond\phi} \quad \begin{matrix}[@_c\phi]\ [@_a\Diamond c]\\ \vdots\,\pi_3\\ \theta\end{matrix}}{\theta} \quad \rightsquigarrow \quad \begin{matrix}\begin{matrix}\vdots\,\pi_1\\ @_e\phi\end{matrix} \quad \begin{matrix}\vdots\,\pi_2\\ @_a\Diamond e\end{matrix}\\ \vdots\,\pi_3[e/c]\\ \theta\end{matrix}$$

It turns out that we need further reduction rules in connection with the derivation rules $(\bot E)$, $(\vee E)$, $(\Diamond E)$, and (R_θ). A *permutable formula* in a derivation is a formula occurrence that is both the conclusion of $(\bot E)$, $(\vee E)$, $(\Diamond E)$, or (R_θ) and the major premise of an elimination rule. Permutable formulas in a derivation can be removed by applying *permutative reductions*. The rules for permutative reductions are as follows in the case where the elimination rule has two premises. We have omitted the reduction rule where (R_θ) is followed by an elimination which can be found in Section 2.2.4.

$(\bot E)$ followed by a two-premise elimination

$$\dfrac{\dfrac{\begin{matrix}\vdots\,\pi_1\\ @_a\bot\end{matrix}}{\theta} \quad \begin{matrix}\vdots\,\pi\\ \chi\end{matrix}}{\xi} \quad \rightsquigarrow \quad \dfrac{\begin{matrix}\vdots\,\pi_1\\ @_a\bot\end{matrix}}{\xi}$$

$(\vee E)$ followed by a two-premise elimination

$$\dfrac{\dfrac{\begin{array}{c}\vdots\,\pi_1\\ @_a(\phi\vee\psi)\end{array}\quad \begin{array}{c}[@_a\phi]\\ \vdots\,\pi_2\\ \theta\end{array}\quad \begin{array}{c}[@_a\psi]\\ \vdots\,\pi_3\\ \theta\end{array}}{\theta}\quad \begin{array}{c}\vdots\,\pi\\ \chi\end{array}}{\xi} \quad\rightsquigarrow\quad \dfrac{\begin{array}{c}\vdots\,\pi_1\\ @_a(\phi\vee\psi)\end{array}\quad \dfrac{\begin{array}{c}[@_a\phi]\\ \vdots\,\pi_2\\ \theta\end{array}\quad \begin{array}{c}\vdots\,\pi\\ \chi\end{array}}{\xi}\quad \dfrac{\begin{array}{c}[@_a\psi]\\ \vdots\,\pi_3\\ \theta\end{array}\quad \begin{array}{c}\vdots\,\pi\\ \chi\end{array}}{\xi}}{\xi}$$

$(\Diamond E)$ followed by a two-premise elimination

$$\dfrac{\dfrac{\begin{array}{c}\vdots\,\pi_1\\ @_a\Diamond\phi\end{array}\quad \begin{array}{c}[@_c\phi]\;[@_a\Diamond c]\\ \vdots\,\pi_2\\ \theta\end{array}}{\theta}\quad \begin{array}{c}\vdots\,\pi\\ \chi\end{array}}{\xi} \quad\rightsquigarrow\quad \dfrac{\begin{array}{c}\vdots\,\pi_1\\ @_a\Diamond\phi\end{array}\quad \dfrac{\begin{array}{c}[@_b\phi]\;[@_a\Diamond b]\\ \vdots\,\pi_2[b/c]\\ \theta\end{array}\quad \begin{array}{c}\vdots\,\pi\\ \chi\end{array}}{\xi}}{\xi}$$

The cases where the elimination rule has one or three premises are obtained by deleting or adding derivations as appropriate.

A derivation is *normal* if it contains no maximum or permutable formula. In what follows we shall prove a normalization theorem which says that any derivation can be rewritten to a normal derivation by repeated applications of reductions. To this end we need a number of definitions and lemmas.

Definition 8.6. The $\Diamond$-*graph* of a derivation π is the binary relation on the set of formula occurrences in π of the form $@_a\Diamond c$ which is defined as follows. A pair of formula occurrences (ϕ, ψ) is an element of the $\Diamond$-graph of π if and only if it satisfies one of the following conditions.

1. ϕ is the relational premise of an instance of $(\Diamond I)$ which has ψ as the conclusion.
2. ϕ is the major premise of an instance of $(\Diamond E)$ at which ψ is relationally discharged.
3. ϕ is a parametric premise of an instance of $(\vee E)$, $(\Diamond E)$, or (R_θ) which has ψ as the conclusion.

Note that the $\Diamond$-graph of π is a relation on the set of formula occurrences of π; not the set of formulas occurring in π. Also, note that every formula occurrence in a $\Diamond$-graph is of the form $@_a\Diamond c$.

Lemma 8.4. *The $\Diamond$-graph of a derivation π does not contain cycles.*

Proof. Induction on the structure of π.

Definition 8.7. The *potential* of a chain in the $\Diamond$-graph of π is the number of formula occurrences in the chain which are major premises of instances of $(\Diamond E)$. A *stubborn formula* in a derivation π is a maximum or permutable formula of the form $@_a\Diamond c$ and the *stubbornness* of a stubborn formula in π is the maximal potential of a chain in the $\Diamond$-graph of π that contains the stubborn formula.

Note that the notion of potential in the definition above is well-defined as Lemma 8.4 implies that the number of formula occurrences of a chain in a $\Diamond$-graph is bounded.

Lemma 8.5. *Let π be a derivation where all stubborn maximum formulas have stubbornness less than or equal to d and all stubborn permutable formulas have stubbornness less than d. Assume that ϕ is a stubborn maximum formula with stubbornness d such that no formula occurrence above ϕ is a stubborn maximum formula with stubbornness d. Let π' be the derivation obtained by applying the reduction such that ϕ is removed. Then all stubborn maximum formulas in π' have stubbornness less than or equal to d and all stubborn permutable formulas in π' have stubbornness less than d, and moreover, the number of stubborn maximum formulas with stubbornness d in π' is less than the number of stubborn maximum formulas with stubbornness d in π.*

Proof. The derivations π and π' have the forms below.

$$\begin{array}{c}\dfrac{\dfrac{\begin{array}{c}\vdots\,\pi_1\\ @_e d\end{array} \qquad \begin{array}{c}\vdots\,\pi_2\\ @_a\Diamond e\end{array}}{@_a\Diamond d} \qquad \begin{array}{c}[@_c d]\ \ [@_a\Diamond c]\\ \vdots\,\pi_3\\ \theta\end{array}}{\theta}\\ \vdots\,\tau\\ \psi\end{array} \qquad\qquad \begin{array}{c}\begin{array}{cc}\vdots\,\pi_1 & \vdots\,\pi_2\\ @_e d & @_a\Diamond e\end{array}\\ \vdots\,\pi_3[e/c]\\ \theta\\ \vdots\,\tau\\ \psi\end{array}$$

Note that any formula occurrence in π', except the indicated occurrences of $@_e d$, $@_a\Diamond e$, and θ, in an obvious way can be mapped to a formula occurrence in π. Let f be the map thus defined (note that f need not be injective as the instance of $(\Diamond E)$ in π might discharge more than one occurrence of $@_a\Diamond c$). Using the map f, a map from the $\Diamond$-graph of π' to the $\Diamond$-graph of π is defined as follows. There are a number of cases to consider. Case 1: An element (ξ,χ) of the $\Diamond$-graph of π', where the formula occurrences ξ and χ both are in the domain of f, is mapped to $(f(\xi),f(\chi))$, which straightforwardly can be shown to be an element of the $\Diamond$-graph of π (observe that no assumption in π_1 or π_2 is discharged at a rule-instance in $\pi_3[e/c]$). Case 2: An element (ξ,χ), where χ is one of the indicated occurrences of $@_a\Diamond e$ (and ξ therefore is in the domain of f), is mapped to $(f(\xi),\chi')$, where χ' is the relational premise of the instance of $(\Diamond I)$. Case 3: An element (ξ,χ), where ξ is the indicated occurrence of θ (and χ therefore is in the domain of f), is mapped to $(\xi',f(\chi))$, where ξ' is the conclusion of the instance of $(\Diamond E)$. Case 4: An element (ξ,χ), where ξ is one of the indicated occurrences of $@_a\Diamond e$, ξ is different from the indicated occurrence of θ, and χ is in the domain of f, is mapped to $(\xi',f(\chi))$, where ξ' is the assumption in π_3 discharged by the instance of $(\Diamond E)$ corresponding to the occurrence of $@_a\Diamond e$ in question. Case 5: An element (ξ,χ), where χ is the indicated occurrence of θ, χ is different from each of the indicated occurrences of $@_a\Diamond e$, and ξ is in the domain of f, is mapped to $(f(\xi),\chi')$, where χ' is the parametric premise of the instance of $(\Diamond I)$. Case 6: An element (ξ,χ), where ξ is one of the indicated occurrences of $@_a\Diamond e$ and χ is the indicated occurrence of θ, is mapped to (ξ',χ'), where ξ' is the assumption in π_3 discharged by the instance of $(\Diamond E)$ corresponding to the occurrence of $@_a\Diamond e$ in question and χ' is the parametric premise of the instance of $(\Diamond E)$. By using the map from the $\Diamond$-graph of π' to the $\Diamond$-

graph of π, any chain in the $\Diamond$-graph of π' that does not contain any of the indicated occurrences of $@_a\Diamond e$ can in an obvious way be mapped to a chain in the $\Diamond$-graph of π with the same potential and which does not contain the indicated occurrences of $@_a\Diamond e$, $@_a\Diamond d$, and $@_a\Diamond c$, and similarly, any chain in the $\Diamond$-graph of π' that contains one of the indicated occurrences of $@_a\Diamond e$ can in an obvious way be mapped to a chain in the $\Diamond$-graph of π with greater potential and which contains the mentioned formula occurrences. The conclusions of the lemma follow straightforwardly.

Definition 8.8. A *segment* in a derivation π is a non-empty list $\phi_1, \ldots, \phi_n$ of formula occurrences in π with the following properties.

1. ϕ_1 is not the conclusion of an instance of $(\vee E)$, an instance of $(\Diamond E)$, or an instance of (R_θ) with more than zero parametric premises.
2. For each $i < n$, ϕ_i is a parametric premise of an instance of $(\vee E)$, $(\Diamond E)$, or (R_θ) which has ϕ_{i+1} as the conclusion.
3. ϕ_n is not a parametric premise of an instance of $(\vee E)$, $(\Diamond E)$, or (R_θ).

The *length* of a segment is the number of formula occurrences in the segment. A segment σ_1 *stands above* a segment σ_2 if and only if the last formula occurrence in σ_1 stands above the first formula occurrence in σ_2. A *maximum segment* (*permutable segment*) is a segment in which the last formula occurrence is a maximum formula (permutable formula). A *stubborn segment* is a maximum or permutable segment where the formula that occurs in the segment is of the form $@_a\Diamond c$. The *degree* of a segment is the degree of the formula that occurs in the segment.

The following lemma is along the lines of a similar result for ordinary intuitionistic first-order logic given in Prawitz (1965).

Lemma 8.6. *Any derivation π can be rewritten to a derivation π' that does not contain permutable formulas or non-stubborn maximum formulas, by repeated applications of permutative reductions applied to permutable formulas and proper reductions applied to non-stubborn maximum formulas.*

Proof. To any derivation π we assign the pair (d,k) of non-negative integers where d is the maximal degree of a permutable or non-stubborn maximum segment in π or 0 if there is no such segment and k is the sum of the lengths of permutable and non-stubborn maximum segments in π of degree d (note that a list of formula occurrences with only one element is a segment if the one and only formula occurrence in the list is a maximum formula). The proof is by induction on such pairs equipped with the lexicographic order. Let π be a derivation to which a pair (d,k) is assigned such that $d > 0$. It is straightforward to check that there exists a permutable or non-stubborn maximum segment σ of degree d in π such that there is i) no permutable or non-stubborn maximum segment with degree d that stands above σ and ii) no permutable or non-stubborn maximum segment with degree d that stands above or contains a minor, relational, or parametric premise of the rule instance of which the last formula occurrence in σ is the major premise. Let π' be the derivation obtained by applying the appropriate reduction rule such that the last formula occurrence in σ is removed. Then it is straightforward to check that the pair (d',k') assigned to π' is less than (d,k) in the lexicographic order

We are now ready to prove the normalization theorem.

Theorem 8.3. *(Normalization) Any derivation can be rewritten to a normal derivation by repeated applications of proper and permutative reductions.*

Proof. By Lemma 8.6 we just need to consider derivations that do not contain permutable formulas or non-stubborn maximum formulas. To any such derivation π we assign the non-negative integer d where d is the maximal stubbornness of a stubborn maximum formula in π or 0 if there is no stubborn maximum formula. Let π be a derivation to which an integer d is assigned such that $d > 0$. It is straightforward that there exists a stubborn maximum formula ϕ with stubbornness d such that no formula occurrence above ϕ is a stubborn maximum formula with stubbornness d. Let π' be the derivation obtained by applying the reduction such that ϕ is removed. Then by inspecting the involved reduction rule it is trivial to check that all maximum or permutable formulas in π' are stubborn, and moreover, by Lemma 8.5 all stubborn maximum formulas in π' have stubbornness less than or equal to d and all stubborn permutable formulas in π' have stubbornness less than d, and furthermore, the number of stubborn maximum formulas with stubbornness d in π' is less than the number of stubborn maximum formulas with stubbornness d in π. By repeated applications of this procedure a derivation is obtained in which all maximum or permutable formulas are stubborn with stubbornness less than d. By application of Lemma 8.6 a derivation π'' is obtained that does not contain permutable formulas or non-stubborn maximum formulas. If all maximum or permutable formulas in a derivation τ are stubborn with stubbornness less than d, then it is trivial to check by inspecting the involved reduction rules that all maximum or permutable formulas in the derivation τ' obtained by applying a permutative reduction are stubborn, and moreover, it can be proved in a way similar to the way in which Lemma 8.5 is proved, that all stubborn formulas in τ' have stubbornness less than d. Thus, all maximum formulas in π'' are stubborn with stubbornness less than d. We are therefore done by induction.

8.2.5 The Form of Normal Derivations

Below we adapt an important definition from Prawitz (1965) to intuitionistic hybrid logic.

Definition 8.9. A *path* in a derivation π is a non-empty list $\phi_1, \ldots, \phi_n$ of formula occurrences in π with the following properties.

1. ϕ_1 is a relational conclusion, or the conclusion of a (R_θ) rule with zero parametric premises, or an assumption that is not non-relationally discharged by an instance of $(\vee E)$ or $(\Diamond E)$.
2. For each $i < n$, ϕ_i is not a minor or relational premise and either
   a. ϕ_i is not the major premise of an instance of $(\vee E)$ or $(\Diamond E)$ and ϕ_i stands immediately above ϕ_{i+1}, or

   b. ϕ_i is the major premise of an instance r of $(\vee E)$ or $(\Diamond E)$ and ϕ_{i+1} is an assumption non-relationally discharged by r.

3. ϕ_n is either the end-formula of π, or a minor or relational premise, or the major premise of an instance of $(\vee E)$ or $(\Diamond E)$ that does not non-relationally discharge any assumptions.

Note that ϕ_1 in the definition above might be a discharged assumption.

Lemma 8.7. *Any formula occurrence in a derivation π belongs to some path in π.*

Proof. Induction on the structure of π.

The definition of a path leads us to the lemma below. The lemma says that a path in a normal derivation can be split up into three parts: An analytical part in which formulas are broken down in their components by successive applications of the elimination rules, a minimum part in which an instance of the rule $(\bot E)$ may occur, and a synthetical part in which formulas are put together by successive applications of the introduction rules. See Prawitz (1971).

Lemma 8.8. *Let $\beta = \phi_1, \ldots, \phi_n$ be a path in a normal derivation. Then there exists a formula occurrence ϕ_i in β, called the* minimum formula *in β, such that*

1. *for each $j < i$, ϕ_j is a major or parametric premise or the non-relational premise of an instance of $(Nom1)$;*
2. *if $i \neq n$, then ϕ_i is a non-relational premise of an introduction rule or the premise of an instance of $(\bot E)$; and*
3. *for each j, where $i < j < n$, ϕ_j is a non-relational premise of an introduction rule, a parametric premise, or the non-relational premise of an instance of $(Nom1)$.*

Proof. Let ϕ_i be the first formula occurrence in β which is not the non-relational premise of an instance of $(Nom1)$, and is not a parametric premise, and is not the major premise of an elimination rule, save possibly the major premise of an instance of $(\vee E)$ or $(\Diamond E)$ that does not non-relationally discharge any assumptions (such a formula occurrence exists in β as ϕ_n satisfies the mentioned criterium). We are done if $i = n$. Otherwise ϕ_i is a non-relational premise of an introduction rule or the premise of an instance of $(\bot E)$ (by inspection of the rules and the definition of a path). If ϕ_i is the premise of an instance of $(\bot E)$, then each ϕ_j, where $i < j < n$, is a non-relational premise of an introduction rule, or the non-relational premise of an instance of $(Nom1)$, or a parametric premise (by inspection of the rules, the definition of a branch, and normality of π). Similarly, if ϕ_i is a non-relational premise of an introduction rule, then each ϕ_j, where $i < j < n$, is a non-relational premise of an introduction rule or a parametric premise.

In what follows we shall consider the form of normal derivations. To this end we give the following definition.

Definition 8.10. The notion of a *subformula* is defined by the conventions that

- ϕ is a subformula of ϕ;

- if $\psi \wedge \theta$, $\psi \vee \theta$, or $\psi \rightarrow \theta$ is a subformula of ϕ, then so are ψ and θ; and
- if $@_a\psi$, $\Diamond\psi$, or $\Box\psi$ is a subformula of ϕ, then so is ψ.

A formula $@_a\phi$ is a *quasi-subformula* of a formula $@_c\psi$ if and only if ϕ is a subformula of ψ.

Now we state the theorem which says that normal derivations satisfy a version of the subformula property.

Theorem 8.4. *(Quasi-subformula property) Let π be a normal derivation of ϕ from a set of satisfaction statements Γ. Any formula occurrence θ in π is a quasi-subformula of ϕ, or of some satisfaction statement in Γ, or of some relational premise, or of some relational conclusion, or of some relationally discharged assumption.*

Proof. First a convention: The *order* of a path in π is the number of formula occurrences in π which stand below the last formula occurrence of the path. Now consider a path $\beta = \phi_1, \ldots, \phi_n$ in π of order p. By induction we can assume that the theorem holds for all formula occurrences in paths of order less than p. Note that it follows from Lemma 8.8 that any formula occurrence ϕ_j such that $j \leq i$, where ϕ_i is the minimum formula in β, is a quasi-subformula of ϕ_1, and similarly, any ϕ_j such that $j > i$, is a quasi-subformula of ϕ_n.

We first consider ϕ_n. We are done if ϕ_n is the end-formula ϕ or a relational premise. If ϕ_n is the minor premise of an instance of $(\rightarrow E)$, then we are done by induction as the major premise belongs to a path of order less than p. If ϕ_n is the major premise of an instance of $(\vee E)$ or $(\Diamond E)$ that does not non-relationally discharge any assumptions, then ϕ_n is the minimum formula and hence a quasi-subformula of ϕ_1. Now, we are done if ϕ_1 is a relational conclusion, or an undischarged assumption, or a relationally discharged assumption. Otherwise ϕ_1 is discharged by an instance of $(\rightarrow I)$ with a conclusion that belongs to some branch of order less than p (note that due to normality of π, ϕ_1 is not the conclusion of a (R_θ) rule with zero parametric premises).

We now consider ϕ_1. We are done if ϕ_1 is a relational conclusion, or an undischarged assumption, or a relationally discharged assumption. If ϕ_1 is the conclusion of a (R_θ) rule with zero parametric premises, then ϕ_1 has the same form as the minimum formula which is a quasi-subformula of ϕ_n. Otherwise ϕ_1 is discharged by an instance of $(\rightarrow I)$ with a conclusion that belongs to β or to some path of order less than p.

Note that a consequence of the theorem is that θ is a quasi-subformula of ϕ, or of some formula in Γ, or of a formula of the form $@_a c$ or $@_a\Diamond c$ (since relational premises, relational conclusions, and relationally discharged assumptions are of the form $@_a c$ or $@_a\Diamond c$).

8.3 Axiom Systems for Intuitionistic Hybrid Logic

In this section we shall give a sound and complete Hilbert-style axiom system for the intuitionistic hybrid logic $\mathscr{H}^{\mathscr{I}}$. The axiom system is comprised of all instances of theorems of intuitionistic propositional logic together with the axioms and rules in Figure 8.4. The system will be denoted $\mathbf{A}_{\mathscr{H}\mathscr{I}}$. Note that the intuitionistic axiom system $\mathbf{A}_{\mathscr{H}\mathscr{I}}$ can be seen as obtained from the classical axiom system $\mathbf{A}_{\mathscr{H}}$ given in Figure 2.11 of Section 2.5 by changing the surrounding propositional logic from classical to intuitionistic and by adding axioms and rules for the connectives $\wedge$, $\vee$, and $\Diamond$ (which are not intuitionistically definable in terms of the other connectives).

$(Distr_{\wedge})$	$@_a(\phi \wedge \psi) \leftrightarrow (@_a\phi \wedge @_a\psi)$
$(Distr_{\vee})$	$@_a(\phi \vee \psi) \leftrightarrow (@_a\phi \vee @_a\psi)$
$(Distr_{\rightarrow})$	$@_a(\phi \rightarrow \psi) \leftrightarrow (@_a\phi \rightarrow @_a\psi)$
$(\bot)$	$@_a\bot \rightarrow \bot$
$(Scope)$	$@_a@_b\phi \leftrightarrow @_b\phi$
(Ref)	$@_aa$
$(Intro)$	$(a \wedge \phi) \rightarrow @_a\phi$
$(\Diamond I)$	$(@_e\phi \wedge \Diamond e) \rightarrow \Diamond\phi$
$(\Box E)$	$(\Box\phi \wedge \Diamond e) \rightarrow @_e\phi$

$$\frac{\phi \rightarrow \psi \quad \phi}{\psi}\ (MP) \qquad \frac{\phi}{@_a\phi}\ (N_{@}) \qquad \frac{@_a\phi}{\phi}\ (Name)^{*}$$

$$\frac{(\psi \rightarrow \Diamond\phi) \wedge ((\psi \wedge @_c\phi \wedge \Diamond c) \rightarrow \theta)}{\psi \rightarrow \theta}\ (\Diamond E)^{\star} \qquad \frac{(\psi \wedge \Diamond c) \rightarrow @_c\phi}{\psi \rightarrow \Box\phi}\ (\Box I)^{\dagger}$$

∗ a does not occur in ϕ.
⋆ c does not occur in ϕ, θ, or ψ.
† c does not occur in ϕ or ψ.

Fig. 8.4 Axioms and rules for intuitionistic hybrid logic

The axiom system $\mathbf{A}_{\mathscr{H}\mathscr{I}}$ is sound and complete with respect to the intuitionistic semantics given in Section 8.1. Soundness is a straightforward induction proof analogous to the classical Theorem 2.7 where we make use of Proposition 8.1 and Proposition 8.2. Completeness is analogous to the proof of the classical Theorem 2.8 but here we make use of the complete natural deduction system $\mathbf{N}_{\mathscr{H}\mathscr{I}}$ for intuitionistic hybrid logic given in Section 8.2.

It is straightforward to modify soundness and completeness to encompass rules corresponding to a basic geometric theory $\mathbf{T}$. First, a Hilbert-style rule (R_θ) is associated with each formula θ in the basic geometric theory $\mathbf{T}$, exactly as in the classical case described in Section 2.5. Second, an intuitionistic frame, cf. Definition 8.1 in Section 8.1, is called a $\mathbf{T}$-*frame* if and only if for every model $\mathfrak{M}$ for intuitionistic hybrid logic which is based on the frame in question and every formula $\theta \in \mathbf{T}$, it is the case that $\mathfrak{M}^* \models \theta$. The notion of validity is then relativised to the class of intuitionistic **T**-frames and the axiom system $\mathbf{A}_{\mathscr{H}\mathscr{I}}$ is extended with the set of rules $\{(R_\theta) \mid \theta \in \mathbf{T}\}$.

8.4 Axiom Systems for a Paraconsistent Hybrid Logic

In this section we shall give a sound and complete Hilbert-style axiom system for a hybridized version of D. Nelson's constructive and paraconsistent logic N4 (strictly speaking, our logical basis is a variant of N4 extended with the falsum constant $\perp$, but for simplicity we just denote it N4). Completeness of the axiom system is proved by reduction to completeness of the axiom system $\mathbf{A}_{\mathscr{H}\mathscr{I}}$ for intuitionistic hybrid logic given in the previous section.

Now, the basic notion in the standard Kripke semantics for intuitionistic logic is that of *known truth*, that is, at any state of knowledge either it is known that a given propositional symbol is true or it is not known that it is true. It is very natural to extend this semantics with a symmetric notion of *known falsity*. Of course, it is required that if a propositional symbol is known to be false at some state of knowledge, then this knowledge is preserved by the epistemic partial order. D. Nelson's constructive logics N3 and N4 are based on such symmetric notions of known truth and known falsity. The logic N3 is equipped with the restriction that at no state of knowledge there is a propositional symbol known to be both true and false. This restriction is not imposed on N4, and therefore that logic is paraconsistent. In the semantics of N3 and N4, the notions of known truth and known falsity are generalized from propositional symbols to arbitrary formulas by defining two parallelly defined forcing relations, and the languages of the two logics involve a new kind of negation called *strong negation* which switches between the forcing relations. Strong negation, denoted $\sim$, has the property that if the formula $\sim(\phi \wedge \psi)$ is valid, then $\sim\phi$ or $\sim\psi$ is valid, which is a desirable property from an intuitionistic point of view (note that ordinary intuitionistic negation does not have this property).

The logic N4 has been suggested as the basis of intuitionistic versions of description logics, see Wansing and Odintsov (2003). That paper gives a tableau system for an N4 version of the description logic $\mathscr{A}\mathscr{L}\mathscr{C}$ which is a notational variant of multi-modal K (whereas we here give an axiom system for an N4 version of hybrid logic).

Note that extending our intuitionistic hybrid logic with N4 machinery makes the reasoning capability more powerful since more can be known about the subject-matter, namely the falsity of propositions, but this additional reasoning capability is distinguished from the subject-matter. Thus, the extension with N4 machinery is in line with distinguishing between the way of reasoning and what the reasoning is about.

We shall now be more formal. As indicated above, the language of paraconsistent hybrid logic is obtained by extending the language of intuitionistic hybrid logic with a new unary connective $\sim$. Below we give the semantics.

Definition 8.11. A *model* for paraconsistent hybrid logic is a tuple

$$(W, \leq, \{D_w\}_{w\in W}, \{\sim_w\}_{w\in W}, \{R_w\}_{w\in W}, \{V_w^+\}_{w\in W}, \{V_w^-\}_{w\in W})$$

where

1. W is a non-empty set partially ordered by $\leq$;
2. for each w, D_w is a non-empty set such that $w \leq v$ implies $D_w \subseteq D_v$;
3. for each w, $\sim_w$ is an equivalence relation on D_w such that $w \leq v$ implies $\sim_w \subseteq \sim_v$;
4. for each w, R_w is a binary relation on D_w such that $w \leq v$ implies $R_w \subseteq R_v$; and
5. for each w, V_w^+ is a function that to each ordinary propositional symbol p assigns a subset of D_w such that $w \leq v$ implies $V_w^+(p) \subseteq V_v^+(p)$ (and the same for V_w^-).

It is assumed that if $d \sim_w d'$, $e \sim_w e'$, and dR_we, then $d'R_we'$, and similarly, if $d \sim_w d'$ and $d \in V_w^+(p)$, then $d' \in V_w^+(p)$ (and the same for V_w^-). The model is said to be *based* on the intuitionistic frame $(W, \leq, \{D_w\}_{w\in W}, \{\sim_w\}_{w\in W}, \{R_w\}_{w\in W})$.

Thus, a model for paraconsistent hybrid logic is a model for intuitionistic hybrid logic, cf. Definition 8.1 in Section 8.1, where V_w has been renamed to V_w^+ and where a second function V_w^- has been added satisfying the same requirements as V_w^+.

Let $\mathfrak{M} = (W, \leq, \{D_w\}_{w\in W}, \{\sim_w\}_{w\in W}, \{R_w\}_{w\in W}, \{V_w^+\}_{w\in W}, \{V_w^-\}_{w\in W})$ be a model for paraconsistent hybrid logic. The relations $\mathfrak{M}, g, w, d \models^+ \phi$ and $\mathfrak{M}, g, w, d \models^- \phi$ are defined by mutual induction, where w is an element of W, g is a w-assignment, d is an element of D_w, and ϕ is a formula. The clauses in the definition are obtained from the clauses for the plain intuitionistic relation $\models$ given in Section 8.1 by renaming $\models$ to $\models^+$ and V_w to V_w^+, and moreover, by adding the clauses below.

$$\begin{array}{rl}
\mathfrak{M}, g, w, d \models^- p & \text{iff } d \in V_w^-(p) \\
\mathfrak{M}, g, w, d \models^- a & \text{iff for all } v \geq w, \text{ not } d \sim_v g(a) \\
\mathfrak{M}, g, w, d \models^- \phi \wedge \psi & \text{iff } \mathfrak{M}, g, w, d \models^- \phi \text{ or } \mathfrak{M}, g, w, d \models^- \psi \\
\mathfrak{M}, g, w, d \models^- \phi \vee \psi & \text{iff } \mathfrak{M}, g, w, d \models^- \phi \text{ and } \mathfrak{M}, g, w, d \models^- \psi \\
\mathfrak{M}, g, w, d \models^- \phi \rightarrow \psi & \text{iff } \mathfrak{M}, g, w, d \models^+ \phi \text{ and } \mathfrak{M}, g, w, d \models^- \psi \\
\mathfrak{M}, g, w, d \models^- \bot & \text{iff verum} \\
\mathfrak{M}, g, w, d \models^- \Box\phi & \text{iff for some } e \in D_w, dR_we \text{ and } \mathfrak{M}, g, w, e \models^- \phi \\
\mathfrak{M}, g, w, d \models^- \Diamond\phi & \text{iff for all } v \geq w, \text{ for all } e \in D_v, \\
& \quad dR_ve \text{ implies } \mathfrak{M}, g, v, e \models^- \phi \\
\mathfrak{M}, g, w, d \models^- @_a\phi & \text{iff } \mathfrak{M}, g, w, g(a) \models^- \phi \\
\mathfrak{M}, g, w, d \models^+ {\sim}\phi & \text{iff } \mathfrak{M}, g, w, d \models^- \phi \\
\mathfrak{M}, g, w, d \models^- {\sim}\phi & \text{iff } \mathfrak{M}, g, w, d \models^+ \phi
\end{array}$$

By convention $\mathfrak{M} \models^+ \phi$ means $\mathfrak{M}, g, w, d \models^+ \phi$ for every element w of W, every w-assignment g, and every element d of D_w. A formula ϕ is *valid* in a frame if and only if $\mathfrak{M} \models^+ \phi$ for any model $\mathfrak{M}$ that is based on the frame. A formula ϕ is *valid* in a class of frames if and only if ϕ is valid in every frame in the class in question. A formula ϕ is *valid* if and only if ϕ is valid in the class of all frames.

Note that defining two parallel two-valued evaluation procedures for formulas corresponds to defining one four-valued evaluation procedure, wherefore N4 is sometimes called a four-valued logic. The four truth-values are naturally taken to be the subsets of a set $\{T, F\}$ where T stands for known truth and F stands for known falsity. Hence, the truth-values, that is, the subsets of $\{T,F\}$, are interpreted as knowledge about a state of affairs in the following way.

$\emptyset$ Nothing is known about the state of affairs.
$\{T\}$ It is known that the state of affairs obtains.
$\{F\}$ It is known that the state of affairs fails.
$\{T,F\}$ It is known that the state of affairs obtains as well as fails.

Of course, in the last case the knowledge is conflicting (which is allowed since we are in a paraconsistent setting). The set of truth-values equipped with the inclusion ordering, $\subseteq$, constitutes a lattice. The inclusion ordering can be considered an ordering involving the information embodied in a truth-value since an increase means that more is known about truth or more is known about falsity. The truth-values can be equipped with a second ordering, denoted $\leq$, defined by letting $x \leq y$ if and only if $T \in x$ implies $T \in y$ and $F \notin x$ implies $F \notin y$. The second ordering can be considered an ordering involving the degree of truth embodied in a truth-value in the sense that an increase means an increase in knowledge about truth or a decrease in knowledge about falsity. The set of truth-values equipped with the ordering $\leq$ constitutes a lattice. The truth-values together with the two orderings can be depicted as

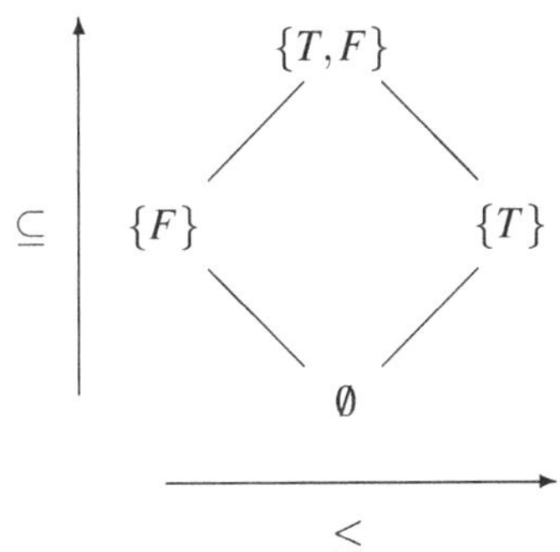

where the up-down direction represents the $\subseteq$ ordering and the left-right direction represents the $\leq$ ordering.[3] The four-valued space of truth-values depicted above has its origin in Nuel D. Belnap's work, see Belnap (1976).

We let $\mathscr{H}^{\mathsf{N4}}$ denote our hybridized version of the paraconsistent logic N4. Note that the notion of validity for $\mathscr{H}^{\mathsf{N4}}$ as defined above has the property that if the formula $\sim(\phi \wedge \psi)$ is valid, then $\sim\phi$ or $\sim\psi$ is valid. The following important propositions are satisfied.

Proposition 8.5. *(Monotonicity) If* $\mathfrak{M}, g, w, d \models^{+} \phi$ *and* $w \leq v$, *then* $\mathfrak{M}, g, v, d \models^{+} \phi$ *(and the same for the relation* $\models^{-}$*).*

Proof. Induction on the structure of ϕ.

[3] It turns out that the set of truth-values equipped with both of the orderings $\subseteq$ and $\leq$ constitutes what is called a bilattice. See Fitting (2006a) where it is demonstrated that bilattices naturally generalize a number of truth-value spaces. It would be interesting to investigate whether the results of the present section can be generalized to other bilattices than the described four-valued bilattice.

Proposition 8.6. *If* $\mathfrak{M}, g, w, d \models^+ \phi$ *and* $d \sim_w d'$, *then* $\mathfrak{M}, g, w, d' \models^+ \phi$ *(and the same for the relation* $\models^-$*).*

Proof. induction on the structure of ϕ.

We shall now use our axiom system for the intuitionistic hybrid logic $\mathscr{H}^{\mathscr{I}}$ to give an axiom system for $\mathscr{H}^{\mathsf{N4}}$. The axiom system is obtained by extending the axiom system for intuitionistic hybrid logic $\mathbf{A}_{\mathscr{H}^{\mathscr{I}}}$ given in the previous section with the axioms in Figure 8.5. The axiom system thus obtained will be denoted $\mathbf{A}_{\mathscr{H}^{\mathsf{N4}}}$. The axioms $(A1), \ldots, (A4)$ in Figure 8.5 are well-known from the literature on the logic N4, see for example Wansing and Odintsov (2003). Of course, the axiom $(A5)$ is a nullary version of $(A2)$.

$(A1)$	$\sim\sim\phi \leftrightarrow \phi$
$(A2)$	$\sim(\phi \vee \psi) \leftrightarrow (\sim\phi \wedge \sim\psi)$
$(A3)$	$\sim(\phi \wedge \psi) \leftrightarrow (\sim\phi \vee \sim\psi)$
$(A4)$	$\sim(\phi \rightarrow \psi) \leftrightarrow (\phi \wedge \sim\psi)$
$(A5)$	$\sim\bot \leftrightarrow \top$
$(A6)$	$\sim a \leftrightarrow \neg a$
$(A7)$	$\sim @_a\phi \leftrightarrow @_a \sim\phi$
$(A8)$	$\sim\Box\phi \leftrightarrow \Diamond\sim\phi$
$(A9)$	$\sim\Diamond\phi \leftrightarrow \Box\sim\phi$

Fig. 8.5 Axioms for paraconsistent hybrid logic

8.4.1 Soundness and Completeness

The axiom system $\mathbf{A}_{\mathscr{H}^{\mathsf{N4}}}$ is sound and complete with respect to the semantics given above.

Theorem 8.5. *(Soundness) If a formula* ϕ *is derivable in* $\mathbf{A}_{\mathscr{H}^{\mathsf{N4}}}$, *then* ϕ *is valid.*

Proof. Induction on the structure of the derivation of ϕ where we make use of Proposition 8.5 and Proposition 8.6.

Completeness is proved by reduction to completeness of our axiom system $\mathscr{H}^{\mathscr{I}}$ for intuitionistic hybrid logic. In comparison, the tableau system for the N4 version of $\mathscr{ALC}$ given in Wansing and Odintsov (2003) is shown to be complete by reduction to completeness of an axiom system for intuitionistic first-order logic. Like the completeness proof in Wansing and Odintsov (2003), our completeness proof makes use of a translation into negation normal form.

We say that a formula of $\mathscr{H}^{\mathsf{N4}}$ is in *negation normal form* if strong negations only occur as prefixes of ordinary propositional symbols. Below we give a translation that translates any formula of $\mathscr{H}^{\mathsf{N4}}$ into a formula in negation normal form.

$$\begin{array}{rclcrcl}
\overline{p} &=& p & \qquad & \overline{\sim p} &=& \sim p \\
\overline{a} &=& a & & \overline{\sim a} &=& \neg a \\
\overline{\phi \wedge \psi} &=& \overline{\phi} \wedge \overline{\psi} & & \overline{\sim(\phi \wedge \psi)} &=& \overline{\sim\phi} \vee \overline{\sim\psi} \\
\overline{\phi \vee \psi} &=& \overline{\phi} \vee \overline{\psi} & & \overline{\sim(\phi \vee \psi)} &=& \overline{\sim\phi} \wedge \overline{\sim\psi} \\
\overline{\phi \rightarrow \psi} &=& \overline{\phi} \rightarrow \overline{\psi} & & \overline{\sim(\phi \rightarrow \psi)} &=& \overline{\phi} \wedge \overline{\sim\psi} \\
\overline{\bot} &=& \bot & & \overline{\sim\bot} &=& \top \\
\overline{\Box\phi} &=& \Box\overline{\phi} & & \overline{\sim\Box\phi} &=& \Diamond\overline{\sim\phi} \\
\overline{\Diamond\phi} &=& \Diamond\overline{\phi} & & \overline{\sim\Diamond\phi} &=& \Box\overline{\sim\phi} \\
\overline{@_a\phi} &=& @_a\overline{\phi} & & \overline{\sim @_a\phi} &=& @_a\overline{\sim\phi} \\
& & & & \overline{\sim\sim\phi} &=& \overline{\phi}
\end{array}$$

The translation is a straightforward extension of the usual translation of modal-logical formulas into negation normal form where classical negation has been replaced by strong negation.

Proposition 8.7. *For any formula ϕ, the formula $\phi \leftrightarrow \overline{\phi}$ is derivable in $\mathbf{A}_{\mathscr{H}\mathrm{N4}}$.*

Proof. It is proved by induction on the structure of ϕ that each of the formulas $\phi \leftrightarrow \overline{\phi}$ and $\sim\phi \leftrightarrow \overline{\sim\phi}$ are derivable.

For any formula ϕ in negation normal form, we define ϕ^* to be the formula obtained by replacing each subformula of the form $\sim p$ by a new ordinary propositional symbol which we shall denote by $p^\sim$. Obviously, the formula ϕ^* is a formula of intuitionistic hybrid logic.

Proposition 8.8. *A formula ϕ is derivable in $\mathbf{A}_{\mathscr{H}\mathrm{N4}}$ if and only if $\overline{\phi}^*$ is derivable in $\mathbf{A}_{\mathscr{H}\mathscr{I}}$.*

Proof. If ϕ is derivable in $\mathbf{A}_{\mathscr{H}\mathrm{N4}}$, then it follows by induction on the structure of the derivation that $\overline{\phi}^*$ is derivable in $\mathbf{A}_{\mathscr{H}\mathscr{I}}$. Conversely, if $\overline{\phi}^*$ is derivable in $\mathbf{A}_{\mathscr{H}\mathscr{I}}$, then a derivation of $\overline{\phi}$ in $\mathbf{A}_{\mathscr{H}\mathrm{N4}}$ can be obtained by replacing each propositional symbol $p^\sim$ by $\sim p$. It follows from Proposition 8.7 that ϕ is derivable in $\mathbf{A}_{\mathscr{H}\mathrm{N4}}$.

We now prove the completeness theorem.

Theorem 8.6. *(Completeness) If a formula ϕ is valid, then ϕ is derivable in $\mathbf{A}_{\mathscr{H}\mathrm{N4}}$.*

Proof. Assume that ϕ is valid but ϕ is not derivable in $\mathbf{A}_{\mathscr{H}\mathrm{N4}}$. Then by Proposition 8.8, the formula $\overline{\phi}^*$ is not derivable in $\mathbf{A}_{\mathscr{H}\mathscr{I}}$. Thus, by completeness of $\mathbf{A}_{\mathscr{H}\mathscr{I}}$ there exists a model $\mathfrak{M} = (W, \leq, \{D_w\}_{w\in W}, \{\sim_w\}_{w\in W}, \{R_w\}_{w\in W}, \{V_w\}_{w\in W})$ for intuitionistic hybrid logic and an element w of W, a w-assignment g, and an element d of D_w such that $\mathfrak{M}, g, w, d \models \overline{\phi}^*$ does not hold. It is now straightforward to extend $\mathfrak{M}$ to a model $\mathfrak{M}'$ for paraconsistent hybrid logic such that $\mathfrak{M}', g, w, d \models^+ \overline{\phi}$ does not hold. Therefore by Proposition 8.7 and soundness of $\mathbf{A}_{\mathscr{H}\mathrm{N4}}$, it is not the case that $\mathfrak{M}', g, w, d \models^+ \phi$ holds, which contradicts the validity of ϕ.

It is straightforward to modify the results above to encompass rules corresponding to geometric conditions on the accessibility relation: Relativise validity to the class of intuitionistic $\mathbf{T}$-frames and extend the axiom system with the rules $\{(R_\theta) \mid \theta \in \mathbf{T}\}$, exactly as in the previous section.

8.5 A Curry-Howard Interpretation of Intuitionistic Hybrid Logic

A natural deduction system for an intuitionistic version of S5 extended with satisfaction operators has been proposed in Jia and Walker (2004), the aim being to provide a foundation for distributed functional programming languages where satisfaction operators are used for reasoning about the distribution of resources at different locations.[4] To this end Jia and Walker (2004) gave a Curry-Howard interpretation of the natural deduction formulation of intuitionistic S5 with satisfaction operators.

According to the standard Curry-Howard interpretation for intuitionistic logic, formulas correspond to types, natural deduction proofs correspond to terms of the lambda calculus (that is, programs), and normalization corresponds to reduction (that is, evaluation), in particular:

- A program of type $\phi \wedge \psi$ evaluates to a pair (t, u) where t is of type ϕ and u is of type ψ.
- A program of type $\phi \vee \psi$ evaluates either to $\mathtt{inl}(t)$ where t is of type ϕ or to $\mathtt{inr}(u)$ where u is of type ψ.
- A program of type $\phi \rightarrow \psi$ evaluates to an abstraction $\lambda x.t$ where the variable x is of type ϕ and t is of type ψ.

See Girard et al. (1989) for a general introduction to the Curry-Howard interpretation. In Jia and Walker (2004) the standard Curry-Howard interpretation for intuitionistic logic is extended with modal operators and satisfaction operators. Nominals represent locations in a network and proofs correspond to distributed programs. Roughly, the extension is as follows.

- A program of type $\Box\phi$ evaluates to a program of type ϕ which can be run at any location in the network.
- A program of type $\Diamond\phi$ evaluates to a program of type ϕ which can be run at some unspecified location in the network.
- A program of type $@_a\phi$ evaluates to a program of type ϕ which can be run at location a.

Mainly being interested in proof-theoretic aspects of intuitionistic hybrid logic, Jia and Walker (2004) did not give a Kripke semantics. A Kripke semantics for the intuitionistic hybrid logic of Jia and Walker (2004), that is, intuitionistic S5 with satisfaction operators, was given in Chadha et al. (2006). The Kripke semantics of Chadha et al. (2006) is similar to the one given in Braüner and de Paiva (2003, 2006), and also considered previously in the present chapter. In Chadha et al. (2006) it is proved that the natural deduction system given in Jia and Walker (2004) is sound and complete relative to the Kripke semantics. In Chadha et al. (2006) it is also proved that the natural deduction system is sound and complete relative to a birelational semantics.

[4] For a paper also having the aim being of providing a foundation for distributed functional programming languages, but using a natural deduction system for intuitionistic S5 without satisfaction operators, see Murphy et al. (2004).

Galmiche and Salhi (2008) give a sequent system for intuitionistic S5 with satisfaction operators corresponding to the natural deduction system considered in Jia and Walker (2004) and Chadha et al. (2006). Galmiche and Salhi (2008) demonstrated how to turn this sequent system into a terminating sequent system for the $\Box$-free fragment such that if a sequent is not provable, then a finite Kripke model falsifying the sequent can be defined. From this it follows that intuitionistic S5 with satisfaction operators does satisfy the finite model property with respect to the Kripke semantics if the $\Box$ modality is disregarded. A decision procedure for full intuitionistic hybrid logic will be described in a forthcoming paper (Galmiche and Salhi 2011).[5]

[5] Announced by D. Galmiche (personal communication).

Chapter 9
Labelled Versus Internalized Natural Deduction

In this chapter we compare the hybrid-logical natural deduction system given in Section 2.2 to a labelled natural deduction system for modal logic. The chapter is structured as follows. In the first section of the chapter we describe the labelled natural deduction system under consideration and in the second section we define a translation from this system to the hybrid-logical natural deduction system given in Section 2.2. In the third section we compare reductions in the two systems. The material in this chapter is taken from Braüner (2007).

9.1 A Labelled Natural Deduction System for Modal Logic

A *labelled* natural deduction, Gentzen, or tableau system for modal logic is a system where formulas involved in the rules are metalingustic formulas obtained by attaching labels to ordinary modal-logical formulas. The labels of labelled systems represent possible worlds of the usual Kripke semantics. Labelled systems often also involve an explicit representation of the accessibility relation of the Kripke semantics. Thus, rules in labelled systems are rules for reasoning directly about the Kripke models.

In this section we describe a labelled natural deduction system for modal logic. In what follows it is assumed that a countably infinite set of labels is given. The metavariables a, b, c, ... range over labels. The derivation rules of the labelled system for modal logic are given in Figure 9.1. In this system a distinction is made between the language of modal logic and a metalanguage involving two sorts of formulas: Formulas of the first sort are atomic first-order formulas of the form $R(a,c)$ and formulas of the second sort are formulas of the form (a,ϕ) where ϕ is a modal-logical formula. All formulas in the rules are such metalinguistic formulas. The first sort is used for relational reasoning and the second sort is used for propositional reasoning, relative to worlds. The modal-logical natural deduction system is a slightly modified version of a natural deduction system given in Basin et al. (1997). See also Viganò (2000). The system is sound and complete in the appropriate sense,

T. Braüner, *Hybrid Logic and its Proof-Theory*, Applied Logic Series 37,
DOI 10.1007/978-94-007-0002-4_9, © Springer Science+Business Media B.V. 2011

see Basin et al. (1997). The system can be seen as a classical version of a natural deduction system for intuitionistic modal logic given in Simpson (1994).

$$\frac{(a,\phi) \quad (a,\psi)}{(a,\phi\wedge\psi)}\ (\wedge I) \qquad \frac{(a,\phi\wedge\psi)}{(a,\phi)}\ (\wedge E1) \qquad \frac{(a,\phi\wedge\psi)}{(a,\psi)}\ (\wedge E2)$$

$$\frac{\begin{matrix}[(a,\phi)]\\ \vdots \\ (a,\psi)\end{matrix}}{(a,\phi\rightarrow\psi)}\ (\rightarrow I) \qquad \frac{(a,\phi\rightarrow\psi) \quad (a,\phi)}{(a,\psi)}\ (\rightarrow E)$$

$$\frac{\begin{matrix}[(a,\neg\phi)]\\ \vdots \\ (a,\bot)\end{matrix}}{(a,\phi)}\ (\bot 1)^{*} \qquad \frac{(a,\bot)}{(c,\bot)}\ (\bot 2)$$

$$\frac{\begin{matrix}[R(a,c)]\\ \vdots \\ (c,\phi)\end{matrix}}{(a,\Box\phi)}\ (\Box I)^{\star} \qquad \frac{(a,\Box\phi) \quad R(a,e)}{(e,\phi)}\ (\Box E)$$

$*$ ϕ is a propositional symbol.
$\star$ c does not occur in $(a,\Box\phi)$ or in any undischarged assumptions other than the specified occurrences of $R(a,c)$.

Fig. 9.1 Labelled natural deduction rules for modal logic

It is instructive to compare the rules for the labelled system given in Figure 9.1 with the rules for the system $\mathbf{N}_{\mathscr{H}}$ which are given in Figures 2.2 and 2.3 of Section 2.2 (the rules for binders are disregarded). First note that contrary to the labelled system, all formulas in the rules for $\mathbf{N}_{\mathscr{H}}$ are formulas of the object language, thus, in this sense the system $\mathbf{N}_{\mathscr{H}}$ is internalized. Also, note that contrary to the labelled system, relational reasoning is not separated from propositional reasoning in the system $\mathbf{N}_{\mathscr{H}}$.

9.2 The Internalization Translation

It is straightforward to translate formulas and derivations of the labelled natural deduction system for modal logic into formulas and derivations of the hybrid-logical natural deduction system $\mathbf{N}_{\mathscr{H}}$ of Section 2.2. We shall call this translation the *internalization* translation and denote it I. A metalinguistic formula ϕ of the first system is translated to a hybrid-logical satisfaction statement $I(\phi)$ by letting $I((a,\phi)) = @_a\phi$ and $I(R(a,c)) = @_a\Diamond c$. Obviously, the internalization translation preserves the semantics, where the modal-logical semantics is defined in an appropriate way, taking the metalinguistic machinery into account (details are left to the reader).

Having translated formulas, we translate derivations. In the next section we show that the translation preserves reductions and this involves a small lemma saying that the translation commutes with substitution of derivations for parcels of undischarged assumptions, and since a derivation is substituted for each undischarged assumption in a specified parcel, we need to be able to keep track of the identity of parcels when translating a derivation. To this end we introduce a few further conventions (similar conventions were introduced for similar reasons in Section 4.3): A set of annotated formulas will be called a *context* and the metavariables $\Phi, \Psi, \Omega, \ldots$ will range over contexts. Moreover, a derivation π is a *derivation from* a context Φ if each undischarged assumption in π is an occurrence of an annotated formula in Φ. Note that, in this book, we have most often considered derivations as being derivations from sets of formulas, that is, we have ignored numbers annotating undischarged assumptions. Keeping the numbers (as in Sections 4.3 and 4.4) enables us to keep track of the identity of parcels of undischarged assumptions when translating a derivation. The above translation of metalinguistic formulas is extended to contexts in the obvious way, namely by letting $I(\Phi) = \{(I(\theta))^r \mid \theta^r \in \Phi\}$.

Definition 9.1. A derivation π of ϕ from Φ in the modal-logical natural deduction system is translated to a derivation $I(\pi)$ of $I(\phi)$ from $I(\Phi)$ in the hybrid-logical natural deduction system $\mathbf{N}_{\mathscr{H}}$ by replacing each formula occurrence ψ in π by $I(\psi)$.

Note that the hybrid-logical introduction and elimination rules for the connectives $\wedge$, $\rightarrow$, and $\Box$ can be seen as obtained by taking the image under the translation I of the labelled modal-logical rules for these connectives.

9.3 Reductions

In this section we show that the internalization translation preserves reductions. Before doing so, we need to fix reduction rules for the natural deduction systems under consideration, that is, the labelled natural deduction system for modal logic given earlier in the present chapter and the internalized natural deduction system $\mathbf{N}_{\mathscr{H}}$. We have already given reduction rules for $\mathbf{N}_{\mathscr{H}}$ in Section 2.2.4, so we just need to give reduction rules for the modal-logical natural deduction system. First some conventions in connection with the modal-logical derivation rules. The premise of the form $R(a,e)$ in the rule $(\Box E)$ is called the *relational premise* and the premise of the form (a,ϕ) in the rule $(\rightarrow E)$ is called the *minor premise*. A premise of an elimination rule that is neither minor nor relational is called *major* . Of course, these conventions are analogous to the conventions for $\mathbf{N}_{\mathscr{H}}$ given in Section 2.2.4.

As usual for natural deduction systems, a *maximum formula* in a derivation is a formula occurrence that is both the conclusion of an introduction rule and the major premise of an elimination rule. Maximum formulas can be removed by applying *reduction rules*. The rules for reductions of the modal-logical system are as follows.

$(\wedge I)$ followed by $(\wedge E1)$ (analogously in the case of $(\wedge E2)$)

$$\dfrac{\dfrac{\begin{matrix}\vdots\ \pi_1\\(a,\phi)\end{matrix}\qquad\begin{matrix}\vdots\ \pi_2\\(a,\psi)\end{matrix}}{(a,\phi\wedge\psi)}}{(a,\phi)}\qquad\rightsquigarrow\qquad\begin{matrix}\vdots\ \pi_1\\(a,\phi)\end{matrix}$$

$(\rightarrow I)$ followed by $(\rightarrow E)$

$$\dfrac{\dfrac{\begin{matrix}[(a,\phi)]\\\vdots\ \pi_1\\(a,\psi)\end{matrix}}{(a,\phi\rightarrow\psi)}\qquad\begin{matrix}\vdots\ \pi_2\\(a,\phi)\end{matrix}}{(a,\psi)}\qquad\rightsquigarrow\qquad\begin{matrix}\vdots\ \pi_2\\(a,\phi)\\\vdots\ \pi_1\\(a,\psi)\end{matrix}$$

$(\Box I)$ followed by $(\Box E)$

$$\dfrac{\dfrac{\begin{matrix}[R(a,c)]\\\vdots\ \pi_1\\(c,\phi)\end{matrix}}{(a,\Box\phi)}\qquad\begin{matrix}\vdots\ \pi_2\\R(a,e)\end{matrix}}{(e,\phi)}\qquad\rightsquigarrow\qquad\begin{matrix}\vdots\ \pi_2\\R(a,e)\\\vdots\ \pi_1[e/c]\\(e,\phi)\end{matrix}$$

Note that the reduction rules for $\mathbf{N}_{\mathcal{H}}$ given in Section 2.2.4 can be seen as obtained by applying the internalization translation to all formulas displayed in the modal-logical reduction rules above, and adding a reduction rule for satisfaction operators. The reduction rules above can also be found in Basin et al. (1997) and Simpson (1994).

As usual for natural deduction systems, a derivation is *normal* if it contains no maximum formula. Using a variation of a standard technique originally given in Prawitz (1965), in Basin et al. (1997) it is proved that the modal-logical system satisfies a *normalization* theorem, that is, any derivation can be rewritten to a normal derivation by repeated applications of reductions (this is also the technique used in Lemma 2.10 of Section 2.2.4). The strategy of this technique is to select reductions such that a reduction of a maximum formula only generates new maximum formulas having fewer connectives than the original one. In Basin et al. (1997) it is also proved that every normal derivation satisfies a version of the subformula property.

Before proving that the internalization translation preserves reductions, we give a small lemma which says that the internalization translation commutes with substitution of derivations for undischarged assumptions in derivations.

Lemma 9.1. *Let τ and π be modal-logical derivations such that τ is a derivation of ϕ from Φ and π is a derivation from $\{\phi^r\} \cup \Phi \cup \Psi$ where $\phi^r \notin \Phi \cup \Psi$ and $\Phi \cap \Psi = \emptyset$. Moreover, let κ be the derivation obtained by substituting $I(\tau)$ for $(I(\phi))^r$ in $I(\pi)$ and let λ be the derivation obtained by substituting τ for ϕ^r in π. Then $\kappa = I(\lambda)$.*

Proof. Induction on the structure of the derivation of π.

We can now give the theorem which says that the internalization translation preserves reductions.

Theorem 9.1. *(Preservation of reductions) Let π be a modal-logical derivation. If $\pi \rightsquigarrow \tau$, then $I(\pi) \rightsquigarrow I(\tau)$.*

Proof. By induction on the structure of the derivation of π. If the end-formula of π is the conclusion of an elimination rule that is involved in a reduction, then in the cases $\rightarrow$ and $\Box$ we use Lemma 9.1, and in the latter case we furthermore use the observation that the translation commutes with substitution of nominals for nominals.

Preservation of reductions is a desirable property since the application of a reduction rule to a derivation is supposed to leave the identity of the proof represented by the derivation unchanged. Rather, the application of a reduction rule just removes a "detour" in the derivation. See the discussion in Prawitz (1971, p. 257).

9.4 Comparison of Reductions

From the above we conclude that formulas and derivations of the modal-logical natural deduction system correspond in a natural way to formulas and derivations of the hybrid-logical natural deduction system via the internalization translation, and moreover, reductions of the modal-logical system correspond naturally to reductions of the hybrid-logical system, as was shown in Theorem 9.1.

On the other hand, not all formulas and derivations of the hybrid-logical system correspond to formulas and derivations of the modal-logical system, which of course is no surprise since the hybrid-logical language involves the more expressive nominals and satisfaction operators, and associated derivation rules. But if we disregard the reduction rules for satisfaction operators, then the reduction rules of the hybrid-logical system only involve introductions followed by eliminations of connectives in the common language, namely introductions followed by eliminations of the connectives $\wedge$, $\rightarrow$, and $\Box$, hence, one might think that all sequences of such reductions in the hybrid-logical system correspond to sequences of reductions in the modal-logical system. However, this is not the case, in the following we shall explain why.

First note that the hybrid-logical introduction rule for the modal operator (given in Figure 2.2 of Section 2.2) not only exhibits a modal operator in the conclusion, but also in the discharged assumptions. To be more specific, the formula $@_a\Diamond c$

displayed in the hybrid-logical introduction rule for the modal operator, is a non-atomic formula of the object language containing a modal operator. This is also the case with the formula $@_a\Diamond e$ displayed in the hybrid-logical elimination rule for the modal operator (also in Figure 2.2). Thus, these formulas, as well as their subformulas, can be introduced and eliminated exactly as other formulas of the object language can be introduced and eliminated. This can simply not happen in the modal-logical system where the corresponding formulas $R(a,c)$ and $R(a,e)$ in the introduction and elimination rules for the modal operator, see Figure 9.1, are formulas of the metalanguage.

This difference between the systems is what gives rise to reduction sequences in the hybrid-logical system which do not correspond to any reduction sequences in the modal-logical system. As an example, consider the reduction sequence below where maximum formulas are indicated by putting frames around them.

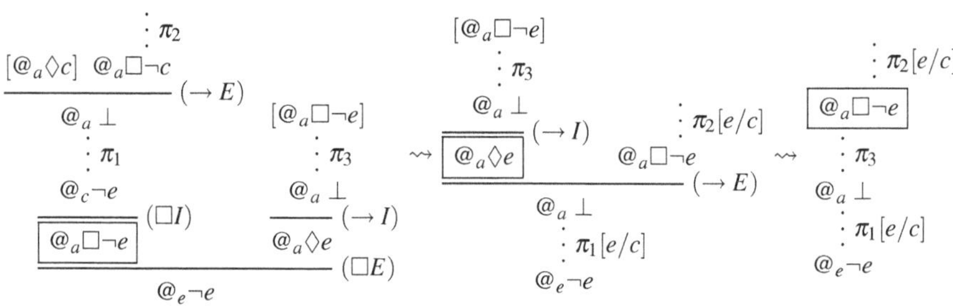

(Recall that $@_a\Diamond c$ is an abbreviation for the formula $@_a\neg\Box\neg c$ that in turn is an abbreviation for the formula $@_a(\Box\neg c \rightarrow \bot)$ which has $\rightarrow$ as the top-most connective, ignoring the satisfaction operator, meaning that this formula is introduced and eliminated by the rules for the $\rightarrow$ connective.) It is straightforward to define normal derivations π_1, π_2, and π_3 having the displayed forms such that the side-condition on the rule $(\Box I)$ is satisfied, that is, such that the nominal c does not occur in any undischarged assumptions of π_1 and π_2. It is assumed that the end-formula $@_a\Box\neg c$ in the derivation π_2 is the conclusion of an introduction rule and the specified assumption $@_a\Box\neg e$ in the derivation π_3 is the major premise of an elimination rule.

Now, the three derivations in the above reduction sequence each have exactly one maximum formula, and the first reduction removes the maximum formula $@_a\Box\neg e$ in the left-hand-side derivation, and this generates the new maximum formula $@_a\Diamond e$ in the middle derivation, which is removed by the second reduction, but this generates the new maximum formula $@_a\Box\neg e$ in the right-hand-side derivation, having exactly the same form as the maximum formula in the left-hand-side derivation. This reduction sequence does not correspond to any reduction sequence in the modal-logical natural deduction system, the reason being that in the modal-logical system the maximum formulas in such a deterministic reduction sequence will step by step become simpler, that is, they will have fewer and fewer connectives.

We mentioned in the previous section (in the remarks following the modal-logical reduction rules) that the technique used in the normalization proof for the modal-logical natural deduction system is a variation of a standard technique where reductions are selected such that the reduction of a maximum formula only generates new maximum formulas with fewer connectives than the original one. Clearly, this standard technique does not work in the example reduction sequence above where the reduction of a maximum formula of the form $@_a\Box\neg e$ generates a new maximum formula of the form $@_a\Diamond e$, the reason being that $@_a\Diamond e$ does not have fewer connectives than $@_a\Box\neg e$. However, the hybrid-logical natural deduction system does satisfy normalization, which was proved in Section 2.2.4 where the problem is solved by using what we have called the $\Box$-graph of a derivation to systematically control the application of reductions to new maximum formulas like $@_a\Diamond e$.[1]

9.4.1 A Remark on Normalization

As mentioned above, the hybrid-logical natural deduction system satisfies normalization. It is in this connection remarkable that there are connectives with introduction and elimination rules, which are similar to the hybrid-logical introduction and elimination rules for $\Box$, but which allow derivations that cannot be normalized. Consider for example the following introduction and elimination rules for a new unary connective $\sharp$.

$$\dfrac{\begin{array}{c}[@_a\sharp c]\\ \vdots \\ @_c\phi\end{array}}{@_a\sharp\phi}\,(\sharp I) \qquad\qquad \dfrac{@_a\sharp\phi \quad @_a\sharp e}{@_e\phi}\,(\sharp E)$$

The introduction rule for $\sharp$ is equipped with the side-condition that the nominal c does not occur in $@_a\sharp\phi$ or in any undischarged assumptions other than the specified occurrences of $@_a\sharp c$.[2] The reduction rule for the new connective $\sharp$ can be read off from the associated introduction and elimination rules (in fact, this is generally the case if the introduction and elimination rules have a certain form, see Sections 5.3.1 and 5.3.2 for an elaboration).

[1] This is actually a generally occurring problem (with a generally applicable solution) since the same problem crops up (and is solved in the same way) in connection with normalization for intuitionistic hybrid logic, cf. Section 8.2.4, and normalization for first-order hybrid logic, cf. Section 6.2.4. In the first case the reduction rule for the connective $\Diamond$, which in intuitionistic hybrid logic is primitive, not defined, might generate new maximum formulas on the form $@_a\Diamond e$, and in the second case, the reduction rule for $\forall$ might generate new maximum formulas on the form $@_aE(t)$ where $E(t)$, called the existence predicate, is an abbreviation for $\exists y(y=t)$ which in turn is an abbreviation for $\neg\forall y\neg(y=t)$.

[2] Melvin Fitting has pointed out that the rules for $\sharp$ are sound with respect to models based on the integers ordered by the successor relation, where $\sharp\phi$ is true at a point if and only if ϕ is true at the next point. This is a case of discrete, linear time. Note that the operator $\sharp$ is self-dual as the dual operator $\neg\sharp\neg$ gets the same interpretation as $\sharp$. Actually, in such models the standard modal operators $\Box$ and $\Diamond$ collapse and get the same interpretation as $\sharp$.

$$\dfrac{\dfrac{\begin{array}{c}[@_a\sharp c]\\ \vdots\,\pi_1\\ @_c\phi\end{array}}{@_a\sharp\phi} \qquad \begin{array}{c}\vdots\,\pi_2\\ @_a\sharp e\end{array}}{@_e\phi} \quad\rightsquigarrow\quad \begin{array}{c}\vdots\,\pi_2\\ @_a\sharp e\\ \vdots\,\pi_1[e/c]\\ @_e\phi\end{array}$$

The example reduction sequence below shows that there are derivations involving the new connective $\sharp$ that cannot be normalized.

$$\dfrac{\dfrac{\begin{array}{c}\dfrac{[@_a\sharp c]\quad[@_a\sharp c]}{@_c c}\,(\sharp E)\\ \vdots\,\pi\\ @_c e\end{array}}{@_a\sharp e}\,(\sharp I) \qquad \dfrac{\begin{array}{c}\dfrac{[@_a\sharp c]\quad[@_a\sharp c]}{@_c c}\,(\sharp E)\\ \vdots\,\pi\\ @_c e\end{array}}{@_a\sharp e}\,(\sharp I)}{@_e e}\,(\sharp E) \quad\rightsquigarrow\quad \begin{array}{c}\dfrac{\dfrac{\begin{array}{c}\dfrac{[@_a\sharp c]\quad[@_a\sharp c]}{@_c c}\,(\sharp E)\\ \vdots\,\pi\\ @_c e\end{array}}{@_a\sharp e}\,(\sharp I) \qquad \dfrac{\begin{array}{c}\dfrac{[@_a\sharp c]\quad[@_a\sharp c]}{@_c c}\,(\sharp E)\\ \vdots\,\pi\\ @_c e\end{array}}{@_a\sharp e}\,(\sharp I)}{@_e e}\,(\sharp E)\\ \vdots\,\pi[e/c]\\ @_e e\end{array}$$

It is straightforward to define a normal derivation π having the displayed form such that the side-condition on the rule $(\sharp I)$ is satisfied, that is, such that the nominal c does not occur in any undischarged assumptions of π. Note that there is a copy of the left-hand-side derivation in the right-hand-side derivation, thus, even though the reduction removes the maximum formula $@_a\sharp e$ in the left-hand-side derivation, the right-hand-side derivation contains a new copy of this maximum formula. This shows that normalization of the left-hand-side derivation is not possible.

Chapter 10
Why Does the Proof-Theory of Hybrid Logic Behave So Well?

The material in this chapter is primarily of a conceptual nature, the goal of the chapter being to put into perspective hybrid logic and the proof-theory of hybrid logic. The chapter is structured as follows. In the first section we explicate what we mean when we say that a proof-system is well-behaved. In the second section we shall try to give an answer to the following question: Why does the proof-theory of hybrid logic behave so well compared to the proof-theory of ordinary modal logic? In the third section we make some remarks in relation to proof-systems for classical propositional logic. In the fourth section we make some concluding philosophical remarks. The material in this chapter is mainly taken from Braüner (2007).

10.1 The Success Criteria

In this section we shall explicate what we mean when we say that a proof-system is well-behaved. That is, we shall describe the success criteria behind our claim that the proof-theory of hybrid logic behaves well compared to the proof-theory of ordinary modal logic. We state our success criteria in terms of natural deduction systems (but similar criteria can be given for Gentzen systems).

1. The introduction and elimination rules associated with each connective satisfy Prawitz' inversion principle.
2. The system satisfies normalization such that normal derivations satisfy a version of the subformula property.
3. Conditions on the accessibility relation can be incorporated into the system in a uniform way, that is, by just adding appropriate rules.

T. Braüner, *Hybrid Logic and its Proof-Theory*, Applied Logic Series 37,
DOI 10.1007/978-94-007-0002-4_10, © Springer Science+Business Media B.V. 2011

We find that these thee criteria are absolutely central.[1] We presented and discussed Prawitz' inversion principle earlier in Section 2.1. The inversion principle is also discussed in Section 5.3. As we explained in Section 2.1, it follows from the inversion principle that by rewriting a derivation, it is possible to remove a formula occurrence that is both introduced by an introduction rule and eliminated by an elimination rule, that is, a maximum formula. This rewrite process is formalized in proper reduction rules, hence, the inversion principle can be seen as a prerequisite for formulating a normalization theorem since such a theorem is relative to a set of reduction rules.[2] Note that the inversion principle is a property of the set of rules associated with a particular connective, whereas normalization is a property of a natural deduction system as a whole, relative to a set of reduction rules.

Now, roughly, there are two different kinds of natural deduction, Gentzen, and tableau systems for modal logic. The first kind of systems are the labelled systems described in the previous chapter where formulas involved in the rules are metalingustic formulas, and the second kind of systems are systems where formulas involved in the rules are formulas of the object language, that is, ordinary modal-logical formulas.[3] Systems of the second kind will be called *standard* systems. In the following two sections, we discuss these two kinds of systems.

[1] Also other criteria could be considered, one important example being interpolation, that is, the criterion that a proof system should lend itself to the calculation of interpolants, perhaps after being enhanced with further machinery, like the tableau system for first-order hybrid logic which in Blackburn and Marx (2003) is used as the basis of an algorithm that calculates interpolants. See the remarks on interpolation in Section 1.4. Note that there are two steps: The first step is the requirement of a logic (which here is a formal language together with a semantics) that it satisfies interpolation. This might be proved semantically, independent of any proof systems. If the logic does satisfy interpolation, then the second step is the requirement of a proof system for the logic that the proof system in question can be used as the basis for calculating interpolants.

[2] But the inversion principle does not imply that the normalization theorem actually holds, cf. the introduction and elimination rules for the connective $\sharp$ given in Section 9.4.1. These rules satisfy the inversion principle, thus, reduction rules for the connective $\sharp$ can be given, but if the natural deduction system $\mathbf{N}_{\mathcal{H}}$ for hybrid logic given in Section 2.2 is extended with the introduction and elimination rules for $\sharp$, and the set of reduction rules is extended with the reduction rule for $\sharp$, then the system no longer satisfies normalization.

[3] It should be mentioned that there are a number of natural deduction and Gentzen style formulations for modal logic that do not fit this categorization well. Notable here are formulations in terms of Nuel Belnap's display logic and Kosta Dŏsen's higher-level sequents. However, these formulations differ considerably from Gentzen's original natural deduction and sequent systems and they are more complicated from a technical point of view. (Although it has to be acknowledged that display sequents as well as higher-level sequents were introduced as natural generalisations of Gentzen's notion of a sequent, intended to allow a uniform sequent-style formulation of many different logics.) An overview can be found in Wansing (1994). Also notable are modal hypersequent systems, see Avron (1996) as well as the handbook chapter (Fitting 2007).

10.1.1 Standard Systems for Modal Logic

No known standard natural deduction systems for modal logic satisfy all three success criteria given above. The first two criteria are satisfied by a number of systems, but when a standard modal-logical natural deduction, Gentzen, or tableau system is given, it is usually for one particular modal logic. This is for example the case with the natural deduction systems for the modal logics S4 and S5 given in Prawitz (1965).[4] With reference to Prawitz's systems for S4 and S5, Robert Bull and Krister Segerberg note the following in their survey paper on modal logic in *Handbook of Philosophical Logic*.

> However, it has proved difficult to extend this sort of analysis to the great multitude of other systems of modal logic. It seems fair to say that a deductive treatment congenial to modal logic is yet to be found, for Hilbert systems are not suited for the purpose of actual deduction, ... (Bull and Segerberg 2001, p. 25)

Bull and Segerberg continue.

> ... only exceptional systems would seem to be characterizable in terms of reasonably simple rules. (Bull and Segerberg 2001, p. 27)

This view was also expressed in the quotation by Heinrich Wansing in the preface of the present book. In the quoted passage, which we recapitulate below, Wansing gives a succint summary of the status of modal-logical proof-theory, pointing out the lack of general results.

> Compared with the multitude of not only existing but also interesting axiomatically presentable normal modal propositional logics, the number of systems for which sequent calculus presentations (of some sort) are known is disappointingly small. In contrast to the axiomatic approach, the standard sequent-style proof theory for normal modal logics fails to be 'modular', and the very mechanism behind the small range of known possible variations is not very clear. (Wansing 1994, p. 128)

We find that the lack of uniformity described above is a major deficiency of standard modal-logical proof-theory.

10.1.2 Labelled Systems for Modal Logic

Contrary to standard natural deduction systems for modal logic, labelled systems usually satisfy all three criteria given above. Thus, rules for reasoning directly about the Kripke models are proof-theoretically well-behaved. This is for example the case with Viganò's natural deduction system for classical modal logic given in Figure 9.1 of Section 9.1. Another example is Simpson's natural deduction system for

[4] In fact, Prawitz' systems for S4 and S5 deviate from most standard systems since his introduction rules for $\Box$ make use of "non-local" side-conditions, that is, side-conditions that do not just refer to the premises of the rules and to undischarged assumptions, but to the whole derivations of the premises.

intuitionistic modal logic given in Simpson (1994). Both of these systems involve metalinguistic formulas of two sorts, namely atomic first-order formulas of the form $R(a,c)$ and labelled formulas of the form (a,ϕ) where ϕ is a modal-logical formula. This metalinguistic machinery enables the formulation of the following introduction and elimination rules for the modal operator (cf. Figure 9.1).

$$\dfrac{\begin{array}{c}[R(a,c)]\\ \vdots\\ (c,\phi)\end{array}}{(a,\Box\phi)}\ (\Box I) \qquad\qquad \dfrac{(a,\Box\phi) \quad R(a,c)}{(c,\phi)}\ (\Box E)$$

The introduction rule, that is, the rule $(\Box I)$, is equipped with the side-condition that the label c does not occur in $(a,\Box\phi)$ or in any undischarged assumptions other than the specified occurrences of $R(a,c)$. Compare these introduction and elimination rules to the truth-condition for the modal operator in the Kripke semantics.

$$\Box\phi \text{ is true at } a \text{ iff for any } c \text{ such that } aRc,\ \phi \text{ is true at } c$$

Clearly, the introduction rule for the modal operator can be read off from the right-to-left direction of the truth-condition and the elimination rule can be read off from the left-to-right direction. Thus, the rules can be read off from the modal operator's truth-condition in the Kripke semantics. We shall come back to this in Section 10.4. Beside the above introduction and elimination rules for the modal operator, the metalinguistic machinery enables the formulation of rules for first-order conditions on the accessibility relation, in Viganò's case conditions expressed by Horn clause theories, and in Simpson's case, conditions expressed by geometric theories.[5]

For other works on labelled proof systems for modal logic, see Fitting (1983) as well as later publications by Fitting. Moreover, see Gabbay (1996) on labelled deductive systems. The fundamental idea of Gabbay's labelled deductive systems is to prefix formulas in derivations by labels with the aim of regulating the proof process. In fact, labelled deductive systems are proposed as a systematic way of giving proof systems to many different logics. See also the discussion in Section 2.2.6.

10.2 Why Hybrid-Logical Proof-Theory Behaves So Well

To sum up, standard natural deduction systems for modal logic do not satisfy all three criteria given earlier, whereas labelled systems usually do satisfy the criteria, but at the expense of making use of metalinguistic machinery. As has been demon-

[5] Recall from Section 2.2.1 that a first-order formula is geometric if it is built out of atomic formulas of the forms $R(a,c)$ and $a=c$ using only the connectives $\bot$, $\wedge$, $\vee$, and $\exists$. A geometric theory is a finite set of closed first-order formulas, each having the form $\forall\bar{a}(\phi\rightarrow\psi)$, where the formulas ϕ and ψ are geometric. See Section 2.2.1 for more details.

strated in Section 2.2, with the propositional hybrid-logical natural deduction system $\mathbf{N}_{\mathscr{H}}$, this deficiency can be remedied by hybridization, that is, hybridization of modal logic enables the formulation of a natural deduction system such that the criteria all are satisfied without involving metalinguistic machinery, in particular, rules can be added to the system corresponding to first-order conditions on the accessibility relation expressed by geometric theories.[6]

Which features of hybrid logic have enabled us to formulate natural deduction systems satisfying the three criteria without involving metalinguistic machinery? In technical terms, the answer is that hybrid-logic has the following two features.[7]

- We can express that a formula ϕ is true at a world a, that is, the formula $@_a\phi$ is true.
- We can express that a world a is R-related to a world c, that is, the formula $@_a\Diamond c$ is true.

In Section 9.2 these features enabled us to define the internalization translation I which translates metalinguistic formulas (a,ϕ) and $R(a,c)$ into hybrid-logical formulas by letting $I((a,\phi)) = @_a\phi$ and $I(R(a,c)) = @_a\Diamond c$. Furthermore, by applying the translation I to all formulas in a derivation, the translation was extended such that it translates a derivation of the labelled modal-logical natural deduction system to a derivation of the hybrid-logical natural deduction system $\mathbf{N}_{\mathscr{H}}$. By considering this translation, we can see why we can formulate a hybrid-logical natural deduction system, namely $\mathbf{N}_{\mathscr{H}}$, that satisfies the three criteria. We consider each of the criteria in turn.

As pointed out in Section 9.2, the hybrid-logical introduction and elimination rules for the connectives $\wedge$, $\rightarrow$, and $\Box$ can be seen as obtained by taking the image under the translation I of the labelled modal-logical rules for these connectives, that is, rules for reasoning directly about the Kripke models. In the case of the modal operator, this results in the following introduction and elimination rules (cf. Figure 2.2 of Section 2.2).

$$\dfrac{\begin{matrix}[@_a\Diamond c] \\ \vdots \\ @_c\phi\end{matrix}}{@_a\Box\phi}\ (\Box I) \qquad\qquad \dfrac{@_a\Box\phi \quad @_a\Diamond c}{@_c\phi}\ (\Box E)$$

The introduction rule is equipped with the side-condition that the nominal c does not occur in $@_a\Box\phi$ or in any undischarged assumptions other than the specified occur-

[6] See Sections 8.2 and 6.2 for the cases of intuitionistic and first-order hybrid logic. In the latter case the accessibility relation as well as the quantifier domains are subject to first-order conditions expressed by geometric theories.

[7] These two features also enable the formulation of natural deduction systems for intuitionistic hybrid logics satisfying the criteria, cf. Section 8.2, but in that case the features are interpreted intuitionistically, that is, they are interpreted as statements in intuitionistic first-order logic and intuitionistic hybrid logic. In the case of first-order hybrid logic, cf. Section 6.2, we can furthermore express that an individual t exists at a world a, that is, the formula $@_aE(t)$ is true, which enables the formulation of natural deduction systems for first-order hybrid logics satisfying the criteria.

rences of $@_a\Diamond c$. For each of the connectives $\wedge$, $\rightarrow$, and $\Box$, the hybrid-logical rules satisfy the inversion principle as the labelled rules satisfy it. Besides these connectives, it is straightforward to give introduction and elimination rules for satisfaction operators which satisfy the inversion principle.

The inversion principle gives rise to hybrid-logical reduction rules such that normalization is satisfied and such that normal derivations satisfy a version of the subformula property, see Sections 2.2.4 and 2.2.5. Reductions in the labelled modal-logical system correspond to reductions in the hybrid-logical system, as the translation I preserves reductions, cf. Section 9.3, but there are reduction sequences in the hybrid-logical system involving the modal operator that do not correspond to any reduction sequences in the labelled modal-logical system, as explained in Section 9.4. That such reduction sequences are possible follows from the fact that the hybrid-logical introduction rule for the modal operator not only exhibits a modal operator in the conclusion, but also in the discharged assumptions (that is, in the formula $@_a\Diamond c$ displayed in the introduction rule).[8] The proof that normalization is satisfied involves controlling such reduction sequences in a systematic way, cf. Section 9.4.

Conditions on the accessibility relation can be incorporated into the system by adding hybrid-logical rules obtained by taking the image under the translation I of labelled modal-logical natural deduction rules given in Simpson (1994) (strictly speaking, extended with atomic first-order formulas of the form $a = c$ which are translated to $@_a c$). The conditions on the accessibility relation are first-order conditions expressed by geometric theories. This actually requires the addition of further reduction rules with the aim of removing permutable formulas in a derivation, see Section 2.2.4 for more on such permutative reductions.

In conclusion, what has happened is that the metalinguistic formulas and rules of the labelled modal-logical natural deduction system have been internalized as hybrid-logical formulas and rules via the translation I, which has enabled the formulation of an internalized hybrid-logical natural deduction system involving only object language formulas such that the three criteria are satisfied.[9] In other words, we have provided a proof-theoretic analysis demonstrating that the good proof-theoretic behaviour of the labelled rules for reasoning directly about models is preserved by internalization. We are now in position to give an answer to the question why the proof-theory of hybrid logic behaves so well. The answer is that internalization of

[8] According to Prawitz' terminology, cf. Section 5.3.1, a natural deduction introduction rule for a connective is called explicit if the connective in question is exhibited exactly once, namely in the conclusion of the rule. Thus, according to this terminology, the hybrid-logical introduction rule for the modal operator is not explicit.

[9] Kai Brünnler (2006) compares labelled and unlabelled Gentzen systems for modal logic. A system of the latter kind is a system that does not use labels, which he makes more precise by calling a Gentzen system *pure* if each sequent has an equivalent object language formula. Clearly, what we here call standard proof systems for modal logic are pure: In natural deduction terminology, a derivation of a modal-logical formula ϕ from a set of modal-logical formulas Γ is equivalent to the modal-logical formula $\bigwedge\Gamma \rightarrow \phi$. On the other hand, labelled natural deduction systems for modal logic are clearly not pure, but it is remarkable that hybrid-logical natural deduction systems actually are pure in Brünnler's sense.

metalinguistic model-theoretic machinery in the object language enables us to give well-behaved proof-theory for hybrid logic.

The issue of internalizing a metalanguage in a hybrid-logical object language is also discussed in a range of papers by Patrick Blackburn, see Blackburn (2000a,b). See also the discussion in Section 2.2.6.

Moreover, internalizing a metalanguage in an object language is the subject of Seligman (2001). The approach in that paper is, however, different from the approach taken here: In Seligman (2001) a Gentzen system for hybrid logic is developed from a Gentzen system for first-order predicate logic by a series of transformations which step by step internalizes the (first-order) semantic theory of hybrid logic.

10.3 Comparison to Internalization of Bivalent Semantics

It is instructive to compare the internalization process described in the previous section with the internalization of the standard bivalent semantics of classical propositional logic in tableau systems. We shall do this in the present section. To this end we shall make use of a language for propositional logic where the connectives negation and conjunction are both taken as primitive, like in Chapter 3.

It is well-known that there are two different types of tableau systems for classical propositional logic, namely signed and unsigned tableau systems. An unsigned system was introduced already in Figure 3.1 of Section 3.1. For the convenience of the reader the unsigned system is reproduced in Figure 10.1. A signed tableau system is introduced in Figure 10.2. In the signed tableau system, the truth-values true and false of the standard bivalent semantics are explicitly represented in the metalanguage by the signs T and F,[10] whereas in unsigned tableau system the truth-values are not represented explicitly.

$$\frac{\neg\neg\phi}{\phi}\,(\neg\neg)$$

$$\frac{\phi \wedge \psi}{\phi, \psi}\,(\wedge) \qquad \frac{\neg(\phi \wedge \psi)}{\neg\phi \mid \neg\psi}\,(\neg\wedge)$$

Fig. 10.1 Unsigned tableau rules for propositional logic

However, the metalinguistic signs T and F in the signed system can be simulated in the unsigned system by using an object language connective, namely negation,

[10] This is different from signed tableau systems for intuitionistic propositional logic where the signs T and F do not represent the truth-values true and false in the usual classical sense, but rather *known* and *not known*

$$\frac{\mathtt{T}\neg\phi}{\mathtt{F}\phi}\,(\mathtt{T}\neg) \qquad \frac{\mathtt{F}\neg\phi}{\mathtt{T}\phi}\,(\mathtt{F}\neg)$$

$$\frac{\mathtt{T}(\phi\wedge\psi)}{\mathtt{T}\phi,\mathtt{T}\psi}\,(\mathtt{T}\wedge) \qquad \frac{\mathtt{F}(\phi\wedge\psi)}{\mathtt{F}\phi\mid\mathtt{F}\psi}\,(\mathtt{F}\wedge)$$

Fig. 10.2 Signed tableau rules for propositional logic

that is, a formula $\mathtt{F}\phi$ in the signed system can be simulated by $\neg\phi$ in the unsigned system, and a formula $\mathtt{T}\psi$ can simply be simulated by ψ. To be precise, a tableau in the signed system can be translated into a tableau in the unsigned system by replacing all occurrences of $\mathtt{F}$ by $\neg$, by deleting all occurrences of $\mathtt{T}$, and by deleting all instances of the rule $(\mathtt{T}\neg)$. Conversely, a tableau in the unsigned system can be translated into a tableau in the signed system by replacing each formula of the form $\neg\phi$ by $\mathtt{F}\phi$, by replacing each formula ψ not of the form $\neg\phi$ by $\mathtt{T}\psi$, and by inserting instances of the rule $(\mathtt{T}\neg)$ where relevant. Of course, both of these translations preserve open and closed branches.

In the sense above, the standard bivalent semantics which is explicitly represented in the signed tableau system has been internalized in the unsigned tableau system. This has simplified the metalanguage in the sense that signed formulas have been replaced by ordinary unsigned formulas, but the unsigned rule corresponding to a signed rule involves more than one connective, that is, besides the connective involved in the original signed rule, the unsigned rule involves negation (the signed rule $(\mathtt{F}\wedge)$ only involves conjunction whereas the unsigned rule $(\neg\wedge)$ involves conjunction as well as negation). Thus, the metalanguage representation of a feature of the semantics (bivalence) has been replaced by the representation of the same feature in the object language (negation).[11]

Now, the feature of the Kripke semantics that truth is relative to possible worlds is exactly what is explicitly represented by the metalinguistic labels of the labelled natural deduction system for modal logic given in Figure 9.1. In hybrid logic, the feature that truth is relative to possible worlds is represented in the object language by the satisfaction operators. This means that the metalinguistic labels a, c, e, ... in the rules of Figure 9.1 can be simulated by satisfaction operators, to be precise, the metalinguistic labels can simply be replaced by satisfaction operators prefixing all formulas. This is indeed what is done by the internalization translation I defined in Section 9.2. This is analogous to the case with ordinary propositional logic where the explicit metalanguage representation of bivalence in the signed rules of Figure 10.2 has been replaced by the object language representation of bivalence in the unsigned rules of Figure 10.1. Moreover, the accessibility relation of the Kripke

[11] The above comparison of signed and unsigned tableau systems is analogous to two-sided Gentzen systems in comparison to one-sided Gentzen systems. In two-sided Gentzen systems, the truth-values true and false are represented in the metalanguage by the separation between the left and right hand sides of a sequent, whereas in one-sided Gentzen systems the truth-values are not represented.

semantics is represented by metalinguistic first-order formulas of the form $R(a,c)$ in the labelled natural deduction system. In hybrid logic, the accessibility relation can be represented by object language formulas of the form $@_a\Diamond c$, thus, formulas of the form $R(a,c)$ in the rules of Figure 9.1 can be simulated by hybrid-logical formulas of the form $@_a\Diamond c$. Again, this is what is done by the internalization translation.

Note the general pattern: A rule in which some feature of the semantics (bivalence, relativisation of truth to possible worlds, the accessibility relation) is explicitly represented can be simulated by a rule in which the feature in question is only represented in the object language (as negation, satisfaction operators, formulas of the form $@_a\Diamond c$) and the new rule involves this object language representation besides the connective which was originally involved in the old rule.

10.4 Some Concluding Philosophical Remarks

Recall that in Section 10.1.2 we pointed out that the labelled introduction and elimination rules for the modal operator can be read off from the modal operator's truth-condition in the Kripke semantics. This also applies to the hybrid-logical, internalized introduction and elimination rules for the modal operator discussed in Section 10.2. Thus, we have justified the derivation rules for the modal operator by an antecedent understanding of the modal operator's meaning, namely by its truth-condition in the Kripke semantics. Some remarks should be made in this connection.

First note that our justification of the derivation rules for the modal operator is in terms of model-theoretic semantics, namely in terms of the Kripke semantics. This is related to a distinction Jaako Hintikka draws between two different traditions in viewing the relation of logic to reality, namely "language as the universal medium" (or "the universalist tradition") and "language as calculus" (or "the model-theoretical tradition"), see Hintikka (1988). According to the first tradition, one cannot step outside the language and one cannot theorize about changes in the interpretation of the language. Contrary to this, the second tradition includes meta-logical considerations and the tradition takes as a cornerstone the relation between formulas and models defined by Tarski-style truth-conditions. Clearly, our justification of the introduction and elimination rules for the modal operator presupposes the second view since the justification is in terms of a model-theoretic semantics.

An alternative to justifying derivation rules for logical connectives in terms of model-theoretic semantics, is to explain the meaning of the connectives in terms of the roles the connectives play in derivations, independently of model-theoretic notions. It is arguable that both kinds of explanation are legitimate, cf. the following long quotation by Nuel D. Belnap (1962).

> It seems plain that throughout the whole texture of philosophy one can distinguish between two modes of explanation: the analytic mode, which tends to explain wholes in terms of parts, and the synthetic mode, which explains parts in terms of the wholes or contexts in which they occur. In logic, the analytic mode would be represented by Aristotle, who commences with terms as the ultimate atoms, explains propositions or judgements by means

> of these, syllogisms by means of the propositions which go to make them up, and finally ends with the notion of science as a tissue of syllogisms. The analytic mode is also represented by the contemporary logician who first explains the meaning of complex sentences, by means of truth-tables, as a function of their parts, and then proceeds to give an account of correct inference in terms of the sentences occurring therein. ... Among formal logicians, use of the synthetic mode in logic is illustrated by Kneale and Popper ..., as well as by Jaskowski, Gentzen, Fitch, and Curry, all of these treating the meaning of connectives as arising from the role they play in the context of formal inference. It is equally well illustrated, I think, by aspects of Wittgenstein and those who learned from him to treat the meanings of words as arising from the role they play in the context of discourse. It seems to me nearly self-evident that employment of both modes of explanation is important and useful. (Belnap 1962, pp. 130–131)[12]

The programme of explaining the meaning of logical connectives in terms of the roles they play in derivations has developed into a separate branch of logic called *proof-theoretic semantics*. To be more precise, proof-theoretic semantics is based on the idea of explaining the meaning of a logical connective in terms of a set of derivation rules.[13] In certain respects, proof-theoretic semantics is in line with the first tradition identified by Hintikka. We shall not go into further details of proof-theoretic semantics. See Wansing (2000) for a presentation of a number of different semantic paradigms.

[12] Belnap (1962) is a response to Prior (1960) in which Prior raises doubt as to whether the meaning of logical connectives can be explained in terms of derivation rules along the lines of natural deduction introduction and elimination rules. In his paper, Prior introduces a logical connective *tonk* with introduction rules similar to the standard natural deduction introduction rules for disjunction and elimination rules similar to the standard natural deduction elimination rules for conjunction. An effect of extending a formal system with *tonk* together with the mentioned rules is that any formula becomes derivable, which obviously is absurd. In his response to Prior's paper, Belnap suggests imposing certain restrictions on derivation rules, thereby excluding Prior's rules for *tonk* from the permissible rules for a connective. According to Belnap, the crucial restriction is *conservativity*: When a formal system is extended with a new logical connective together with a set of derivation rules, then for any formula built using only the original connectives, it is required that if the formula is derivable in the extended system, then it is also derivable in the original system, that is, it is derivable without using the derivation rules for the new connective.

[13] A frequently discussed issue in proof-theoretic semantics is which restrictions to impose on the derivation rules for a connective, that is, which sets of derivation rules to take as permissible. This discussion can be traced back to Prior (1960) and Belnap (1962), cf. the previous footnote. A number of restrictions on derivation rules have been proposed, one proposal being conservativity, cf. the previous footnote, another proposal being the inversion principle, cf. Prawitz (1971).

References

R. Alur and T.A. Henzinger. A really temporal logic. In *Proceedings of the 30th Annual Symposium on Foundations of Computer Science (FOCS)*, pages 164–169. IEEE Computer Society Press, 1989.

R. Alur and T.A. Henzinger. Logics and models of real time: A survey. In *Real Time: Theory in Practice*, volume 600 of *Lecture Notes in Computer Science*, pages 74–106. Springer Berlin, 1992.

A.R. Anderson and N.D. Belnap. *Entailment: The Logic of Relevance and Necessity*, volume 1. Princeton University Press, 1975.

C. Areces. *Logic Engineering. The Case of Description and Hybrid Logics*. PhD thesis, Institute for Logic, Language and Computation, University of Amsterdam, 2000.

C. Areces and J. Heguiabehere. HyLoRes 1.0: Direct resolution for hybrid logics. In A. Voronkov, editor, *Proceedings of the 18th International Conference on Automated Deduction*, volume 2392 of *Lecture Notes in Computer Science*, pages 156–160, 2002.

C. Areces and B. ten Cate. Hybrid logics. In P. Blackburn, J. van Benthem, and F. Wolter, editors, *Handbook of Modal Logic*, pages 821–868. Elsevier, 2007.

C. Areces, P. Blackburn, and M. Marx. The computational complexity of hybrid temporal logics. *Logic Journal of the IGPL*, 8:653–679, 1999.

C. Areces, P. Blackburn, and M. Marx. Hybrid logics: Characterization, interpolation and complexity. *Journal of Symbolic Logic*, 66:977–1010, 2001.

C. Areces, P. Blackburn, and M. Marx. Repairing the interpolation theorem in quantified modal logic. *Annals of Pure and Applied Logic*, 124:287–299, 2003.

C. Areces, M. de Rijke, and H. de Nivelle. Resolution in modal, description and hybrid logic. *Journal of Logic and Computation*, 11:717–736, 2001.

A. Avron. The method of hypersequents in the proof theory of propositional non-classical logics. In W. Hodges, M. Hyland, C. Steinhorn, and J. Truss, editors, *Logic: Foundations to Applications*, pages 1–32. Oxford Science Publications, 1996.

R.C. Barcan. A functional calculus of first order based on strict implication. *Journal of Symbolic Logic*, 11:1–16, 1946.

D. Basin, S. Matthews, and L. Viganò. Labelled propositional modal logics: Theory and practice. *Journal of Logic and Computation*, 7:685–717, 1997.

N.D. Belnap. Tonk, plonk and plink. *Analysis*, 22:130–134, 1962.

N.D. Belnap. How a computer should think. In G. Ryle, editor, *Contemporary Aspects of Philosophy*, pages 30–55. Oriel Press, 1976.

E. Bencivenga. Free logics. In D.M. Gabbay and F. Guenthner, editors, *Handbook of Philosophical Logic, 2nd Edition*, volume 5, pages 147–196. Kluwer, Dordrecht 2002.

N. Benton, G.M. Bierman, V. de Paiva, and M. Hyland. Term assignment for intuitionistic linear logic. Technical Report 262, Computer Laboratory, University of Cambridge, 1992.

T. Braüner, *Hybrid Logic and its Proof-Theory*, Applied Logic Series 37,
DOI 10.1007/978-94-007-0002-4,

N. Benton, G.M. Bierman, V. de Paiva, and M. Hyland. A term calculus for intuitionistic linear logic. In M. Bezem and J.F. Groote, editors, *Proceedings of First International Conference on Typed Lambda Calculi and Applications*, volume 664 of *Lecture Notes in Computer Science*, pages 75–90. Springer Berlin, 1993.

G.M. Bierman and V. de Paiva. On an intuitionistic modal logic. *Studia Logica*, 65:383–416, 2000.

P. Blackburn. Nominal tense logic. *Notre Dame Journal of Formal Logic*, 14:56–83, 1993.

P. Blackburn. Internalizing labelled deduction. *Journal of Logic and Computation*, 10:137–168, 2000a.

P. Blackburn. Representation, reasoning, and relational structures: A hybrid logic manifesto. *Logic Journal of the IGPL*, 8:339–365, 2000b.

P. Blackburn. Arthur Prior and hybrid logic. *Synthese*, 150:329–372, 2006. Special issue edited by T. Braüner, P. Hasle, and P. Øhrstrøm.

P. Blackburn and M. Marx. Tableaux for quantified hybrid logic. In U. Egly and C. Fermüller, editors, *Automated Reasoning with Analytic Tableaux and Related Methods, TABLEAUX 2002*, volume 2381 of *Lecture Notes in Artificial Intelligence*, pages 38–52. Springer Berlin, 2002.

P. Blackburn and M. Marx. Constructive interpolation in hybrid logic. *Journal of Symbolic Logic*, 68:463–480, 2003.

P. Blackburn and J. Seligman. Hybrid languages. *Journal of Logic, Language and Information*, 4:251–272, 1995.

P. Blackburn and B. ten Cate. Pure extensions, proof rules, and hybrid axiomatics. *Studia Logica*, 84:277–322, 2006.

P. Blackburn and M. Tzakova. Hybridizing concept languages. *Annals of Mathematics and Artificial Intelligence*, 24:23–49, 1998.

P. Blackburn and M. Tzakova. Hybrid languages and temporal logic. *Logic Journal of the IGPL*, 7:27–54, 1999.

P. Blackburn, M. de Rijke, and Y. Venema. *Modal Logic*, volume 53 of *Cambridge Tracts in Theoretical Computer Science*. Cambridge University Press, Cambridge, 2001.

G. Boolos. Don't eliminate cut. *Journal of Philosophical Logic*, 13:373–378, 1984.

T. Bolander and P. Blackburn. Termination for hybrid tableaus. *Journal of Logic and Computation*, 17:517–554, 2007.

T. Bolander and P. Blackburn. Terminating tableau calculi for hybrid logics extending K. In C. Areces and S. Demri, editors, *Proceedings of Methods for Modalities 5*, volume 231 of *Electronic Notes in Theoretical Computer Science*, pages 21–39. Elsevier, 2009.

T. Bolander and T. Braüner. Two tableau-based decision procedures for hybrid logic. In H. Schlingloff, editor, *4th Workshop "Methods for Modalities" (M4M), Informatik-Bericht Nr. 194*, pages 79–96. Humboldt-Universität zu Berlin, 2005.

T. Bolander and T. Braüner. Tableau-based decision procedures for hybrid logic. *Journal of Logic and Computation*, 16:737–763, 2006. Revised and extended version of Bolander and Braüner (2005).

L. Bozzato, M. Ferrari, and P. Villa. A note on constructive semantics for description logics. In *Proceedings of CILC09, 24-esimo Convegno Italiano di Logica Computazionale*, 2009.

T. Braüner. *An Axiomatic Approach to Adequacy*. PhD thesis, Department of Computer Science, University of Aarhus, 1996. 168 pages. Published as Technical Report BRICS-DS-96-4.

T. Braüner. A cut-free Gentzen formulation of the modal logic S5. In R. de Queiroz and W. Carnielli, editors, *Proceedings of 6th Workshop on Logic, Language, Information and Computation*, pages 79–88, 1999.

T. Braüner. A cut-free Gentzen formulation of the modal logic S5. *Logic Journal of the IGPL*, 8:629–643, 2000. Full version of Braüner (1999).

T. Braüner. Natural deduction for hybrid logic (extended abstract). In C. Areces and M. de Rijke, editors, *Workshop Proceedings of Methods for Modalities 2*. ILLC Amsterdam, Netherlands, 2001.

T. Braüner. Homophonic theory of truth for tense logic. In F. Wolter, H. Wansing, M. de Rijke, and M. Zakharyaschev, editors, *Advances in Modal Logic*, Volume 3, pages 59–72. World Scientific, 2002a.

T. Braüner. Modal logic, truth, and the master modality. *Journal of Philosophical Logic*, 31:359–386, 2002b. Revised and extended version of Braüner (2002a).

T. Braüner. Natural deduction for first-order hybrid logic. In C. Areces, P. Blackburn, M. Marx, and U. Sattler, editors, *Workshop Proceedings of Fourth Workshop on Hybrid Logics*, pages 37–51, 2002c. Affiliated to *IEEE Symposium on Logic in Computer Science*.

T. Braüner. Functional completeness for a natural deduction formulation of hybridized S5. In P. Balbiani, N.-Y. Suzuki, F. Wolter, and M. Zakharyaschev, editors, *Advances in Modal Logic, Volume 4*, pages 31–49. King's College Publications, 2003.

T. Braüner. Natural deduction for hybrid logic. *Journal of Logic and Computation*, 14:329–353, 2004a. Revised and extended version of Braüner (2001).

T. Braüner. Two natural deduction systems for hybrid logic: A comparison. *Journal of Logic, Language and Information*, 13:1–23, 2004b.

T. Braüner. Hybrid logic. In E.N. Zalta, editor, *The Stanford Encyclopedia of Philosophy*. Stanford University, 2005a. On-line encyclopedia article available at http://plato.stanford.edu/entries/logic-hybrid.

T. Braüner. Natural deduction for first-order hybrid logic. *Journal of Logic, Language and Information*, 14:173–198, 2005b. Revised and extended version of Braüner (2002c).

T. Braüner. Proof-theoretic functional completeness for the hybrid logics of everywhere and elsewhere. *Studia Logica*, 81:191–226, 2005c. Revised and extended version of Braüner (2003).

T. Braüner. Axioms for classical, intuitionistic, and paraconsistent hybrid logic. *Journal of Logic, Language and Information*, 15:179–194, 2006.

T. Braüner. Why does the proof-theory of hybrid logic work so well? *Journal of Applied Non-Classical Logics*, 17:521–543, 2007.

T. Braüner. Adding intensional machinery to hybrid logic. *Journal of Logic and Computation*, 18:631–648, 2008.

T. Braüner. Hybrid logic and its proof-theory. Computer Science Research Report 124, Roskilde University, Denmark, 2009. Thesis accepted in fulfillment of the requirements for the Danish higher doctorate dr.scient. (doctor scientiarum).

T. Braüner. Hybrid logic. In D.M. Gabbay and F. Guenthner, editors, *Handbook of Philosophical Logic, 2nd Edition*, volume 17. Springer, Berlin, 2011. To appear.

T. Braüner and V. de Paiva. Towards constructive hybrid logic (extended abstract). In C. Areces and P. Blackburn, editors, *Workshop Proceedings of Methods for Modalities 3*, 15 pages. 2003.

T. Braüner and V. de Paiva. Intuitionistic hybrid logic. *Journal of Applied Logic*, 4:231–255, 2006. Revised and extended version of Braüner and de Paiva (2003).

T. Braüner and S. Ghilardi. First-order modal logic. In P. Blackburn, J. van Benthem, and F. Wolter, editors, *Handbook of Modal Logic*, pages 549–620. Elsevier, 2007.

K. Broda, M. Finger, and A. Russo. Labelled natural deduction for substructural logics. *Logic Journal of the IGPL*, 7:283–318, 1999.

K. Brünnler. Deep sequent systems for modal logic. In G. Governatori, I. Hodkinson, and Y. Venema, editors, *Advances in Modal Logic, Volume 6*, pages 117–119. College Publications, 2006.

R.A. Bull. MIPC as the formalisation of an intuitionistic concept of modality. *Journal of Symbolic Logic*, 31:609–616, 1966.

R.A. Bull. An approach to tense logic. *Theoria*, 36:282–300, 1970.

R.A. Bull and K. Segerberg. Basic modal logic. In D.M. Gabbay and F. Guenthner, editors, *Handbook of Philosophical Logic, 2nd Edition*, volume 3, pages 1–81. Kluwer, 2001.

S. Cerrito and M. Cialdea. Nominal substitution at work with the global and converse modalities. In L. Beklemishev, V. Goranko, and V. Shehtman, editors, *Advances in Modal Logic*, Volume 8:57–74, 2010. To appear.

J. Copeland, editor. *Logic and Reality: Essays in the Legacy of Arthur Prior*. Oxford University Press/Clarendon Press, 1996.

J. Copeland. Arthur Prior. In E.N. Zalta, editor, *The Stanford Encyclopedia of Philosophy*. Stanford University, 2007. On-line encyclopedia article available at http://plato.stanford.edu/entries/prior.

R. Chadha, D. Macedonio, and V. Sassone. A hybrid intuitionistic logic: Semantics and decidability. *Journal of Logic and Computation*, 16:27–59, 2006.

C.C. Chang and H.J. Keisler. *Model Theory, 3rd Edition*. Elsevier, 1990.

M. D'Agostino and M. Mondadori. The taming of the cut. Classical refutations with analytical cut. *Journal of Logic and Computation*, 4:285–319, 1994.

M. D'Agostino, D.M. Gabbay, R. Hähnle, and J. Posegga, editors. *Handbook of Tableau Methods*. Springer, Berlin 1999.

V. de Paiva. Constructive description logics: What, why and how. Extended draft presented at *Context Representation and Reasoning*. Available at V. de Paiva's homepage http://www.cs.bham.ac.uk/~vdp, 2006. Accessed 21 Sep 2010.

M. de Rijke. The modal logic of inequality. *Journal of Symbolic Logic*, 57:566–584, 1992.

F.M. Donini and F. Massacci. Exptime tableaux for $\mathscr{ALC}$. *Artificial Intelligence*, 124:87–138, 2000.

W.B. Ewald. Intuitionistic tense and modal logic. *Journal of Symbolic Logic*, 51:166–179, 1986.

M. Fitting. Tableau methods of proof for modal logics. *Notre Dame Journal of Formal Logic*, 13:237–247, 1972a.

M. Fitting. ε-calculus based axiom systems for some propositional modal logics. *Notre Dame Journal of Formal Logic*, 13:381–384, 1972b.

M. Fitting. *Proof Methods for Modal and Intuitionistic Logic*. Reidel, 1983.

M. Fitting. Linear reasoning in modal logic. *Journal of Symbolic Logic*, 49:1363–1378, 1984.

M. Fitting. Many-valued modal logics. *Fundamenta Informaticae*, 15:235–254, 1992a.

M. Fitting. Many-valued modal logics, II. *Fundamenta Informaticae*, 17:55–73, 1992b.

M. Fitting. Basic modal logic. In D.M. Gabbay et al., editors, *Handbook of Logic in Artificial Intelligence and Logic Programming*, volume 1, pages 365–448. Oxford University Press, Oxford, 1993.

M. Fitting. Tableaus for many-valued modal logic. *Studia Logica*, 55:63–87, 1995.

M. Fitting. A modal Herbrand theorem. *Fundamenta Informaticae*, 28:101–122, 1996.

M. Fitting. Herbrand's theorem for a modal logic. In A. Cantini, E. Casari, and P. Minari, editors, *Logic and Foundations of Mathematics*, pages 219–225. Kluwer, 1999.

M. Fitting. *Types, Tableaus, and Gödel's God*. Kluwer, 2002.

M. Fitting. First-order intensional logic. *Annals of Pure and Applied Logic*, 127:171–193, 2004. Essays in the memory of Alfred Tarski. Parts IV, V and VI.

M. Fitting. Bilattices are nice things. In T. Bolander, V.F. Hendricks, and S.A. Pedersen, editors, *Self-Reference*, pages 53–77. CSLI Publications, 2006a.

M. Fitting. FOIL axiomatized. *Studia Logica*, 84:1–22, 2006b.

M. Fitting. Modal proof theory. In P. Blackburn, J. van Benthem, and F. Wolter, editors, *Handbook of Modal Logic*, pages 85–138. Elsevier, 2007.

M. Fitting and R.L. Mendelsohn. *First-Order Modal Logic*. Kluwer, 1998.

K. Fine and A.N. Prior. *Worlds, Times and Selves*. Duckworth, London, 1977. Based on manuscripts by Prior with a preface and a postscript by K. Fine.

M. Franceschet and M. de Rijke. Model checking hybrid logics (with an application to semistructured data). *Journal of Applied Logic*, 4:279–304, 2006.

D.M. Gabbay. *Labelled Deductive Systems*. Oxford University Press, Oxford 1996.

D.M. Gabbay and G. Malod. Naming worlds in modal and temporal logic. *Journal of Logic, Language and Information*, 11:29–65, 2002.

D.M. Gabbay, A. Kurucz, F. Wolter, and M. Zakharyaschev. *Many-Dimensional Modal Logics: Theory and Applications*. Elsevier, 2003.

D. Gallin. *Intensional and Higher-order Modal Logic*. North-Holland, Amsterdam, 1975.

A. Galton. Operators vs. arguments: The ins and outs of reification. *Synthese*, 150:415–441, 2006. Special issue edited by T. Braüner, P. Hasle, and P. Øhrstrøm.

D. Galmiche and Y. Salhi. A family of Gödel hybrid logics. *Journal of Applied Logic*, 2010. To appear.

D. Galmiche and Y. Salhi. Calculi for an intuitionistic hybrid modal logic. In V. de Paiva and A. Nanevski, editors, *Proceedings of the Workshop on Intuitionistic Modal Logic and Applications (IMLA'08)*, 2008. Technical Report MSR-TR-2008-90, Microsoft Research.

D. Galmiche and Y. Salhi. Sequent calculi and decidability for intuitionistic hybrid logic. Information and Computation, 2011. To appear.

G. Gargov and V. Goranko. Modal logic with names. *Journal of Philosophical Logic*, 22:607–636, 1993.

J.W. Garson. Quantification in modal logic. In D.M. Gabbay and F. Guenthner, editors, *Handbook of Philosophical Logic, 2nd Edition*, volume 3, pages 267–323. Kluwer, 2001.

G. Gentzen. Investigations into logical deduction. In M.E. Szabo, editor, *The Collected Papers of Gerhard Gentzen*, pages 68–131. North-Holland, 1969.

J.-Y. Girard, Y. Lafont, and P. Taylor. *Proofs and Types*, volume 7 of *Cambridge Tracts in Theoretical Computer Science*. Cambridge University Press, Cambridge 1989.

V. Goranko. Temporal logic with reference pointers. In D. Gabbay and H. -J. Ohlbach, editors, *Proceedings of the 1st International Conference on Temporal Logic*, volume 827 of *Lecture Notes in Artificial Intelligence*, pages 133–148. Springer Berlin, 1994.

V. Goranko. Hierarchies of modal and temporal logics with reference pointers. *Journal of Logic, Language, and Information*, 5:1–24, 1996.

V. Goranko. Sorting and hybrid modal logics. Manuscript, 2000.

V. Goranko and S. Passy. Using the universal modality: Gains and questions. *Journal of Logic and Computation*, 2:5–30, 1992.

R. Goré. Chapter 6: Tableau methods for modal and temporal logics. In M. D'Agostino, D.M. Gabbay, R. Hähnle, and J. Posegga, editors, *Handbook of Tableau Methods*, pages 297–396. Kluwer, 1999.

S. Görnemann. A logic stronger than intuitionism. *Journal of Symbolic Logic*, 36:249–261, 1971.

J.U. Hansen. A tableau system for a first-order hybrid logic. In J. Villadsen, T. Bolander, and T. Braüner, editors, *Proceedings of the International Workshop on Hybrid Logic 2007*, pages 32–40. *19th European Summerschool in Logic, Language and Information*, 2007.

J.U. Hansen. Terminating tableaux for dynamic epistemic logics. In T. Bolander and T. Braüner, editors, *Proceedings of the 6th Workshop on Methods for Modalities (M4M-6 2009)*, volume 262 of *Electronic Notes in Theoretical Computer Science*, pages 141–156. Elsevier, 2010.

J.U. Hansen, T. Bolander, and T. Braüner. Many-valued hybrid logic. In C. Areces and R. Goldblatt, editors, *Advances in Modal Logic, Volume 7*, pages 111–132. College Publications, 2008.

M. Hardt and G. Smolka. Higher-order syntax and saturation algorithms for hybrid logic. In P. Blackburn, T. Bolander, T. Braüner, V. de Paiva, and J. Villadsen, editors, *Proceedings of the International Workshop on Hybrid Logic 2006*, volume 174 of *Electronic Notes in Theoretical Computer Science*, pages 15–27. Elsevier, 2007.

P. Hasle, P. Øhrstrøm, T. Braüner, and J. Copeland, editors. *Revised and Expanded Edition of Arthur N. Prior: Papers on Time and Tense*. Oxford University Press, Oxford 342 pages. 2003.

A. Hazen. Expressive completeness in modal language. *Journal of Philosophical Logic*, 5:25–46, 1976.

J. Hintikka. Form and content in quantification theory. *Acta Philosophica Fennica*, 8:8–55, 1955.

J. Hintikka. On the development of the model-theoretic viewpoint in logical theory. *Synthese*, 77:1–36, 1988.

H.T. Hodes. Some theorems on the expressive limitations of modal languages. *Journal of Philosophical Logic*, 13:13–26, 1984.

G. Hoffmann and C. Areces. HTab: A terminating tableaux system for hybrid logic. In C. Areces and S. Demri, editors, *Workshop Proceedings of Methods for Modalities 5*, pages 255–268. École Normale Supérieure de Cachan, France, 2007.

I. Horrocks and U. Sattler. A tableaux decision procedure for $\mathcal{SHOIQ}$. In *Proceedings of the 19th International Joint Conference on Artificial Intelligence (IJCAI 2005)*, pages 448–453, 2005.

I. Horrocks, U. Sattler, and S. Tobies. Practical reasoning for expressive description logics. In *Proceedings of LPAR'99*, volume 1705 of *Lecture Notes in Artificial Intelligence*, pages 161–180. Springer Berlin, 1999.

I. Horrocks, U. Hustadt, U. Sattler, and R. Schmidt. Computational modal logic. In P. Blackburn, J. van Benthem, and F. Wolter, editors, *Handbook of Modal Logic*, pages 181–245. Elsevier, 2007.
G.E. Hughes and M.J. Cresswell. *An Introduction to Modal Logic*. Methuen, 1968.
G.E. Hughes and M.J. Cresswell. *A New Introduction to Modal Logic*. Routledge, 1996.
T. Jager. An actualist semantics for quantified modal logic. *Notre Dame Journal of Formal Logic*, 23:335–349, 1982.
L. Jia and D. Walker. Modal proofs as distributed programs (extended abstract). In D. Schmidt, editor, *European Symposium on Programming*, volume 2986 of *Lecture Notes in Computer Science*, pages 219–233. Springer Berlin, 2004.
M. Kaminski and G. Smolka. Hybrid tableaux for the difference modality. In C. Areces and S. Demri, editors, *Workshop Proceedings of Methods for Modalities 5*, pages 269–284. École Normale Supérieure de Cachan, France, 2007.
M. Kaminski and G. Smolka. Terminating tableau systems for hybrid logic with difference and converse. *Journal of Logic, Language and Information*, 18:437–464, 2009. Special issue edited by T. Braüner and T. Bolander.
A.J.P. Kenny. Arthur Norman Prior (1914–1969). In *Proceedings of the British Academy*, volume LVI, pages 321–349, 1970.
M. Kracht and O. Kutz. The semantics of modal predicate logic I. Counterpart-frames. In F. Wolter, H. Wansing, M. de Rijke, and M. Zakharyaschev, editors, *Advances in Modal Logic*, Volume 3, pages 299–320. World Scientific, 2002.
H. Kushida and M. Okada. A proof-theoretic study of the correspondence of hybrid logic and classical logic. *Journal of Logic, Language and Information*, 16:35–61, 2007.
M. Lange. Model checking for hybrid logic. *Journal of Logic, Language and Information*, 18: 465–491, 2009. Special issue edited by T. Braüner and T. Bolander.
D. Lewis. Counterpart theory and quantified modal logic. *Journal of Philosophy*, 65:113–126, 1968.
D. Lewis. Counterparts of persons and their bodies. *Journal of Philosophy*, 68:203–211, 1971.
S. Lindstrøm and K. Segerberg. Modal logic and philosophy. In P. Blackburn, J. van Benthem, and F. Wolter, editors, *Handbook of Modal Logic*, pages 1149–1214. Elsevier, 2007.
F. Massacci. Single step tableaux for modal logics. *Journal of Automated Reasoning*, 24:319–364, 2000.
D.P. McCullough. Logical connectives for intuitionistic propositional logic. *Journal of Symbolic Logic*, 36:15–20, 1969.
J.M.E. McTaggart. The unreality of time. *Mind*, 187:457–474, 1908.
G. Mints. *A Short Introduction to Modal Logic*. CSLI Publications, 1992.
R. Montague. On the nature of certain philosophical entities. In R.H. Thomason, editor *Formal Philosophy, Selected Papers of Richard Montague*, pages 148–187. Yale University Press, 1974.
T. Müller. Prior's tense-logical universalism. *Logique et Analyse*, 50:223–252, 2007.
T. Murphy, K. Crary, R. Harper, and F. Pfenning. A symmetric modal lambda calculus for distributed computing. In H. Ganzinger, editor, *19th Symposium on Logic in Computer Science, (LICS)*, pages 286–295. IEEE Computer Society Press, 2004.
H.J. Ohlbach, A. Nonnengart, M. de Rijke, and D.M. Gabbay. Encoding two-valued non-classical logics in classical logic. In J.A. Robinson and A. Voronkov, editors, *Handbook of Automated Reasoning*, volume 1, pages 1403–1486. MIT Press, 2001.
M. Ohnishi and K. Matsumoto. Gentzen method in modal calculi. *Osaka Mathematical Journal*, 9:113–130, 1957.
M. Ohnishi and K. Matsumoto. Gentzen method in modal calculi, II. *Osaka Mathematical Journal*, 11:115–120, 1959.
H. Ono and N.-Y. Suzuki. Relations between intuitionistic modal logics and intermediate predicate logics. *Reports on Mathematical Logic*, 22:65–87, 1988.
P. Øhrstrøm and P. Hasle. A.N. Prior's rediscovery of tense logic. *Erkenntnis*, 39:23–50, 1993.
P. Øhrstrøm and P. Hasle. *Temporal Logic: From Ancient Ideas to Artificial Intelligence*. Kluwer, 1995.

P. Øhrstrøm and P. Hasle. A.N. Prior's logic. In D.M. Gabbay and J. Woods, editors, *Logic and the Modalities in the Twentieth Century. The Handbook of the History of Logic*, volume 6. Elsevier, 2005a.

P. Øhrstrøm and P. Hasle. Modern temporal logic: The philosophical background. In D.M. Gabbay and J. Woods, editors, *Logic and the Modalities in the Twentieth Century. The Handbook of the History of Logic*, volume 6. Elsevier, 2005b.

Z. Parks. Investigations into quantified modal logic–I. *Studia Logica*, 35:109–125, 1976.

S. Passy and T. Tinchev. Quantifiers in combinatory PDL: Completeness, definability, incompleteness. In *Fundamentals of Computation Theory FCT 85*, volume 199 of *Lecture Notes in Computer Science*, pages 512–519. Springer Berlin, 1985.

S. Passy and T. Tinchev. An essay in combinatory dynamic logic. *Information and Computation*, 93:263–332, 1991.

A. Plantinga. Actualism and possible worlds. In M. Davidson, editor, *Essays in the Metaphysics of Modality*, pages 103–121. Oxford University Press, 2003. Originally published in *Theoria*, 42:139–160, 1976.

D. Prawitz. *Natural Deduction. A Proof-Theoretical Study*. Almqvist and Wiksell, Stockholm, 1965.

D. Prawitz. Ideas and results in proof theory. In J. E. Fenstad, editor, *Proceedings of the Second Scandinavian Logic Symposium*, volume 63 of *Studies in Logic and The Foundations of Mathematics*, pages 235–307. North-Holland, 1971.

D. Prawitz. Proofs and the meaning and completeness of the logical constants. In J. Hintikka et al., editors, *Essays on Mathematical and Philosophical Logic*, Studies in Logic and The Foundations of Mathematics, pages 25–40. Reidel, 1978.

A. Prior. *Time and Modality*. Clarendon/Oxford University Press, 1957.

A.N. Prior. The runabout inference-ticket. *Analysis*, 21:38–39, 1960.

A. Prior. *Past, Present and Future*. Clarendon/Oxford University Press, 1967.

A. Prior. *Papers on Time and Tense*. Clarendon/Oxford University Press, 1968.

A.N. Prior. Some free thinking about time. In J. Copeland, editor, *Logic and Reality: Essays in the Legacy of Arthur Prior*. Oxford University Press/Clarendon Press, Oxford, 1996. With introduction by P. Øhrstrøm in which the original dating is discussed, pages 43–44 and pages 47–51.

W.V.O. Quine. Three grades of modal involvement. In *The Ways of Paradox and Other Essays*, pages 156–174. Random House, 1953.

J. Reed. Hybridizing a logical framework. In P. Blackburn, T. Bolander, T. Braüner, V. de Paiva, and J. Villadsen, editors, *Proceedings of the International Workshop on Hybrid Logic 2006*, volume 174 of *Electronic Notes in Theoretical Computer Science*, pages 135–148. Elsevier, 2007.

J.A. Robinson. *Logic: Form and Function*. North Holland, 1979.

D.S. Scott. Advice on modal logic. In K. Lambert, editor, *Philosophical Problems in Logic*, pages 143–173. D. Reidel Publishing Company, 1970.

D. Scott, editor. *Notes on the Formalization of Logic*. Sub-faculty of Philosophy, University of Oxford, Oxford 1981.

J. Seligman. The logic of correct description. In M. de Rijke, editor, *Advances in Intensional Logic*, volume 7 of *Applied Logic Series*, pages 107–135. Kluwer, 1997.

J. Seligman. Internalisation: The case of hybrid logics. *Journal of Logic and Computation*, 11: 671–689, 2001. Special Issue on Hybrid Logics. C. Areces and P. Blackburn (eds.).

P. Simons. The logic of location. *Synthese*, 150:443–458, 2006. Special issue edited by T. Braüner, P. Hasle, and P. Øhrstrøm.

A. Simpson. *The Proof Theory and Semantics of Intuitionistic Modal logic*. PhD thesis, University of Edinburgh, 1994.

H. Sturm and F. Wolter. First-order expressivity for S5-models: Modal vs. two-sorted languages. *Journal of Philosophical Logic*, 30:571–591, 2001.

D. Sustretov. Hybrid logics of separation axioms. *Journal of Logic, Language and Information*, 18:541–558, 2009. Special issue edited by T. Braüner and T. Bolander.

R. Sylvan. Other withered stumps of time. In J. Copeland, editor, *Logic and Reality: Essays in the Legacy of Arthur Prior*, pages 111–130. Oxford University Press/Clarendon Press, Oxford, 1996.

A. Tarski. What are logical notions? *History and Philosophy of Logic*, 7:143–154, 1986. The paper is edited by J. Corcoran.

B. ten Cate. *Model Theory for Extended Modal Languages*. PhD thesis, Institute for Logic, Language and Computation, University of Amsterdam, 2004.

B. ten Cate and T. Litak. Topological perspective on the hybrid proof rules. In P. Blackburn, T. Bolander, T. Braüner, V. de Paiva, and J. Villadsen, editors, *Proceedings of the International Workshop on Hybrid Logic 2006*, volume 174 of *Electronic Notes in Theoretical Computer Science*, pages 79–94. Elsevier, 2007.

A.S. Troelstra and H. Schwichtenberg. *Basic Proof Theory*, volume 43 of *Cambridge Tracts in Theoretical Computer Science*. Cambridge University Press, Cambridge 1996.

A. Troelstra and D. van Dalen. *Constructivism in Mathematics: An Introduction*. North-Holland, 1988.

M. Tzakova. Tableaux calculi for hybrid logics. In N.V. Murray, editor, *Automated Reasoning with Analytic Tableaux and Related Methods, TABLEAUX 1999)*, volume 1617 of *Lecture Notes in Artificial Intelligence*, pages 278–292. Springer Berlin, 1999.

J. van Benthem. *Modal Logic and Classical Logic*. Bibliopolis, 1983.

W. van der Hoek and M. de Rijke. Counting objects. *Journal of Logic and Computation*, 5: 325–345, 1995.

S. Vickers. *Topology via Logic*, volume 5 of *Cambridge Tracts in Theoretical Computer Science*. Cambridge University Press, Cambridge 1988.

S. Vickers. Geometric logic in computer science. In G.L. Burn, S.J. Gay, and M.D. Ryan, editors, *Theory and Formal Methods 1993*, pages 37–54. Springer Berlin, 1993.

L. Viganò. *Labelled Non-classical Logics*. Kluwer, 2000.

F. von Kutschera. Die Vollständigkeit des Operatorsystems $\{\neg, \wedge, \vee, \supset\}$ für die intuitionistische Aussagenlogik im Rahmen der Gentzensemantik. *Archiv für Mathematische Logik und Grundlagenforschung*, 11:3–16, 1968.

H. Wansing. Sequent calculi for normal modal propositional logics. *Journal of Logic and Computation*, 4:125–142, 1994.

H. Wansing. A proof-theoretic proof of functional completeness for many modal and tense logics. In H. Wansing, editor, *Proof Theory of Modal Logic*, volume 2 of *Applied Logic Series*, pages 123–136. Kluwer, 1996.

H. Wansing. The idea of a proof-theoretic semantics and the meaning of the logical operations. *Studia Logica*, 64:3–20, 2000.

H. Wansing. Logical connectives for constructive modal logic. *Synthese*, 150:459–482, 2006. Special issue edited by T. Braüner, P. Hasle, and P. Øhrstrøm.

H. Wansing and S.P. Odintsov. Inconsistency-tolerant description logic. Motivation and basic systems. In V. Hendricks and J. Malinowski, editors, *50 Years of Studia Logica*, volume 21 of *Trends in Logic Series*, pages 301–335. Kluwer, 2003.

K. Wenningsted-Torgard. *Theorem Prover for Hybrid Logic*. Roskilde University Digital Archive, 2008. Student project report available at http://hdl.handle.net/1800/3664. Accessed 21 sep 2010.

J. Zucker and R.S. Tragesser. The adequacy problem for inferential logic. *Journal of Philosophical Logic*, 7:501–516, 1978.

Index

FSC
www.fsc.org
MIX
Papier | Fördert gute Waldnutzung
FSC® C083411

Kolibri 360 GmbH
Ferdinand-Jühlke-Straße 7
99095 Erfurt, Deutschland
produktsicherheit@kolibri360.de